Rutland Water - Decade of Change

Developments in Hydrobiology 8

Series editor
H. J. Dumont

DR W. JUNK PUBLISHERS THE HAGUE-BOSTON-LONDON 1982

Rutland Water – Decade of Change

Proceedings of the Conference
held in Leicester, U.K., 1–3 April 1981

Edited by
David M. Harper and John A. Bullock
University of Leicester

Reprinted from *Hydrobiologia,* vol. 88, no 1/2 (1982)

DR W. JUNK PUBLISHERS THE HAGUE-BOSTON-LONDON 1982

Distributors:

for the United States and Cananda

Kluwer Boston, Inc.
190 Old Derby Street
Hingham, MA 02043
U.S.A.

for all other countries

Kluwer Academic Publishers Group
Distribution Center
P.O. Box 322
3300 AH Dordrecht
The Netherlands

ISBN-13: 978-94-009-8008-2 e-ISBN-13: 978-94-009-8006-8
DOI: 10.1007/978-94-009-8006-8

Library of Congress Cataloging in Publication Data
Main entry under title:

Rutland water--decade of change.

(Developments in hydrobiology ; 8)
"Reprinted from Hydrobiologia, vol. 8, no. 1/2, 1982."
1. Reservoirs--England--Rutland--Congresses.
I. Harper, David M. II. Bullock, John A. III. Series.
TD258.R87R87 628.1'32 82-7164
AACR2

Cover design: Max Velthuijs

Rutland Water viewed from the east, over the dam and pumping stations. The circular tanks behind the pumphouse are trout rearing ponds; the limnological tower can be seen in the centre of the photograph.

The south arm of Rutland Water viewed from the west end. The circular clump of trees to the left is Lax Hill, part of the nature reserve, and the reserve car park on the south shore can be seen in the middle right.

Photographs by Tony Fletcher, Anglian Water Authority.

Contents

Introduction – Rutland Water 1970–1981
by D. M. Harper 1

PART ONE: RUTLAND WATER: THE RESOURCE FOR PUBLIC WATER SUPPLY

Rutland Water – from conception to operation
by E. A. Knights 7

A 'regrettable necessity' – the decision to construct
by P. Lawson 19

Escaping from prison: an economic reassessment of Rutland Water
by P. R. Herrington 27

PART TWO: THE RESERVOIR IN THE RUTLAND ENVIRONMENT

The landscape of Rutland Water
by Dame Sylvia Crowe 43

Perimeter forestry and landscape works at Rutland Water
by A. Ford 47

Tree pre-planting at Rutland Water
by C. J. Gill 51

Area structure plans of Leicestershire County and Rutland District Councils
by D. Wightman & P. Burford 52

Traffic and visitor surveys at Rutland Water
by D. Wightman 53

Rutland 2000: educational resources of Rutland Water
by Bryan Waites 55

Archaeological discoveries at Rutland Water
by A. W. Adams, T. H. McK. Clough, M. S. Gorin, N. M. Reynolds & M. Todd 57

PART THREE: WATER QUALITY, ALGAL ECOLOGY AND RESERVOIR MANAGEMENT

Circulation patterns; their investigation and prediction
by E. J. Myers, R. S. Oldham & A. J. D. Ferguson 67

Vertical movements of the surface water and thermal stratification in Rutland Water
by S. L. Maddocks 79

The effect of Langmuir circulation on the distribution and settling of algae and suspended particles
by Tavicha Buranathanitt, David J. Cockrell & Peter H. John 88

Models of annual cycles in lentic water bodies
by B. Henderson-Sellers & P. B. R. Archer 89
Rutland Water raw water quality – problems and management
by N. F. Low 93
The production of public supply drinking water from Rutland Water at the Wing Treatment Works
by D. B. Ford, B. E. Drage & T. J. Roberts 103
Rutland Water phytoplankton: the development of an asset or a nuisance?
by A. J. D. Ferguson & D. M. Harper 117

PART FOUR: ZOOLOGICAL ECOLOGY AND RESERVOIR MANAGEMENT

Zooplankton and their relationships with water quality and fisheries
by D. M. Harper & A. J. D. Ferguson 135
Benthic invertebrates of Rutland Water
by J. A. Bullock, F. Clark & S. Ison 147
The leeches of Rutland Water
by M. Majeed 157
Studies on two species of *Asellus* in Rutland Water
by Y. T. Daoud 158
Bathymetric distribution of some Chironomidae (Diptera) in Rutland Water
by A. E. Brown & R. S. Oldham 159
The protozoan epizooites found on the gills of *Gammarus pulex*
by M. H. Walker & E. Roberts 171
The control of a fly nuisance problem in a woodland area adjacent to Rutland Water
by M. J. Knight & R. S. Oldham 177
Establishing and maintaining the trout fishery at Rutland Water
by D. E. Moore 179
The feeding of trout during the filling phase of Rutland Water
by D. M. Harper 191
Temporal variations in the diet of brown trout (*Salmo trutta* L.) and rainbow trout (*S. gairdneri* R.) in Rutland Water
by A. D. Warlow & R. S. Oldham 199
The population dynamics of eyeflukes *Diplostomum spathaceum* and *Tylodelphys clavata* (Digenea: Diplostomatidae) in rainbow and brown trout in Rutland Water: 1974–1978
by J. Moody & E. Gaten 207
Rutland Water Nature Reserve: concept, design and management
by T. P. Appleton 211
The distribution of micro-organisms through a bunded lagoon in Rutland Water
by J. T. Holah & J. D. McIver 225
Wildlife surveys of the reservoir site prior to inundation 1971–1976
by I. M. Evans 231

Introduction - Rutland Water 1970-1981

D. M. Harper
Department of Adult Education, Leicester University, England, and Hon. Secretary, Rutland Water Research Group

Preamble

Rutland Water, or Empingham Reservoir as it was initially known, came into being officially in May 1970 when Royal Assent was granted to the private Welland and Nene (Empingham Reservoir) and Mid-Northamptonshire Water Act. The end of the first stage in its existence came a decade later. By then, the completed reservoir had reached its top water level and the dam had been inspected; about half its design capacity was being used for water supply and it had received national acclaim for its recreational facilities and international status for its wildfowl populations.

In April 1981 a conference was held at Leicester University to bring together the events of the past decade. The initiative for this conference came from the Rutland Water Research Group, set up in 1971 to coordinate research effort and exchange information amongst those with an active amateur or professional interest in the changes associated with the reservoir (Harper 1979). The Research Group joined with the Institution of Water Engineers and Scientists (Scientific Section) because of the great interest in the primary water supply function of the reservoir, and with the Adult Education Department of Leicester University, which organised the conference.

Rutland Water in its national and regional context

Rutland Water is only ten years old, although plans for such a water supply scheme had been gestating for at least a decade longer. The project itself spans the most interesting period in the history of water management in Britain, and probably of any country in the world (Okun 1977).

Initial planning of future supplies for the Northamptonshire area – where demand was predicted to increase rapidly from expansion of towns like Northampton with 'overspill' population from London and Birmingham – was the responsibility of the Mid-Northamptonshire Water Board, one of the 250 or so independent water supply undertakings in Britain.

In 1963 the Water Resources Act created 29 River Authorities in England and Wales, and gave them responsibilities for water resource development in their catchments in addition to those for land drainage, fisheries and pollution prevention held by their predecessor River Boards. Thus the Welland and Nene River Authority became involved with the planning and ultimately the promotion of this water resource development on a tributary of the river Welland. The area of supply of the new reservoir had widened to include neighbouring water supply undertakings, notably for Peterborough and Leicester, and it was promoted on this basis.

In the four years following the Empingham Reservoir Act, Government proposals for the reorganisation of all aspects of water cycle management in England and Wales were refined and implemented as the 1973 Water Act. This created ten new Regional Water Authorities, charged with the management of water supply, sewage disposal, and the old river authority functions within their new catchment areas (Okun 1977; Parker & Penning-Rowsell 1981). The Anglian Water

Hydrobiologia 88, 1-5 (1982). 0018-8158/82/0881-0001/$01.00.

Fig. 1. England and Wales, showing the boundaries of the Regional Water Authorities, and the position of the catchments of the Welland and Nene rivers within the Anglian Water Authority area (dotted line).

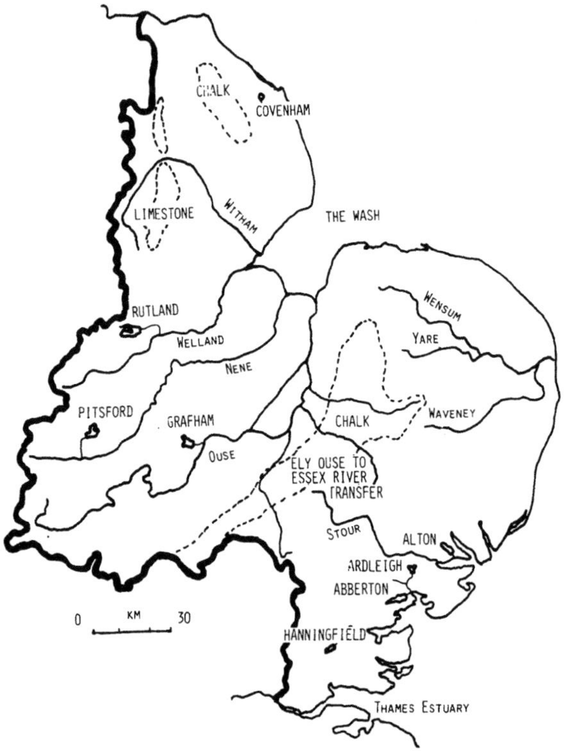

Fig. 2. The Anglian Water Authority showing the major components of the water resources network – reservoirs, groundwater resources, rivers and river transfer schemes. Lincolnshire is the county in the north of the region, Essex in the south adjacent to London.

Authority (Fig. 1) took over responsibility for Rutland Water in the advanced stages of reservoir construction. Six years later it occupies an important role in strategic planning of the water resources in the Anglian region, linking with reservoirs, groundwater resources and river transfers from Lincolnshire down to the outskirts of London (Fig. 2) (A.W.A. 1975).

Rutland Water, like nearly all the major reservoirs of south-eastern England, is a pumped storage reservoir, the largest such scheme in Britain. Pumped storage reservoirs, situated close to the lower reaches of the major rivers from which their supply is taken, are the most practical means of surface water storage in the lowland southeast where demand is high, rainfall low and the supply from impounding reservoirs, groundwater or direct river abstraction insufficient. Thus, the total catchment of Rutland Water is many times larger than its natural catchment (Fig. 3), although the contribution of the natural catchment during the filling phase of the reservoir was greater than anticipated (Harper 1978) because of high rainfall during this period.

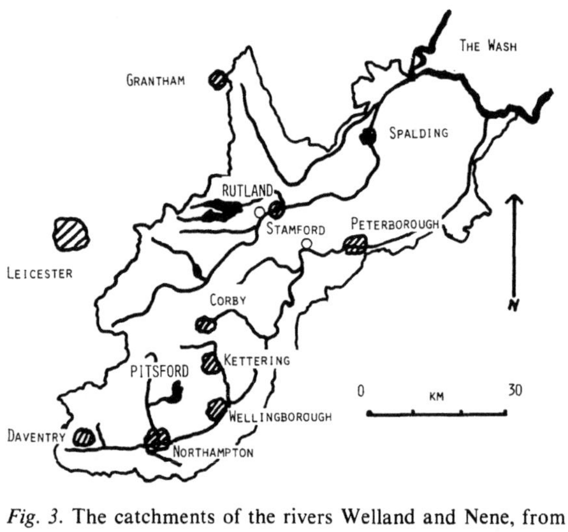

Fig. 3. The catchments of the rivers Welland and Nene, from which water is pumped into Rutland at intakes marked with an open circle. The relative sizes of natural and artificial catchment of Rutland Water can thus be compared. The towns which Rutland supplies are also shown.

The important features of Rutland Water

It is obviously impossible for a conference which was multi-disciplinary in the widest sense of the word and which was held relatively early in the life of the new reservoir, to cover all its aspects. Nevertheless, those features of the reservoir which have given it national, or even international, recognition are recorded in this volume.

The first feature of the reservoir is the controversy which it has generated. When the proposals for the scheme were first made public, considerable local opposition was aroused which was mainly directed at its size and location. The planning aspects of the decision-making process and the activities of the promoters and opposition are reviewed by Lawson (pp. 19–26). Local opposition declined following the Act's Royal Assent; the reasons for local acceptance of the reservoir lay primarily in the thorough consultation procedures adopted by the promoters, outlined by Knights (pp. 7–17) and the major efforts put into recreational and amenity planning outlined in the articles by Knights (pp. 7–17), Crowe (pp. 43–46) and Ford (pp. 47–50). The wider controversy existing a decade ago as to whether a reservoir of this size was justifiable in the national context is re-examined by Herrington (pp. 27–42). Recent events have shown that this controversy still exists (Pearce 1981; Bowyer 1981).

The second feature of this reservoir is its detailed and successful provision for recreation. The foundations for this were laid during the parliamentary debates and ensuing consultations but given greater strength by the re-organisation of the water industry in 1974, which for the first time committed the new authorities to promote the recreational and amenity use of their water resources. The general provision for recreation is outlined by Knights (pp. 7–17) and explained more fully in Langford (1979), and landscaping and amenity provision is covered by Crowe (pp. 43–46) and Ford (pp. 47–50). The two major recreational activities requiring scientific management as the basis for their success – the trout fishery and the nature reserve – are dealt with by Moore (pp. 179–189) and Appleton (pp. 211–224): It has been observed (Parker & Penning-Rowsell 1981) that recreational planning and management is generally accorded a low status and funding within the Regional Water Authorities, with the exception of fisheries, for which separate legislation exists. The success of the planning and integration of all the recreational activities at Rutland Water and in particular the most sensitive – wildlife conservation – provides an excellent example for other water authorities, as well as for Anglian, in other aspects of the water cycle management.

The third feature of the reservoir, but undoubtedly the most important, is its management for water supply. The scientific management of lowland reservoirs is costly and complex because of the high alkalinity and nutrient levels in the impounded water, and the resultant development of planktonic organisms. The Water Research Centre were brought in as consultants to the River Authority during the reservoir's planning and initial construction because of these problems, with the result that the reservoir has a very wide range of management options, summarised in Fig. 4. The uses of these options, the limnology of the reservoir during its first 5 years of water storage, and the operation of the treatment processes are reviewed over pp. 67–209. The management limnology of Rutland can be seen in a wider, regional context when compared with the treatment problems faced by other eutrophic reservoirs in the Anglian Water Authority (Greene & Hayes 1981).

The role of research at Rutland Water

An outstanding feature of the success of the recording of events at Rutland Water in its first ten years has been the co-operation of individuals and institutions with diverse interests, despite the small proportion of external funding. Of course, much of the work described here has been carried out by employees of the Water Authority as part of their duties, but reference is made by Appleton (pp. 211–224) and Evans (pp. 231–232) to the work of volunteers and much of the work carried out by the educational institutes has been internally funded. This has led to gaps in our knowledge of the reservoir; little is known for example about the aquatic plants of reservoir or nature reserve. The Research Group was considerably weakened by the sudden death of Dr. M. Pearson of Nottingham University who was interested in the tolerance of *Phragmites australis* to flooding and its colonisation of the new water's edge.

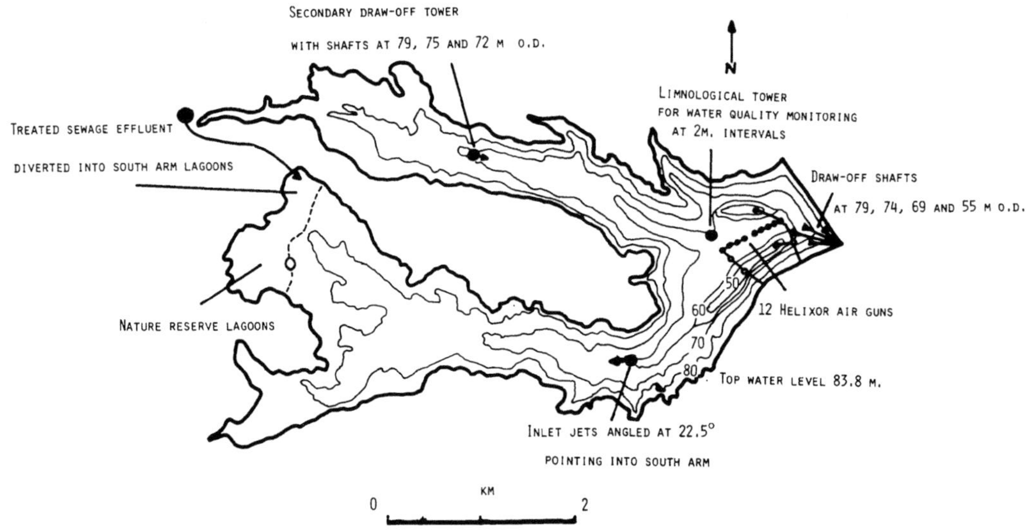

Fig. 4. The main features of Rutland Water of importance in its management limnology. The sewage effluent comes from the town of Oakham (population 6 780), situated on the inflow stream to the north arm about 1 km from the reservoir.

With the exception of a few gaps, the basic groundwork of recording and interpreting changes has been completed and is contained within these pages. The archaeological work is mostly over (see article by Adams *et al.*, pp. 57–66) and as the biological work shows (see articles by Ferguson and Harper, Harper and Ferguson, and Bullock *et al.*, pp. 117–156), many of the changes have now settled down into more or less predictable seasonal patterns.

Rutland Water will continue to be a focal point of interest because of its size and importance, and because of the wealth of background data gathered here. I hope that the members of the Research Group, and other workers, will now direct their enthusiasm and resourcefulness towards solving some of the many problems, both academic and applied, still associated with such an exciting research site.

Acknowledgements

The editors wish to express their gratitude to members of the Rutland Water Research Group who presented papers at the conference which are reproduced here, although some who were involved in the early days of the Group have since left. Mr. Alan Morton, now at Liverpool University, and Peter Nicholson, now the Nature Conservancy Council's Regional Officer for the South West, were instrumental in establishing the group; Barry Gooding when resource planner with the Welland and Nene River Authority and Professor Dix, when head of the Institute of Planning Studies at Nottingham University, were valuable contributors to it. The managers of the River Authority and its successor, the Welland and Nene River Division of A.W.A., particularly the present incumbent Mr. Peter Langford, provided invaluable support to the Group, encouragement for the contributions of their staff and assistance to research workers on site.

The preparation and organisation of the conference was efficiently dealt with by Eileen Sunderland and Barbara Hackney of the Department of Adult Education at Leicester University, and by Andy Elphink and Debora Peckham of the scientific section of the Institution of Water Engineers and Scientists. The Chairmen of sessions – Ken Stroud, John Hellawell and R. I. Hall – together with the opening contributors, helped to produce a lively and stimulating conference. Frank Clark and Steven Brierley recorded the proceedings. Colin Harden, Trout Hatchery Manager, guided delegates around the facilities in his charge after the conference, as did Frank Knights, David Ford and Tim Appleton.

Finally, Mr. Wil Peters, Managing Editor, and Mr. Cliff Morgan, Desk Editor, of Dr. W. Junk Publishers, have my thanks for the speed and efficiency with which they solved all the problems and brought this volume to publication.

References

Anglian Water Authority, 1975. Interim Review of the Long Term Water Plan. A.W.A., Huntingdon.

Bowyer, G., 1981. Letter. New Scientist 92: 332.

Greene, L. A. & Hayes, C. R., 1981. The impact of eutrophication on water treatment and supplies in the Anglian Water Authority. J. Inst. Wat. Engrs Scientists 35: 421–436.

Harper, D. M., 1979. The Rutland Water Research Committee Bull. Br. Ecol. Soc. 10: 37.

Harper, D. M., 1978. The limnology of Rutland Water. Verh. int. Verein. Limnol. 20: 1604–1611.

Langford, P. R., 1979. Leisure and sports facilities at Rutland Water. J. Inst. Wat. Engrs Scientists 33: 117–134.

Okun, D. A., 1977. Regionalization of Water Management. Applied Science Publishers, London.

Parker, D. J. & Penning-Rowsell, E. C., 1981. Water Planning in Britain. George Allen & Unwin, London.

Pearce, F., 1981. Water, water, everywhere. New Scientist 92: 90–93.

Rutland Water – from conception to operation

E. A. Knights
Reservoir Engineer, Anglian Water Authority, Rutland Water, Whitwell, Oakham, Leicestershire, England

Keywords: Rutland Water, reservoir, water supply, demand, design, recreation

Abstract

Rutland Water was planned in the 1960s to meet the water needs of the expanding East Midland region of England. Chosen from 64 possible sites, it was approved in 1970 and construction started in 1971. The dam was closed and the reservoir started to fill from its natural catchment in 1975; pumping started from the rivers Welland and Nene in 1976 and the reservoir was full by March 1979.

Details of the stages in its development are given together with an indication of the ways in which the recreational uses of the reservoir have been integrated with its water supply function.

Introduction

The New Towns Act passed by Parliament in 1965 gave the approval for expansion plans at the towns of Peterborough and Northampton. In addition, 'overspill' population from Birmingham and London was planned to contribute to developments at Wellingborough and Daventry; continued growth was forecast for Corby (Fig. 1).

The size of these expansion areas and their proximity meant that the water supply authorities were faced with the prospect of accelerating demands – from estimates of consumption of 227×10^3 m^3 day^{-1} in 1974 to a projection of 550 by the year 2001. The predicted doubling of population in this region together with the increase in per capita use of water made the future growth of water supply one of the most rapid in Britain.

Forecasts of population growth and demand for water supply were presented at the Parliamentary proceedings for the promotion of the reservoir and these served to indicate the extent of the problem (Fig. 2).

The local water supply authority, the Mid-Northamptonshire Water Board, had realised in the early 1960s that the only way of meeting these major new demands was to rely on a large water storage reservoir supplied with pumped water from the river systems. The region already had a number of small impounding reservoirs in the upper catchments of the Nene and Welland, together with a larger pumped storage reservoir, Pitsford, just outside Northampton. These resources would soon be fully used and a significant quantity of water was being imported into the area from Grafham Water, itself a large new lowland pumped storage reservoir near Huntingdon.

In 1963 the Water Resources Bill became law and created new River Authorities whose task was to oversee the total development of water resources within newly defined hydrometric areas. The two main river catchment areas in the region, the Nene and the Welland, were therefore for the first time administered by one organisation, the Welland and Nene River Authority, which had the responsibility under the Act for assessing water demands and seeing how any deficits in supply could be met during the period to the year 2001.

Investigations began into 64 alternative reservoir sites both inside and outside the Northampton-

Hydrobiologia 88, 7–17 (1982). 0018-8158/82/0881-0007/$02.20.

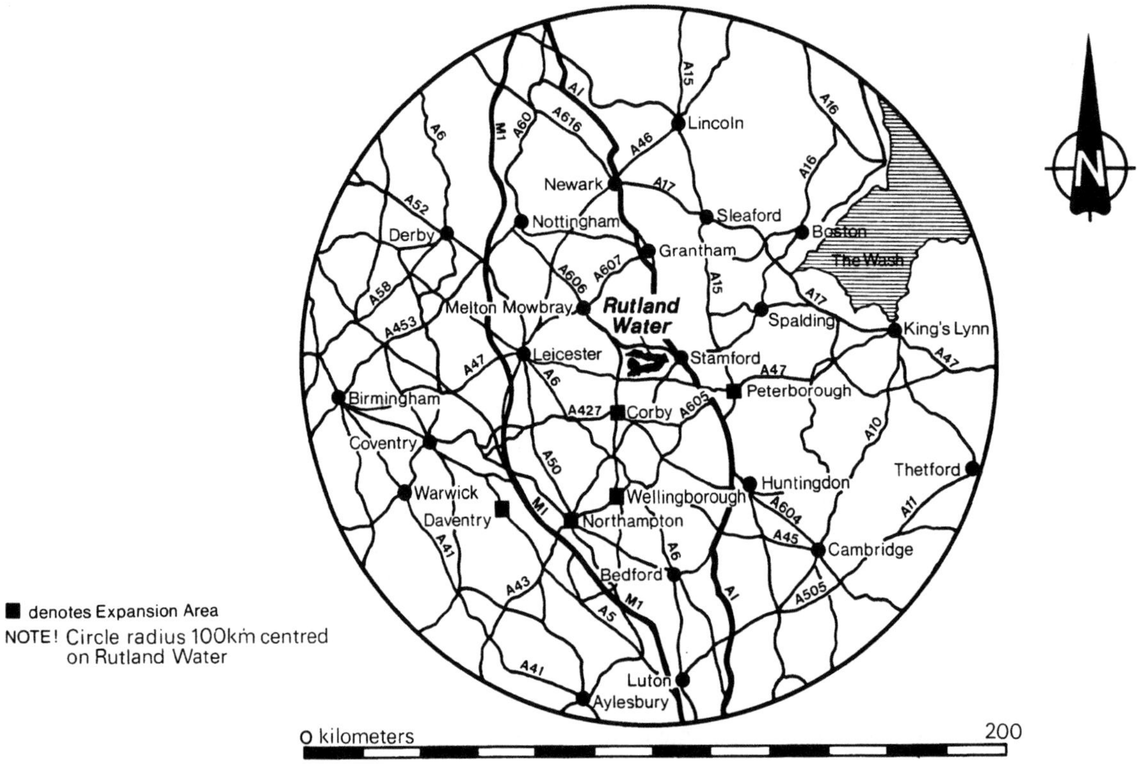

Fig. 1. The location of Rutland Water in the Midlands of England. The circle has a radius of approximately 100 km.

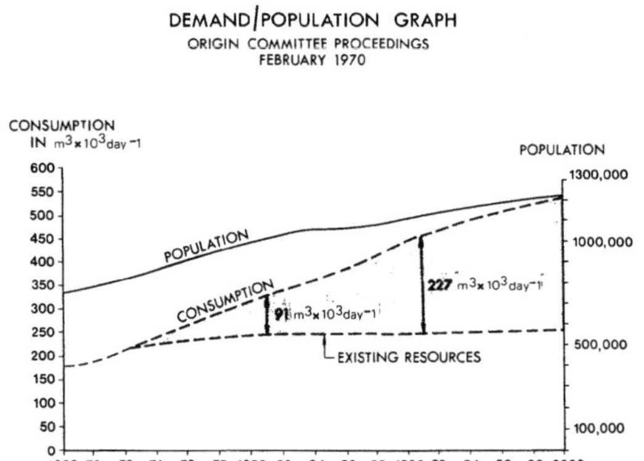

Fig. 2. Forecasts of population increase and demand for water supply presented to Parliamentary committee February 1970. Figures for water in 10^3 m^3 day^{-1}.

Peterborougn supply area, as well as the possible further development of the Lincolnshire limestone groundwater aquifer which provided existing supplies for the eastern part of the area.

The preferred site was one at Empingham, in the old county of Rutland, where a dam across the valley of the river Gwash (a tributary of the Welland) would impound both the northern and southern stream whose confluence lay just above the proposed dam site.

A regional water supply reservoir in the twin valleys of the Gwash had thus been considered for some time before the River Authority and its consultants began their detailed investigations into the desirability of developing this one site to meet demands within the region for the next 30 years.

This Empingham reservoir site offered the following advantages:

1) The scheme would provide a vast storage area occupying the two valleys. The material within the reservoir was suitable for the construction of a single, 35 m high rolled clay dam, the only major earth works required.

2) It was close to points on both the Welland and the Nene where they could be tapped for the water required. The intakes on both rivers could be sited reasonably near their tidal limit so as to maximise the river catchment area contributing to each intake.

Table 1. Reservoir statistics

Cross capacity	124×10^6 m^3
Surface area (max)	1260 ha
Depth (max)	34 m
Top water level	83.8 m O.D.
Catchment area (natural)	7400 ha
Yield	0.33×10^6 m^3 day^{-1}
Compensation flow	4.5×10^3 m^3 day^{-1}
Dam crest level	87 m O.D.
Dam crest length	1200 m
Height above formation level	40 m
Max width at base	810 m
Volume of fill	4.7×10^6 m^3
Outlet shaft no. of draw-offs	4
Overflow capacity	18 m^3 s^{-1}
River abstraction (Wansford)	7.9 m^3 s^{-1}
River abstraction (Tinwell)	5.3 m^3 s^{-1}
Supply aqueducts	21 km
Tunnels	14 km
Pipelines	7 km
Pump capacity:	
Wansford	5 pumps 2250 h.p.
Tinwell	4 pumps 1175 h.p.
Tinwell relift	5 pumps 2250 h.p.
Empingham final lift	8 pumps 8450 h.p.

3) The reservoir would be centrally located for those areas which needed the supply and a considerable amount of water supplied could be returned to the river systems as effluent upstream of the intakes thus increasing the yield of the scheme.

4) The reservoir would be located on the Welland system such that water could, if required, be used to supplement flow downstream. The additional flow could provide for further public water supply (e.g. for Peterborough), future irrigation demands or to combat any sudden pollution.

5) Water storage provided by the reservoir could help alleviate flooding on the downstream catchment.

6) The remaining 63 sites were either of inadequate storage capacity or were geologically unacceptable, apart from an adjacent valley, that of the river Chater at Manton, which would have had a lower yield and higher operating costs.

Representation and public involvement

The Welland and Nene (Empingham Reservoir and Mid-Northamptonshire) Water Act was submitted to Parliament and given the Royal Assent in May 1970 after it had been examined by Select Committees in both Houses. The Committees heard evidence from the River Authority's consultants and advisors as well as from Local Authorities, farming and conservation interests. After consultation and compromise, the proposals were accepted within the following limits:

1) Adequate facilities should be made available to the visiting public so that this burden did not fall unduly upon local ratepayers.

2) Perimeter land should be returned to agricultural use where possible.

3) New roads and services should be constructed as part of the reservoir works to replace those lost in flooded areas.

4) Limits of abstraction were imposed for the two rivers to safeguard their character and downstream functions.

5) Adequate compensation flow was imposed for the river Gwash below the dam.

6) The rural character of the area should be retained and landscape proposals should be implemented to take into account the summer draw-down of the reservoir, which could be extensive at the western end.

The Authority at this stage mounted exhibitions at local village halls, agricultural shows and the like to demonstrate to the public that the proposals were not only necessary but would also provide people with recreation facilities in a controlled manner which would not adversely affect the rural environment. In addition, landowners and farmers in and around the reservoir site were visited and their particular problems discussed. The final result of these endeavours was a high degree of co-operation by those affected by the scheme. This was invaluable when some thirty million pounds sterling in contract works was spread over an area of 1883 ha in an 8 year construction period.

A small Empingham reservoir committee was set up by the River Authority to oversee the project. This committee, advised by full time officers, ensured that positive decisions were made quickly. These involved such diverse matters as environmental problems related to noise levels of heavy plant and early compensation payments to allow those farmers affected to aquire new land on what was then a rapidly rising market. This understanding of everyday problems was invaluable to the smooth running of the project. The farmers affected continued to act as caretakers for the land purchased and maintained its productive use until it was required for contract works or flooding.

A Traffic management committee was formed to monitor traffic flow in the area during construction and when the reservoir first came into recreational use.

A Nature reserve advisory committee was established to co-ordinate the development of the 142 ha nature reserve.

A Reservoir working party was established to maintain liaison. This was made up of engineering staff, landscape consultants, highway authorities, planning officers, Nature Conservancy Council and Leicestershire and Rutland Trust for Nature Conservation staff. It produced its first report in January 1971 and Dame Sylvia Crowe, the Authority's consultant landscape architect, finalised a plan for land use and recreation over the entire area. This study included the re-location of a proliferation of overhead power lines and electrical equipment which would otherwise have marred some of the most beautiful views of the lake. A small grant towards the cost of placing these cables underground was provided by the County Council and the liaison between landscape architect and electricity board was successful in reducing the impact of the re-routed services.

The Empingham reservoir research committee, as it was then called, was formed in 1971 from academics, environmentalists, archaeologists and the staff of bodies involved in the project, with the aim of recording and researching the massive changes about to occur on and around the reservoir site.

Contract and construction

In outline the project consists of the pumped storage reservoir of 124×10^6 m^3 capacity supplied with water pumped from intakes on the rivers Nene (approximately 20 km south-east of the reservoir) and Welland (approximately 10 km) in addition to the small contribution of the natural catchment. The combined abstraction capacity from the rivers is just under 10^6 m^3 day^{-1}, two thirds from the Nene. Water abstracted from the Nene at Wansford (at 20 m O.D.) is pumped to Tinwell (25 m O.D.) and re-lifted together with Welland water to the pumping station at Empingham (50 m O.D.). Here it is pumped up to the reservoir (top water level 84 m O.D.) through inlet jets located on the floor of the south arm facing westwards (Fig. 3).

The potential yield of the reservoir for water supply is presently just under 3×10^5 m^3 day^{-1}. This is expected to increase to over 3.5×10^5 by the year 2001 due to increasing volumes of effluent discharge to the rivers upstream of the abstraction points, together with conjunctive use of the reservoir and other resources. Future strategies may include combination with local groundwater reserves by using reservoir water partly for river regulation rather than wholly for direct supply, and by integrating the operation of Rutland with other sources within the A.W.A. region.

The dam itself is constructed mainly from Upper Lias clay taken from borrow pits in the valley within the reservoir site just upstream of the dam. It is 1.2 km long and a maximum of 0.8 km wide at the base, the width being dictated by the relatively low shear strength of the Lias clay.

The geology of the embankment site which was revealed by the site investigations is as follows:
Lower Lincolnshire limestone
Lower Estuarine series
Northamptonshire Sand ironstone
Upper Lias clay
Marlestone rock bed
Middle Lias clays and silts

The marlstone rock bed is 22 m below the lowest formation level of the embankment and extends under the whole reservoir area. Artesian pressure exists in the rock at the embankment site and relief wells have been installed on the downstream side of the embankment to control uplift pressures.

Underneath the highest part of the embankment, vertical sand drains assist in the consolidation of the clay. Drainage blankets were placed on the clay over the sand drain area and supplemented by gravel strip drains. The under-drainage system discharges to the river Gwash downstream of the tailbay.

Due to the considerable settlement expected in the dam foundation and the necessity for sand drains in the centre of the valley, the outlet tunnel and shaft was positioned at the southern end of the embankment. This discharges into the river Gwash near the downstream toe. These works were completed prior to the construction of the

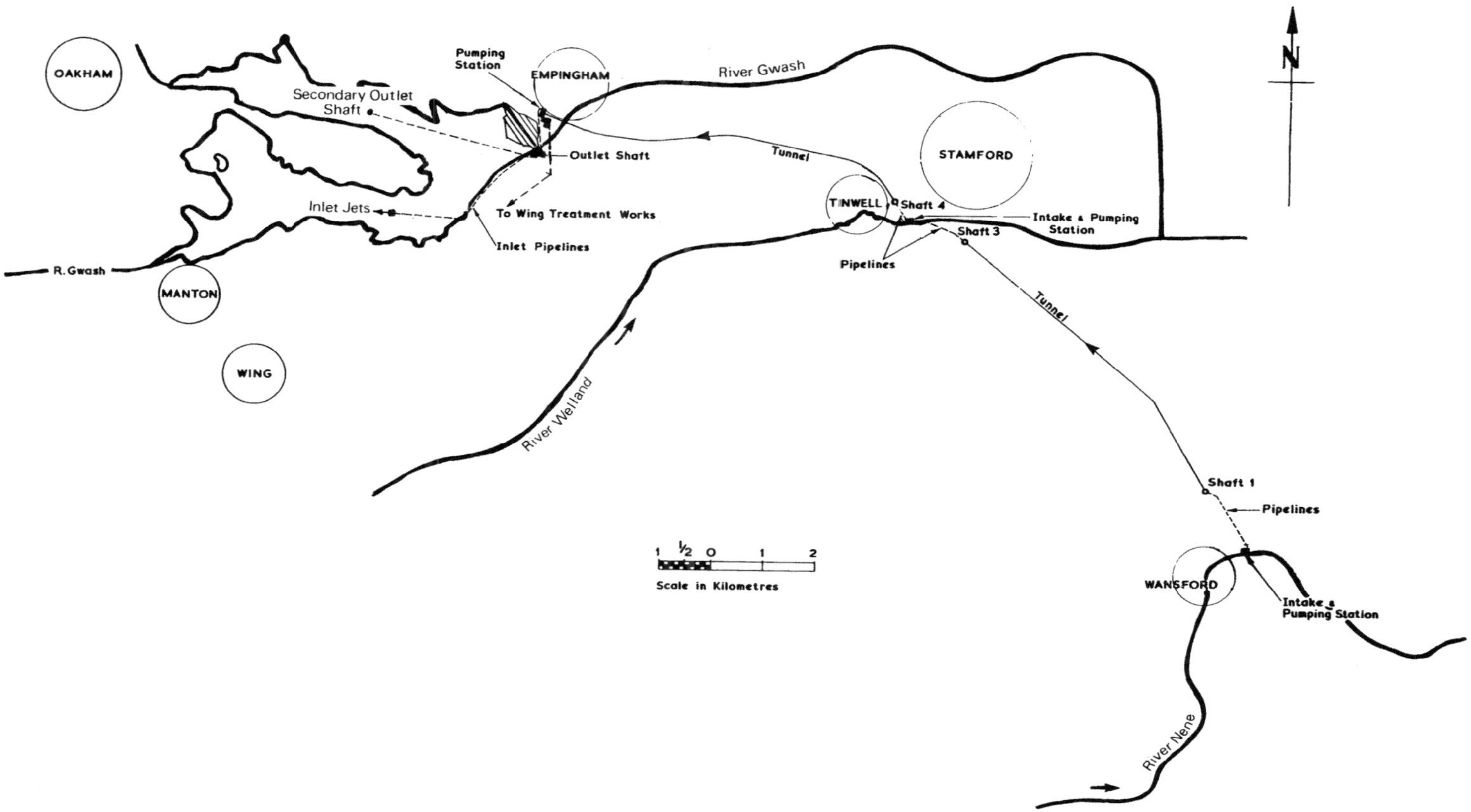

Fig. 3. An outline of the Rutland Water scheme.

embankment, diverting the river flow and allowing foundation stripping in the valley bottom at the start of the embankment contract. This tunnel now serves as the bottom draw-off from the reservoir with three other outlets at higher levels in the form of 1.2 m diameter concrete pipes laid from forebays in the reservoir at its southern corner.

The shape of the reservoir, in the twin drowned valleys of the Gwash, is such that it was considered that water would behave differently in the north arm. A secondary outlet tower and pipeline is located in the north arm, which will allow it to be used as a 'second' reservoir in terms of water quality, as necessary.

A system of compressed air guns is located in the main body of the reservoir, to prevent thermal stratification of the water which would otherwise lead to anaerobic conditions. The guns consist of vertical stack pipes into the base of which compressed air is introduced from a compressor located in Empingham pumping station.

A limnological tower is located in the main body of the reservoir for continuous monitoring of oxygen and temperature, the results being relayed by telemetry to the control room.

The treatment works for Rutland Water are located at Wing, some 8 km to the south, and water is pumped up to the works from a separate set of pumps below the dam.

The network of supply pipelines was originally designed to serve Peterborough, Milton Keynes, most of Northamptonshire and parts of Leicestershire. Further pipelines have been laid subsequently to allow for the future development of supplies to the area of Grantham and north Lincolnshire.

During the construction period, close liaison was maintained between the staff of the Authority and the consulting engineers and much work was carried out by the Authority directly. This included the design and construction of river controls, river intakes and pumping stations, various amenity works, land use surveys and liaison with statutory bodies.

To form an effective buffer between contractors and the farming community over the site of the reservoir, tunnels and pumping stations, a small direct labour unit was set up of local men. This unit helped to:

1) Provide effective control in the site clearance operation.

2) Provide fencing and access to enable the farming operations to continue whilst construction works were proceeding on various sites.

3) Ensure that service diversions were carried out with minimum disruption.

4) Enable rapid response to any problems.

Technical staff were appointed early on in the project to supervise the design, installation and eventual commissioning and operation of the pumping plant, electrical switchgear and electronic control systems.

Environmental aspects of the project were given early consideration; a forester and several workers were appointed by the Authority, and in conjunction with the Leicestershire and Rutland Trust for Nature Conservation a nature reserve warden appointed.

The public began to arrive at the site in large numbers from 1972 onwards to watch the spectacle of 15 giant motor scrapers in their task of moving the 4.7×10^6 m^3 material to construct the embankment. Between May and October in each of the four years of construction, temporary car parks with information boards were opened. The first reservoir wardens were appointed at about this time to control public access throughout the site and to assist with work at the trout hatchery which was under construction.

Most of the landscape and amenity works were completed during the main construction period 1972–1976, taking advantage of the machinery available and the materials quarried on site. A perimeter track was constructed to give access to contractors for removal of power lines, construction of re-routed services and for site clearance. This track was subsequently landscaped and linked to the public roads, enabling the full 40 km of the reservoir perimeter to be traversed.

The architectural designs of the buildings in the scheme were to a common theme approved by both the landscape architect and the Fine Arts Commission. The early consultations with the design engineers ensured that reservoir buildings were located not only from a functional point of view but also to allow the architect to disguise the massive proportions involved. Subsequent landscaping has made these buildings, such as the pumphouse at Empingham, relatively unobtrusive in the rural landscape.

Annual draw-down of the lake had been

anticipated by various landscape proposals. Three km of clay banks ('bunds') retain water at the shallow western end of the reservoir throughout the year. They also receive some catchment inflow and tertiary treated sewage effluent from Oakham. The bunds have independent controls which allow the water to balance the 3 lagoons thus created or to be discharged direct to the reservoir.

Sewage from the settlements surrounding the reservoir has, with the exception of that from Oakham, been fed into a perimeter sewerage system discharging at Empingham downstream of the reservoir. One day storage is provided at each pumping station to prevent any accidental spillage of raw sewage into the reservoir. A number of the villages were connected to a main sewer for the first time in this operation, including the peninsular village of Hambleton.

The dam and tunnelling contractors recruited local labour (within 50 km) throughout the contract period and key staff lived in a temporary caravan site adjacent to the dam. Dust and noise problems were kept to a minimum by the skillful siting of temporary spoil heaps and restriction of working hours, particularly near inhabited areas.

Operation

The programme date for first pumping from the river intakes was delayed 12 months due to engineering, geological and contractual problems and the 'three day week'. This coincided with the severe drought of 1976 which reminded those involved that the reservoir design – based on the river flow statistics of three dry years 1943–1946 – was certainly attainable and that if the scheme had been at full supply a considerable proportion of the stored volume would have been utilised in this year.

The Wing treatment works was not completed at the time of the drought but impounded water was released from the new reservoir down the river system to a site near Peterborough where it was channelled to recharge the gravel strata. Water was re-abstracted a few km distant and received temporary treatment before being supplied to the city to relieve it from the worst aspects of the drought.

This lack of water meant that pumping was prevented until September 1976 when the first rain fell and was followed by a wet winter. This period of drought when no pumping could take place gave an opportunity to both train the control staff and commission the telemetry and computer-aided system for control of pumping. This is a GEC fixed logic unit which has given excellent performance, the river intake and booster pumping stations being unmanned for most of the time. The comprehensive system has a fault finding facility and data storage to enable complete records to be kept. These aids are necessary to assess performance of equipment and operating costs, and prevent wastage of technical staff time. Independent controls exist at river intakes stations to guard against telemetry failure.

A selective filling operation commenced when the pumps became fully operational in late 1976. Top water level was achieved for the first time in March 1979, although the natural catchment had contributed about 30% of the inflow. Initially the water pumped into the reservoir was subjected to certain quality constraints and as much as possible was pumped from the Welland with its better quality water and lower pumping head.

The maintenance of top water level for three months was then necessary in order to test all the valves against the maximum head of water and to monitor the piezometers in and around the dam area.

Gauging stations are now operational on the inflows which enable the contribution of the natural catchment to be assessed more accurately than during the filling phase. A climate station is established at the western end which will facilitate accurate water balance studies from the appropriate hydrological measurements.

The reservoir communications network has now been developed for both rescue and security as well as for normal operational needs. The control centre, manned 24 hours a day, is also used by the Authority now for monitoring regional flood and pollution events and relaying east coast tidal information.

Reservoir perimeter lands are now let annually for sheep grazing to the farmers who previously occupied the land. Income thus derived defrays the high costs of fencing to protect trees and hedges.

Recreation

The Water Act of 1973 gave the Anglian Water Authority the powers to consolidate the wider commitment beyond water supply which the old

River Authority had initiated on this project. Conceding to local opinion, the Authority changed the name from Empingham reservoir to Rutland Water – a perpetuation of the name of the old country, some 4% of whose area is now covered by the reservoir.

Over 250 000 visitors now come to Rutland Water each year. There is a wide range of activities, angling and sailing being the largest ones. The recreation officer has a small wardening force which is made up of anglers and part-time sailing masters, linked by short-wave radios to the control centre and lodge (Fig. 4).

Anglers may purchase day or season tickets for what is now the country's largest and most important trout fishery. These allow them to choise a spot on 29 km of wooded inlets and bays or take one of the 50 powered or 10 pulling fishing dinghies.

The Rutland Sailing Club leases land from the Authority and operates one of the largest inland sailing clubs in the country with 824 ha of water available. In addition, day sailing is available for individuals from the Whitwell spur. Sailing on the reservoir with its two arms is challenging, particularly in the exposed areas. Safety is a priority both to the club who operate their own rescue service and to volunteers who provide additional cover at peak periods. Medical treatment is available for injuries and such things as hypothermia – 216 incidents were recorded in 1980.

In the Whitwell lodge, users of the water have a common room which displays detailed maps of the reservoir, wind speed and barometric pressure. Advice on weather conditions and fishing prospects are also available to both beginners and experts.

The lake can change its mood quickly, so at a wind speed of Force 6 or over, warning lights flash to warn boats to leave the water. Fog has also caused problems so all fishing boats are equipped with weatherproof maps, compass and flares. Life jackets are mandatory on the water; throwing hand lines and lifebelts are available at all picnic sites and access points.

The Nature Reserve is open to both the casual visitor and the serious naturalist through a zoning policy. Observation hides are strategically located and an Interpretive Centre is planned.

Visitors may walk 29 km of reservoir track, and cycling and pony riding is available over parts of it. This track is kept traffic-free, with the exception of the Hambleton peninsular, which is open for a restricted number of anglers' cars only. A number of small, well screened car parks are widely spaced around the perimeter as anglers vary their location with the slightest change of wind to different points around the lake.

Each of the reservoir users has representation on the Consultative Users Panel which acts as the forum for new ideas and monitors for any possible areas of conflict. This is also attended by local planners, police and representatives of farming and conservation interests.

Maintenance

Strong winds across the largest fetches of the 1260 ha lake have caused some erosion of the margins although no damage to the stone facing of the dam. Some lesser protection in the form of local limestone has been placed on the vulnerable parts of the shoreline.

In some bays, particularly where siltation has occurred, there is colonisation by submerged plants. Weedcutting is very limited and selective, leaving sufficient to give cover for fish and their fry. Only the most sheltered bays normally need cutting, mainly for boating activities.

Conclusions

The potential for water supply of Rutland Water is vast – over 3×10^5 m^3 day^{-1} – providing a satisfactory supply for the expanding demands of its region.

Because of its size and location, the development of recreational facilities has been extensive. Problems have been small, however, and with a flexible and sensible approach by both management and users, Rutland Water can provide an excellent example of the successful multi-purpose use of a water supply reservoir.

Acknowledgements

The author wishes to thank the Director of Operations of the Anglian Water Authority for permission to publish this paper. The observations made are the author's own, however, and not necessarily those of the Authority. The author is grateful to Mary Liddington for her assistance with this paper.

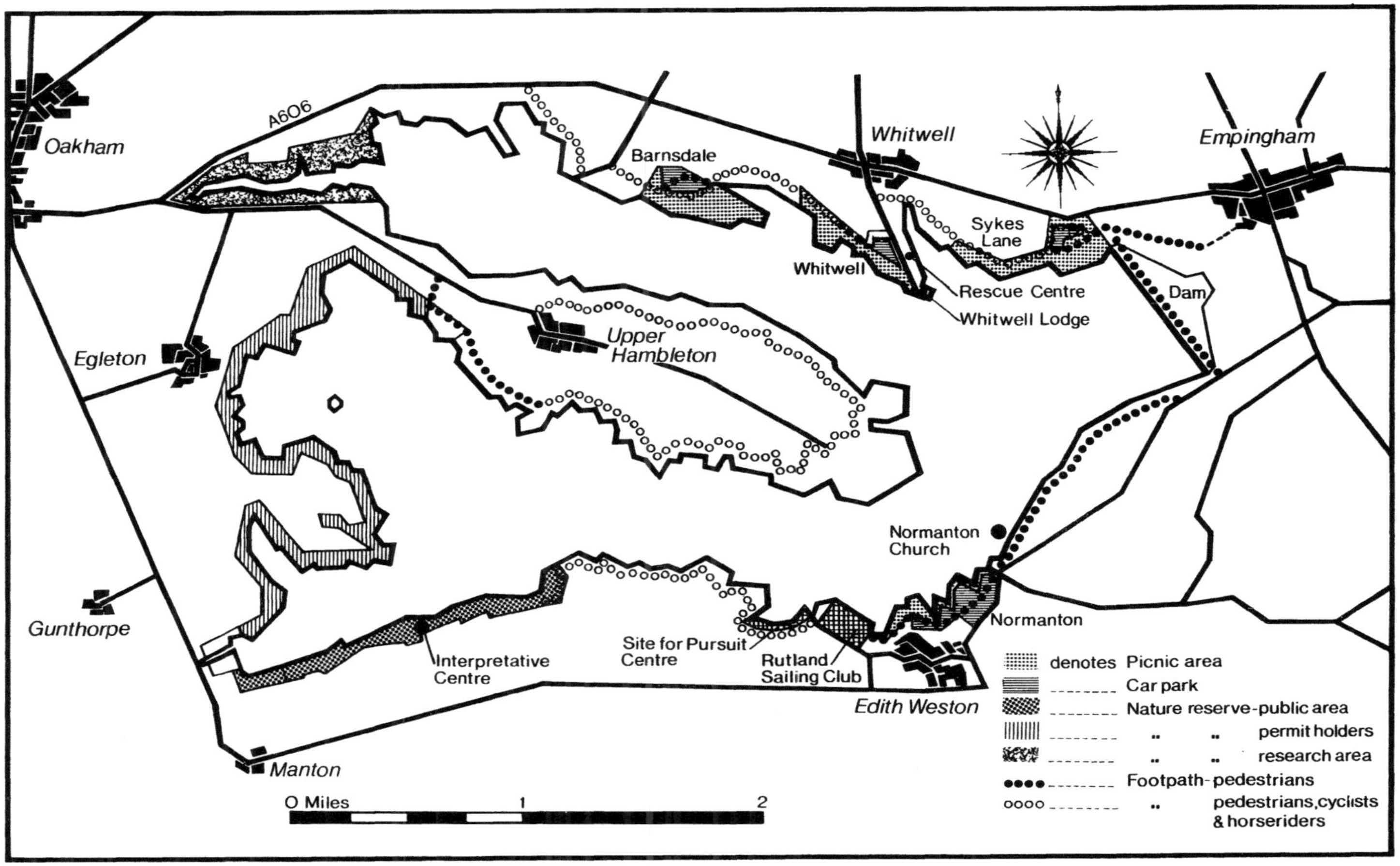

Fig. 4. Map showing the recreational facilities at Rutland Water.

Discussion

The author introduced his paper by referring to a collection of newspaper cuttings concerning the reservoir which had appeared nationally and locally over the past decade. The project had aroused great interest as well as opposition from the outset and its progress had been closely followed in the media. It was appropriate now that the reservoir's water supply functions received analysis and recognition as well as the new standards it had set for amenity provision and environmental consideration. There was a clear need for all aspects of the reservoir to be documented judging by the number of enquiries and visits which the water authority received and the proceedings of the conference would provide this.

Headlines in the early years of the decade expressed local fears of adverse developments, such as caravan sites or hotels. Concern about traffic flow and the local reaction to the first visitors was another headline. Fifteen hundred trout fishermen arriving on the first day of the fishing season caused great consternation – one local magistrate took 2 hours to gain access to her own drive! Such problems were soon resolved at local level.

Subsequent headlines have tended to focus attention on the more sensational and human facets of the reservoir's many uses. For example, two fishing boats containing five men on one occasion spent the night afloat; they were found at 5 a.m. the following morning walking in the village of Egleton some 8 km from the fishing lodge. On another occasion a 16-year-old volunteer was awarded the Royal Life Saving Society award when he rescued two adults and two children from a dinghy in a sudden squall.

Much of the public interest has been focused upon aspects of the reservoir discussed at this conference – such as the dramatic archaeological finds during the course of construction, the success of the trout fishery and nature reserve. Other aspects attracting interest are associated with the landscaping and recreational policy – such as the protection of historic buildings on the reservoir's edge. Normanton church, on the southern shoreline, has been substantially altered with the floor raised 2 m, the windows redesigned and a protective embankment funded by a local trust. More varied recreational uses of the reservoir have developed – wind surfing on the surface, sub-aqua diving below. On the perimeter track walking, limited horse-riding, and since 1981, a cycling hire centre have been developed.

By the late 1970s some of the headlines had begun to draw attention to the primary function of the reservoir – water supply. During the drought in the U.K. in 1976, for example, the headlines dealt with the emergency supply to Peterborough (via the river Gwash) and Grantham (an emergency pipeline laid along the A1 road). More recent water supply developments have not escaped the public notice; the area of supply now includes the local area of Rutland and east Leicestershire. The significance of this cannot be overlooked as the nationally famous Ruddles 'County' beer is now brewed using this supply!

The discussion was opened by Mr. D. Wightman, of Leicestershire County Council. He felt that the Water Authority had made an unprecedented effort to smooth the construction of the reservoir and its subsequent operation, and in general he felt that the transition was relatively free of trauma considering the scale of the project. There had been problems, however, and significant changes in the character of communities surrounding the reservoir. Roads are more crowded at weekends where there was little traffic before; problems of trespass and vandalism have increased, as have the pressures on local authorities to allow commercial developments related to the reservoir. Other smaller problems have occurred – such as plagues of midges invading local villages in the first few summers after the reservoir filled, which fortunately have declined recently.

In contrast to these negative points there have been many positive ones. In particular the quality of the new water recreation facilities provided is very much envied in other counties. The local authorities see Rutland as an area for quiet enjoyment of the facilities provided and would not therefore see the more intrusive sports such as water-skiing acceptable in this location.

Finally the County Council welcomed the deliberate policy of local involvement which had characterised the project and was stressed in Mr. Knights's paper; this was an attitude which could well be more widespread amongst public authorities.

Questions from the floor centred primarily upon

aspects of recreational planning. Mr. Warren, Wessex Water Authority, asked if the basis used for the design of recreational provision was the combined experience of other lowland reservoirs such as Grafham Water and Pitsford, and if this basis had been shown to be substantially correct now that the reservoir was in operation. Mr. Davies, Severn-Trent Water Authority, asked how the estimates for numbers of visitors and of cars had been arrived at, whether the Authority was surprised by the numbers of people arriving during the construction phase, and how it had coped with problems of trespass on the site. Mr. Barnhoorn, Eastbourne Water Company, asked how the capital and revenue costs of the recreational facilities had been met by the Authority.

Mr. Knights confirmed that experience of the recreational provision and its use at reservoirs such as Grafham had been used in planning facilities here; together with estimates of visitor use based upon the experience of the County Council and consideration of the specific location of this water. Within a 100 km radius live approximately 7 million people; in addition it is about half way between major Midlands cities and the North Sea coast.

A large number of visitors did arrive during the construction phase but these were catered for in temporary car parks overlooking the dam site. This confirmed that the planned amount of permanent car parking was realistic (3 000 cars) as well as providing an opportunity to educate visitors about access and trespass. It was an advantage that the boundaries of the reservoir were drawn to existing boundaries which, during the contract stages, were fenced with 5-bar fencing and quickthorn hedging. This, together with clear signposting of the facilities from all major roads has kept unauthorised access around the site to a minimum.

The capital costs of the provision of the recreational facilities were included within the overall budget for the scheme, reduced by grant-aid from such bodies as the Countryside Commission and the Sports Council for particular items. The Authority has tried to make the recreational facilities self-financing through charges at the car parks and realistic pricing policy for the specific activities such as sailing and fishing.

Mr. K. Stroud asked for clarification of the recent press reports that the level of the reservoir was to be allowed to fall in order to save the pumping costs necessary for replenishment. He felt strongly that this would be a mistake to jeopardise the strong goodwil which the Authority had built up. Mr. Knights indicated that this was a proposal which was not now being implemented.

A 'regrettable necessity' – the decision to construct

P. Lawson
North Yorkshire County Council, Northallerton, North Yorkshire, England[1]
[1] *Formerly Institute of Planning Studies, University of Nottingham*

Keywords: regional planning, Parliamentary Bill, petitioners, environmental impact

Abstract

The case for, and opposition to, reservoir development in Rutland needs to be seen in the context of UK regional planning policies and programmes which, in the mid-1960s, were dominated by expectations of rapid economic and population growth. The process of decision making highlighted basic conflicts between urban expansion and rural resources in a way which offered limited scope for compromise. Debate at the local level and in Parliament raised issues which are still unresolved. In the future some of these may be subjected to even more searching technical appraisal.

Introduction

Rutland Water provides a valuable example of decision making processes at work in a case where the implementation of national/regional policies generated major land use and environmental issues and involved a significant impact at the local level. Water resource development was seen to be vital to the realisation of ambitious regional growth plans yet involved reservoir construction in a rural area which had little sympathy with the needs of the towns and cities where the demand for water was concentrated.

Private legislation was used rather than the more common town and country planning procedures. Opponents of the reservoir fought an unequal battle against the compelling case of the promoters.

This non-technical and partly historical review assesses the main arguments and counter-arguments used and the eventual decision to construct. It concludes by bringing the discussion of procedures up to date with observations on current circumstances.

A compelling case

Regional growth plans

Initial consideration of reservoir construction in Rutland coincided with a high point in government attitudes to national and regional planning. In the mid-1960s the National Plan emerged, the economic ideas of George Brown were in vogue and a new generation of Regional Strategies were being prepared by central government. In 1964 the South-East Study emphasised that the availability of water was a critical factor in the implementation of development plans and stressed the need to adopt a sub-regional approach to water supplies. In regional planning terms there was no point in each water undertaking acting as a discrete supply unit and developing its own sources of supply without reference to surrounding areas. It all sounds very obvious but what is accepted practice now was only beginning to emerge as a necessity twenty years ago (McLellan 1963) and to be reflected in the organisational structure of the water industry. The background to the Empingham scheme needs to be seen in terms of the long-term, flexible, regional pro-

Hydrobiologia 88, 19–26 (1982). 0018-8158/82/0881-0019/$01.60.

grammes ushered in by the 1963 Water Resources Act and the subsequent activities of the erstwhile Water Resources Board.

Optimistic population forecasts generally and the growth plans of three New Town Development Corporations and two Expanded Town Schemes in the vicinity, added up to exceptional and accelerating growth prospects for the 1970s. A technical sub-committee of the Water Resources Board was set up charged with forecasting demand and coordinating long term plans for increasing supply. Noting that investments in resource development had a long gestation period the committee concluded that immediate action was needed in order to safeguard the first 10 years during which time a long-term strategy could be developed. In effect this meant completion of works already authorised or being seriously promoted and deferment of further difficult decisions until more was known of the feasibility of innovations like desalination or barrages across estuaries.

A process of elimination

The Gwash Valley was first considered for a pumped storage project over 20 years ago. Indeed in a comparison of 64 possible local reservoir sites it is remarkable how Empingham and nearby Manton emerged as the only satisfactory sites. The others all had deficiencies in terms of capacity, geology or land use. The newly formed Welland and Nene River Authority assumed the major responsibility for promoting the scheme after the reorganisation brought about by the Water Resources Act 1963. In one of several major technical reports produced by the Water Resources Board in the late 1960s (Water Resources Board 1966) it was confirmed that a large new source of supply was required as a matter of urgency if deficiencies were to be avoided. The import of water from other regions, desalination, the Wash Barrage and the use of ground water (perhaps with the artificial recharge of aquifers) were not available as solutions to immediate problems so that a large new surface reservoir emerged as the only feasible option.

The River Authority calculated that even if Empingham sustained a yield of 225×10^6 l/day further deficits would arise by 1991. In other words work would have to start on further sources of supply soon after Empingham was built if supplies were to be ensured into the 21st century. Their answer was to seek powers to construct both Empingham and Manton reservoirs – a step which was seen to be advantageous from a water planning standpoint and, by removing uncertainty blight, from the point of view of local farmers. It is a matter of history that central government took a different view, arguing that because other potential methods of supply were being actively investigated only Empingham should be promoted for the time being.

The decision to construct

A Private Bill

The concept of 'one reservoir at a time' was accepted by the River Authority which, jointly with the Mid-Northamptonshire Water Board, went on to sponsor private legislation (the Welland and Nene (Empingham Reservoir) and Mid-Northamptonshire Water Bill) which was placed before the House of Commons in 1968. The Private Bill procedure has advantages for the promoters of large projects, not the least of which is that it avoids town and country planning procedures such as public inquiries. While the sponsors of the Bill were undoubtedly aware of this there was one compelling reason which, even if the Private Bill procedure had been disadvantageous, would have forced it into use. Because the scheme involved changing the course of a river the consent of Parliament was required and privately sponsored legislation was necessary.

There was little that objectors could do in this situation except to petition against the Bill through the appropriate channels. Petitioning has always been a costly procedure, especially if experts have to be retained. Nevertheless it ensured that there would be detailed consideration by Parliament, including several days in select committee, with the onus of proof that a reservoir at Empingham was necessary falling squarely on the shoulders of the sponsors. Throughout the 18 months or so between deposit of the Bill and Royal Assent in May 1970, its provisions were debated on at least six occasions.

The petitioners had a hard struggle. They expressed a variety of reservations and local objections concerning land use disruption, changes in communications, effect on the local economy, landscape impact, land-take and drawdown effects. Although they tried, they found it hard to refute the technical case supporting the need for a reservoir in this location.

The promoters' case

In contrast the promoters' case could be put simply and convincingly.

i) Regional Planning Policies could not be fulfilled without an increase in supplies;

ii) Per capita consumption was expected to rise at twice the national average rate in the Welland and Nene R.A. area (from 2361/head/day in 1969 to 425 l/head/day in 2001);

iii) Existing sources of supply were stretched and could not cope with dramatic increases in population and per capita consumption. There was a degree of urgency;

iv) Any errors of prediction would tend to underestimate the future demand position and thus reinforce the case for construction;

v) Since Leicester Corporation supplied most of Rutland with water, design forecasts could allocate 18×10^6 l/day to make up for future deficiencies in the Leicester Corporation system. Not only would this be cheap and efficient but it would also have the desirable effect of deferring further resource development in Derbyshire;

vi) Alternatives had been examined but were found wanting or were not available as short term solutions. These included:

Estuary storage in the Wash
Desalination
Artificial recharge of South Lincolnshire aquifer
Use of worked-out clay pits near Peterborough
Conjunctive use of small reservoirs
Increased river abstraction
Development of several small sites on the list of 64
Import of water from North Lincolnshire;

vii) Therefore, taking all these factors into account, more water was needed. It was needed quickly and in quantity and, by a process of technical elimination, was best found by construction at Empingham.

The petitioners' case

The petitioners did their best to present their case through the proper channels in a situation where time, money and technical skills were in short supply. In the event there were four petitions from:

Rutland County Council
Oakham Rural District Council
CLA/NFU/CPRE consortium*
Huntingdon and Peterborough County Council.

The last related to details only. The other three were very similar and argued the following points:

i) The reservoir was not necessary at all. Other solutions were preferable;

ii) Construction in the 1970s would be premature and could not be justified;

iii) Insufficient research had been undertaken on alternative resources and time should be allowed for more acceptable methods of water supply to be proposed;

iv) The limits of deviation were too wide;

v) As much land as possible should be returned to agriculture on completion of construction;

vi) Local life would be excessively disrupted.

These points are a mixture of matters of principle and matters of local impact. To the impartial observer it must have been obvious that the petitioners just could not match the river authority's technical arguments and forecasts while the local details were of a type which were likely to occur at any lowland reservoir site and could partly be overcome by design and by 'concessions' involving relatively little extra expenditure.

The petitioners attempt to question population forecasts failed – the figures used were those adopted by the local authorities in their development plans and had to be broadly accepted by both sides. Their call for greater use of existing sources was countered by the promoters who pointed out that some of these sources were fully developed. Since some would eventually need to be diverted for other purposes the status quo could not be treated as a constant in comparisons of supply and demand at dates in the future.

The petitioners tried to pursue their claim that the River Authority's forecasts of future consumption per capita were unreasonable to the extent that they exceeded the average for SE England as a whole. They produced their own lower predictions

* CLA – Country Landowners' Association.
NFU – National Farmers' Union.
CPRE – Council for the Protection of Rural England.

which had the effect of reducing demand forecasts in the critical years and tried to illustrate that the proposed yield exceeded foreseeable needs.

Finding themselves in a position where they were forced to accept the reservoir solution the only way the petitioners could have averted development at Empingham would have been by proposing a viable alternative. Again time, money and expertise were against them and, in any case, the water industry had been through this process and possessed detailed information which could have been used as counter-argument. To have suggested Manton, generally accepted as the reserve choice if Empingham proposals began to founder, would have merely split the petitioners and shifted the scene of the action. It would have done nothing for the local authorities' case and was not really available as an argument to the CLA/NFU/CPRE consortium which had members in both areas.

So attention focussed on the possibilities of a smaller reservoir involving 'stop dams' to remove one or both of the arms north and south of the Hambleton peninsula or aiming for a lower top water level overall. To their credit the River Authority gave these possibilities serious consideration and did not over-emphasise the very real practical engineering problems involved. It worked out that such truncation would cost at least an extra £1¼ million for each arm and that consequent amendments to the design would introduce a 12 month delay.

Against this there would be a land saving of 1 500 acres and fewer road and electricity pylon diversions – but not enough to offset the extra costs quoted. In net terms double truncation would have added about 7% to the cost of the project.

The petitioners claimed that this was modest and worthwhile. They also argued that the 'stop dams' could be removed at some point in the future if and when the need arose for more storage capacity. Of course the promoters were not so keen and saw problems with such design changes (water circulation, visual amenity, the need for drafting amendments to the Bill, the mechanics of removing 'stop dams' while maintaining supplies).

The petitioners in retreat

Rutland County Council, a small authority with limited financial resources, found it expensive to petition. By the time of the second reading in the Commons when the passage of the Bill still had over a year to go, the County Council had already incurred the equivalent of a 3d rate on opposition. In the debate on the third reading the Member of Parliament (MP) for Rutland and Stamford, Kenneth Lewis, indicated that the County Council may need to withdraw altogether and noted that there was a body of opinion which suggested that the costs of opposition should not fall on local ratepayers. The CLA/NFU/CPRE consortium operated through a local committee which raised substantial but inadequate funds to see the job through. All petitioners complained about the lack of preparation time and the shortage of resources and expertise. They concluded that it was not really possible for farmers and other local interests to successfully oppose a specialised Bill which the promoters had been preparing for many years. Once the Dutch firm of consulting engineers retained by the consortium had withdrawn, the petitioners were unable to obtain adequate technical advice and fell back on the argument that, since there was no real urgency, proceedings should be postponed. Once it became clear that this was getting nowhere the concept of a smaller, truncated reservoir emerged with the object of saving land and minimising local disruption.

When the Commons eventually voted to accept the Bill, in effect regarding the case for Empingham as proven, the petitions were all withdrawn. At this stage local opinion became resigned to the idea of the reservoir and attention switched from opposition to securing the best deal for the local area. Bilateral agreements were struck between the local authorities and the River Authority to construct, at the latter's expense, car parks, a sewage treatment works, sections of new roads and footpaths and extensive landscaping. This was all in keeping with the promoters' stated policy of cooperation, generous compensation and minimising hardship.

Of the 48 agricultural holdings affected it was thought that only 20 would be rendered uneconomic. Compensation was paid to dispossessed owners not only for land losses but also for disturbance, severance and injurious affection. Tenants received more than the statutory minimum compensation of one year's rent for disturbance plus one extra year if loss could be proved. The River Authority offered to alleviate hardship and offset blight by early purchase and then granting a licence back to the occupier on favourable terms and by purchasing land outside the limits of deviation if farmers so wished.

The Bill in Parliament

The Bill did not have an easy passage through the Commons and even when it was finally approved, all parties involved seem to have under-estimated the capabilities of the House of Lords.

Introducing the second reading of the Bill (Hansard 1969) Tom Bradley MP stressed the promoters' arguments (urban growth, rising consumption, lack of alternatives, deferment of Leicester Corporation's further use of Derbyshire resources) and regarded the reservoir scheme as a potential amenity which would 'enhance the attractiveness of and in no way desecrate that delightful county (of Rutland)'. Nevertheless he acknowledged the 'passion and fury' which had built up over this 'Hydrological Stansted'.

Representing that 'passion and fury' Kenneth Lewis (Hansard 1969) voiced the critics' view in a speech which is more memorable for its rhetoric than for its technical content. In arguing against this 'massive and expensive miscalculation' he pointed to rising unemployment and empty houses in Corby, to 'exaggerated demand forecasts', to countryside destruction and to the loss of productive land. He argued in favour of a 'bridging scheme' using worked-out wet gravel pits at Deeping together with groundwater and some smaller reservoirs designed to meet demand in the decade until the Lincolnshire aquifer and desalination could come into use. Despite such positive and constructive ideas the heart of his argument was, of course, the defence of Rutland. 'We have no wish for another Windermere in Rutland.' 'We do not wish to be a kind of towpath round a Lake.' Other rural members supported him. Indeed members seemed to be obsessed with Windermere comparisons and tried to dispose of the concept of the reservoir as a potential amenity and present it as yet another example of the countryside being faced with a 'fait accompli' of benefit to the towns.

The government position was summarised by Kenneth Robinson MP (Hansard 1969). Although the issue 'is by no means a simple one' several of the alternatives are not immediate possibilities. 'The Wash scheme, desalination and the artificial recharge of aquifers were aimed at a later generation.' 'The water engineer who casts this covetous eyes upon our pleasant lakes and valleys does so in response to the demands we press upon him every time we turn the bathroom tap.' Therefore the government was 'regretfully in favour' of construction at Empingham.

Following its second reading in the Commons the Bill spent 9 days in Committee where the promoters' case held up well and Empingham emerged as a 'deplorable and regrettable necessity', 'one of the melancholy consequences of the relentless demands of the urban dweller'. The Committee was however sufficiently impressed by the petitioners' case to call for further investigation of the potential of the Wash as a way of being better prepared 'next time' and supporting its hope that this would be the last great lowland reservoir. It is not known whether this comment was meant only as a palliative for the petitioners or more as a warning shot across the bows of the Water Industry – although later in the same year the Lords were to come out with some very strong words about their likely future attitude to proposals for the development of shallow reservoirs. However it is interesting to note the Commons Private Bills Committee's enthusiasm for the potential of the Wash for water supply. The Committee reported back to the House that there was the prospect of 'the biggest advance since the discovery of oil' if the Wash was used for water supply and that it only requires the need to arise for one more scheme of Empingham proportions for the cost comparisons to begin to look favourable.

The third reading in July 1969 was something of a formality as by that time the crucial decisions had effectively been taken and the government view expressed. The petitioners were running out of steam. In retrospect it is worth noting the government's comments as a pointer to the eventual further reorganisation of the Water Industry. The government spokesman stressed the advantages of a regional approach to supply – demand forecasting and hoped for the flexible use of the reservoir 'to meet the short-term needs not only of the promoters, but also of neighbouring areas'.

Deeply conscious of the financial inability of some petitioners to take their case further, the Lords (House of Lords 1969) subjected the Bill to further scrutiny and deliberately widened their debate to cover the need for the water, alternative possibilities, projections of quantity, programming and the likely local impact. In effect they regarded the Bill as being of more than local significance and took the opportunity to review the national approach to water resources planning, taking the unusual step of setting up a select committee for the purpose. The erstwhile petitioners got value for money in the Lords with all the same points, but no significant new ones, being debated at length. The Lords seemed to be keen to issue a warning that lowland reservoir Bills would be increasingly difficult to pass unless the water industry displayed more urgency in seeking alternatives. Of course the 'not very cordial' reception which the Bill received in Parliament made little difference in the end. The compelling case prevailed and received the Royal Assent but no doubt the debates were instrumental in developing attitudes to the changes in Water Industry organisation which were to follow.

Could history repeat itself?

In a period when Water Authority forecasts of demand are falling it seems unlikely that proposals for major water supply schemes will carry the same urgency in the near future. Gone are the days of optimistic long-term prospects of economic growth. Kenneth Lewis's 1969 reservations about economic development prospects in Corby have now assumed an ironic accuracy. For the time being the position is less concerned with meeting the planners' growth expectations by promoting major short term expansion and more with efficient deployment of the recent (thrifty) investment in what critics may see as surplus capacity based on the optimistic forecasts of a previous age.

There are a number of reasons why the decision-making process on any 'new Empingham' would now be different from the 1960s experience:

i) The changed circumstances of the reorganised water industry. With the 'easy' reservoir sites already developed and ground water not always available, greater attention is being paid to the conjunctive use of resources, to demand management and to promoting efficiency and flexibility by linking resource zones together as a means of satisfying demand in the foreseeable future.

ii) Changing attitudes to resource exploitation generally. Parliament has already warned that it will be more difficult in future to sanction the construction of shallow lowland reservoirs. At local level developers now have to reckon with more open government all round; a participatory democracy, weaned by the planning system onto the consideration of alternatives and an expectation of greater consultation generally. It may be going too far to suggest (but examples such as Belvoir Vale Coal, Windscale Reprocessing and John Tyme's questioning of transportation policy suggest not), that proposals for natural resource development and technological change which interact widely with other interests pose questions about the future shape of society. Should there be more emphasis on the control of demand rather than maintaining an open-ended commitment to increase supply? Environmental pressure groups are adopting an increasingly 'professional' approach. A changed consciousness has become apparent, the origins of which can be traced at least as far back as Colin Buchanan's voice of dissent on the Roskill Commission (Hancock 1980), especially where intangible environmental and social factors are pitted against persuasive arguments based on measurable economic gains and losses.

iii) Growing scepticism about public inquiries. Government has tended to restrict the scope of public inquiries to a consideration of land use factors as they relate to particular sites and has shown some reluctance to extend their scope into discussion of 'need' (for water, for coal, for nuclear energy, for roads) and into the questioning, rather than the interpretation, of policy. Although government is beginning to yield some ground in this area it is doubtful whether the next large-scale reservoir proposal (if it was not handled by Private Bill) will be subjected to the impartial arrangements of a Planning Inquiry Commission rather than the conventional preference for an inquiry where the Department of the Environment gets very close to being both judge and jury. Despite arguments in favour and a statutory capability going back many years, no major development proposal has been subjected to a PIC. Adoption of this approach would enable a two stage approach – policy/principles before site details – to be adopted and would allow the adversarial setting of the public inquiry to give way to a more investigatory approach.

iv) The advent of Environmental Impact Analysis. Practitioners may feel that there is nothing new in EIA. British planners certainly find much of it very familiar but tend to regard it as more vital to countries which lack our well-developed planning system. Nevertheless the UK may be on the verge of having to fall into line with the EEC as a whole and accept mandatory EIA for major projects. A draft directive (EEC 1980) has already been issued. In a way the concept has a lot to commend it especially if, as I have suggested, both public concern and professional worries tend to question the ability of current procedures to maintain and enhance the quality of the environment. The North West Water Authority has already used consultants to carry out an EIA to help in the choice between alternative reservoir sites. 'Post-development audit' research is underway, using many different case studies, to evaluate whether EIA predictions have been borne out in practice. Pressure groups are arguing for EIA to become a formal part of the planning process. Although the government has so far argued against its introduction, the House of Lords European Committee has now broadly endorsed the draft directive. Whether it becomes mandatory or not the broad concept of a balance sheet of impacts may well become a standard feature of the resource development decisions of the future.

Conclusion

Rutland Water is the product of a period of more optimistic growth expectations than could be forecast today. The decision making process was marked by a short term urgency to expand water supplies by conventional means while investigating longer-term innovations which were assumed to be eventual realistic options. Opposition to the reservoir was vociferous and well organised but constrained by a shortage of finance and technical expertise so that, ultimately, it could not rival the compelling case of the promoters. Eventually it declined into a resigned acceptance and now that the reservoir has become a maturing feature of the landscape, is rarely voiced. Nevertheless a review of decisions made over a decade ago and, more significantly, how they were made, is instructive. During the 1970s public attitudes to resource exploitation and environmental change became much more critical, scepticism of the way 'big' decisions were made began to grow and environmental impact analysis began to gain ground as a desirable advisory tool of decision making. Our planning system and utility services need to be adaptable if they are to accommodate these trends.

References

EEC Brussells, 1980. Proposal for a Council Directive concerning the Assessment of the Environmental Effects of certain Public and Private projects C169/14.

Hancock, T., 1980. The Big Decisions. Who Decides? Town and Country Planning Summer School Exeter. Report of Proceedings: 26–33.

Hansard, 1969. 781: 554–618, 785: 823–830, 786: 313–344, 787: 1691–1692.

House of Lords, 1969/70. 305: 563–618 and 307: 1099–1114.

McLellan, A. G., 1963. The growing demand for water and its impact on land use. J. Town Planning Inst. 49: 60–67.

Water Resources Board, 1966. Water Supplies in South East England.

Discussion

Introducing his paper, Mr. Lawson emphasised that it was a non-technical, historical account of the decision-making process placed in the context of the larger scale, regional planning policies which were in vogue during the 1960s. The idea of the giant new reservoir at Empingham was a major part in the concept of 'stepwise' increases in provision to keep water supply ahead of demand – seen as an essential part of the infrastructure necessary for our growing affluence and industrialisation.

Although the reservoir was seen as part of this industrial infrastructure, it became, in its passage through the Houses of Parliament, an example of the changing way of thinking about water supply. Members of Parliament examining the private Parliamentary Bill were alerted to new water supply solutions and expressed the hope in their report that this would be the last great lowland reservoir in Britain.

In addition to the changed ways of looking at water supply there have been considerable changes in the British planning process since the decision to approve the reservoir was made. In particular, we have more open 'participatory democracy' at local level and in the examination of major natural resource problems. Environmental Impact Assessment (E.I.A.) could become a formal requirement. A draft E.E.C. Directive has received the support of the House of Lords and the practice is becoming more common – e.g. the use of E.I.A. by the North West Water Authority in its examination of possible new sources of supply in the future.

The discussion was opened by Mr. Knights, Anglian Water Authority, who complimented Mr. Lawson on his precise and accurate account of the events of a decade ago. He was particularly interested in the changes in the planning process and especially the advent of E.I.A. He asked Mr. Lawson to enlarge upon its practice in the U.K. and to indicate whether, in his opinion, E.I.A. forecasts could pose as many problems as the forecasts of water demand have in a similar 10–12 year time span?

Mr. Lawson pointed out that there was not a great deal of experience of E.I.A. as such in this country but several 'post-development audit' studies were underway in this country which would help to answer the second question. The North West Water Authority had used E.I.A. in its search for new water sources and had looked at four options – including estuarine storage as well as inland sites. It had employed consultants for just over 18 months and had involved a supervisor and several disciplines looking at scientific, recreational and social aspects of the sites – matters which could not always be costed as well as those more conventional aspects of water supply which could.

Mr. Langford, Anglian Water Authority, pointed out that the Empingham project had cost around £33 million and this would almost certainly have been much higher had E.I.A. delayed the scheme – this was something which always had to be borne in mind.

Mr. Herrington, Leicester University, asked Mr. Lawson to expand on the differences between a Public Inquiry and a Planning Inquiry Commission.

A Public Inquiry, which was the usual way of resolving planning and resource development proposals at present, was run by a single Inspector (perhaps supported by a specialist assessor) appointed by the Secretary of State for the Environment. The Inspector reported back to the Secretary of State with his recommendation and the latter then made a decision. On minor matters the power to make the final decision was now increasingly delegated to Inspectors. Usually the Public Inquiry is site-specific and has narrow terms of reference; its procedure has given rise to much criticism in recent years when arguments about Government policy have been brought up but not been accepted.

Planning Inquiry Commissions have been on the Statute Book for some time but have been little used. Such a device, however, could readily lend itself to some of the major resource decisions which Britain faces, without the apparent bias of the Public Inquiry system. A Commission would still be appointed by the Secretary of State, and would report back to him, but there would be three Commissioners. This would give greater confidence of impartiality. The Commission could have a wider, two-stage approach to problems; firstly examining principles, need and policy, and secondly dealing with site specific details. As well as having a wider field of investigation, the Commission would have the advantage of moving away from the quasi-judicial adversarial atmosphere of the Public Inquiry towards a more 'round table' investigatory approach and be in a position to call for specialist studies to be undertaken. It is not clear why Governments have been so reluctant to use Planning Inquiry Commissions but it might well be associated with the delay inherent in such elaborate procedures and the greater weight of the opinions of three experts instead of one where it may be politically expedient to disregard them.

Dr. J. Bullock, Leicester University, asked to what extent the early opposition to the project had influenced the subsequent environmental policies of the Authority? Mr. Lawson felt that, as there was no site-specific detail available at the time of the decision to proceed, the opposition was only able to obtain broad assurances of the policy of the promoters – such as restraint rather than open commercialism, employment of a landscape architect and broad indications of the nature of recreational provision. The reasonable attitude of the promoters once the scheme was approved gave the erstwhile opponents a chance to have a greater influence upon details than might otherwise have been the case.

Escaping from prison: an economic reassessment of Rutland Water

P. R. Herrington
Department of Economics, University of Leicester, England

Keywords: water resources, forecasting, economy, pricing, demand-management

Abstract

Recently in the water industry interest has been growing in solutions to widening demand-supply gaps. These include demand-management elements. The paper examines the 1970 decision by Parliament to build Rutland Water and concludes that, despite over-predictions of water demands, the very few hard facts then available concerning water economy measures would probably not have prevented the reservoir being authorised, had they been taken into account. By 1981, however, dramatic reductions in expected population and per capita public water supply growth had occurred, with much more evidence available concerning the scope of demand-management. It is suggested a 'similar' decision on economic grounds would probably have been a negative one.

Introduction

Despite important changes in the last ten years, the water industry in Britain remains largely insulated from some fundamental issues of resource allocation. Problems of allocative efficiency at the level of the individual industry usually find expression in two forms: decisions about the size of the capacity to produce (the investment problem) and at what output to operate the inherited capacity (the use problem). The information necessary to solve such problems derives from both sides of the economist's familiar scissors framework: from the supply side, where the resource and other social costs of different levels of capacity expansion and operation should be properly identified and wherever possible evaluated, and from the demand side, where the social benefits of altering capacity or the use of existing capacity should ideally be made explicit, most obviously (but not necessarily) through the observation of what consumers are or would be prepared to pay for the particular service under consideration.

'Requirements' and 'supply-fix'

No good reason has been offered as to why the public water supply should be exempt from the profound questions posed by adoption of this framework, questions which rapidly lay bare for analysis the water industry's investment, output and pricing policies. Consider, for example, investment. All long-term public water supply demand forecasts published in Britain in the post-war years have suggested significant growth for twenty or thirty years ahead, even discounting the effect of population growth. The orthodox response has been for the industry to treat the demands as 'requirements' (Warford 1966) and argue that supplies must therefore follow suit, at the very least matching likely demands and hopefully also providing a margin to spare. Over-insurance has been presented as prudent housekeeping, while the statutory water undertakings' secure monopoly position and largely unfettered power to tax have ensured that no problems have arisen in raising the large amounts of finance necessitated by adherence

Hydrobiologia 88, 27–42 (1982). 0018-8158/82/0881-0027/$03.20.

to this 'supply-fix' philosophy (Herrington 1974a).

The dominant engineering tradition in the industry has until very recently perpetuated this approach without serious debate. What has been defended by some as simply the continuing application of well-tried engineering principles has appeared to others, however, to operate as an intellectual straitjacket, inhibiting the industry from responding sensibly to changing economic, financial and environmental circumstances.

To any detached observer interested in wise allocation and conservation of resources, such an automatic adoption of the supply-expansion approach to water resources planning has no rationale to commend it. Certainly the emotional idea that virtually all domestic water use is essential to private and public health cannot be sustained in serious argument; indeed, recent estimates of 'essential' water use in Britain range from 1 litre per head per day (lhd; Cambridgeshire County Council 1980) to 10 lhd for camping (Walker 1977), out of average domestic consumption in England and Wales of 100–120 lhd in 1975 (Central Water Planning Unit 1977).

Solutions to demand-supply gaps

Once the *demand* for water is accepted as a more helpful notion than water *requirements*, then supply-expansion takes its rightful place as just one of the ways in which future supplies and demands may be reconciled. It lies alongside solutions founded on demand-management and those in which supply-expansion and demand-management may be viewed as complementary. The do-nothing response is appropriately regarded as a special case of demand-management, resulting in the longer term in steadily decreasing supply reliability or service quality, or more and more stringent rationing. In any particular situation, the preferred solution will depend on the feasibility, acceptability and costs and benefits of the range of supply- and demand-based alternatives. All these factors may vary over time, space and type of consumer. Mixed solutions might be expected to figure prominently in a low-growth economy valuing increasingly its remaining agricultural land, which characterised Britain up to early in the last decade and probably will do so again in a few years' time.

Looking back over the first decade of Rutland Water conveniently raises both problems identified at the outset. At the beginning of the 1970s Parliament was considering the *investment* problem: should Rutland Water, then in the guise of the Private Bill known as the *Welland and Nene (Empingham Reservoir) and Mid-Northamptonshire Water Bill,* be given the go-ahead? Ten years later, with the reservoir constructed and commissioned, important questions were being raised about the *use* problem: if two areas a short distance apart appear to have too much and too little water respectively, might it be a good idea to join them by pipeline and operate the surplus-area capacity at a higher output? Because of scarcity of time, the remainder of this paper constitutes an attempt to explore only the first of these problems,[1] the investment decision. In particular,

(i) given the information available to Parliament at the time (1969 and 1970), was the decision to pass the Bill the 'correct' one on economic grounds?

and (ii) given the extra information available a decade later, would a 'similar' decision now be resolved differently?

The 1970 decision on Rutland Water

The Private Bill authorising the construction of what was later to be renamed Rutland Water was examined by Select Committees of the House of Commons in April/May 1969 and the House of Lords in February 1970. The more thorough consideration was given in the Commons Committee (where there was opposition, and thus cross-examination of the proposers' witnesses), and the nine volumes of the Minutes of Proceedings constitute the evidence on which much of the following argument is based (House of Commons 1969; henceforth referred to as *Minutes*).

At the end of the 1960s no proposed reservoir had been seriously contested on the grounds that need had not been established. The supply side of the industry was fragmented (in 1969 nearly 250 local water undertakings were still supplying water in England and Wales), engineers called the tune, and accountants picked up the bills, advised on debt repayment and worked out the charges. 'Ignorance is Bliss', killed off by Light Programme planners some twenty years earlier, still characterised the water industry's attitude towards how its product was used, by whom and for what purpose.

Population forecasts

The demand forecasts which lay behind the promotion of Empingham were of a poor standard but no worse than others being used around that time.

National and regional population forecasts in England and Wales were being marked down steadily each year from 1965 onwards; thus the projected 1964–81 increase in the East Midlands standard region was cut from 766 to 506 thousands over the 1965–69 period, while the corresponding increases for a Water-Resources-Board – defined South-East England were 3.19 reduced to 1.90 millions. It is impossible to ascertain the extent to which these rapid revisions had been reflected in the population projections presented by the promoters, the dominant elements of which (representing 88% of the reservoir's 'relevant area' population in 1968) are reproduced in rows (1) and (3) of Table 1. There is evidence, however, of the Peterborough forecasts – part of the South Lincolnshire Water Board (SLWB) area – being reduced in the course of the Commons hearings as a result of Rutland County Council's petition opposing the Bill (*Minutes,* 14 May, p. 33).

Table 1. Population, actual and projected, 1968–2001 (millions).

	Actual		Projected		
	1968	1978	1981		2001
(1) MNWB	0.348	–	0.570	(in 1969)	0.737
(2) AWA(N)	0.462	0.545	0.556	(in 1980)	0.677
(3) SLWB	0.241	–	0.350	(in 1969)	0.430
(4) AWA(P)	0.236	0.279	0.295	(in 1980)	0.381

Code: MNWB = Mid-Northamptonshire Water Board
AWA(N) = Anglian Water Authority, Northampton division
SLWB = South Lincolnshire Water Board
AWA(P) = Anglian Water Authority, Peterborough division.

Sources: rows (1) and (3): *Minutes,* 29 April, various pages, plus author's 1981 estimate for SLWB.
rows (2) and (4): WDU 7 returns to Water Data Unit, and 1978, 1981, 2001 figures kindly provided by Anglian Water Authority.

The likelihood is that the expected population increases had been overestimated, since it is known that the four statutory water boards in the 'relevant area' made their own population forecasts based on figures provided by local planners (*Minutes,* 29 April, pp. 5–6, and 6 May, pp. 15–16). Such a procedure not only suffers from a double lag effect (from regional to local planners' to water boards' revisions); it is also known to have generally resulted in multiple-counting of expected population additions in regions with expected net immigration, since local planners each feared the migration or overspill would be to their particular area. Thackray has drawn attention to these dangers and a dramatic example of this type of overprediction in the West Midlands in which the local authorities' aggregate population growth forecast for 17 years ahead was seven times that predicted by the Office of Population Censuses and Surveys (Thackray 1978). Furthermore, it is difficult to reconcile the 1968–81 population forecasts for the reservoir's 'relevant area' presented to the House of Lords Select Committee in 1970 (House of Lords 1970; p. 5) with the much lower 'official' 1971–81 forecast increase for the East Midlands then current, despite the large London overspill developments expected in the area.[2]

Per capita demand forecasts

The water demand forecasts were crudely presented to the Commons Committee, no distinction being drawn between unmetered and metered consumption and no attempt being made to break down unmetered consumption into its principal components. Table 2 brings together, in rows (1) and (3), the only per capita projections that the *Minutes* record,[3] and also Anglian Water Authority figures for the two divisions that most nearly correspond to the two large water boards in the 'relevant area' in 1968. Abstracting from all post-1968 actual data, the Mid-Northamptonshire Water Board (MNWB) projections to 1981 appear to be very high, the forecast annual rate of growth (in ghd) being one-third greater than that experienced over 1949–68, double that recorded for AWA's Northampton division over 1961–68 and one half as much again as that forecast for SLWB. The 2001 forecast for MNWB was, it is known, derived from steady, but declining, per capita

Table 2. Per capita public water supply, actual and projected, 1949–2001 (gallons per head per day = ghd).

		Actual				Projected		
		1949	1961	1968	1979	1981		2001
(1) MNWB		23	–	51	–	77	(in 1969)	94
(2) AWA(N)	UM		29.7	35.6	45.8	45.1		48.4
	M		12.5	13.9	14.0	15.0		16.2
	Σ		42.2	49.5	59.8	60.1	(in 1980)	64.6
(3) SLWB		–	–	51	–	67.5	(in 1969)	97.5
(4) AWA(P)	UM		29.7	33.5	44.6	45.1		48.4
	M	–	13.9	18.9	17.4	16.4		20.2
	Σ	–	43.6	52.4	62.0	61.5	(in 1980)	68.6

Code: see Table 1. UM = unmetered consumption. M = metered consumption.
Sources: rows (1) and (3): *Minutes*, 29 April, various pages, and 6 May, p. 32.
rows (2) and (4): actuals from WDU 7 returns to Water Data Unit, and projections kindly provided by Anglian Water Authority, from 1978 base and consistent with forecasts in AWA's Annual Plan for 1981.

growth over 1981–2001; how the even higher SLWB forecast was derived is not recorded. All these projections were broadly comparable with those for the Welland and Nene River Authority area in the Water Resources Board's regional study on South-East England (Water Resources Board 1966). Opposition counsel pointed out that they were higher than even the upper limit (88 ghd) of a relatively sophisticated Ministry of Housing and Local Government range forecast for South-East England published as Appendix II of Water Resources Board (1966), and also 6 ghd higher than a planning figure recommended specifically for New Towns in a 1964 Binnie & Partners report on the water resources of the Great Ouse area. The Engineer and Manager of MNWB responded by dismissing such variation as 'simply a difference of opinion between engineering experts' (*Minutes*, 6 May, p. 32), implying there was nothing more to be said on the matter.

The rationale given for such large increases was not very firm, but in the context of 1968 it made some sense. The huge projected overspill and new town developments in Daventry, Northampton, Corby, Wellingborough and Peterborough meant large numbers of new houses, and the association between new dwellings and higher domestic water use was already appreciated, albeit in an unquantified fashion. Counsel for the promoters spoke of four 'chief ways' in which water consumption was expected to increase in the future:

(i) a 'general improvement in living standards, through a ready supply of hot water and an indoor lavatory';

(ii) 'increased use of household machinery for washing clothes and dishes and, e.g., grinding garbage';

(iii) washing cars;

and (iv) water for recreation, from 'gardening to paddling pools' (*Minutes*, 29 April, p. 8).

Although there was not a shred of quantitative analysis to back up these suggestions, they would have had in 1968 a plausible ring about them. Indeed, insofar as the promoters were claiming that, in an environment in which marginal gallons of water had a price to the unmeasured consumer of zero, there was every expectation that unrestricted demands would carry on increasing strongly, it is possible to have much sympathy with the argument. But of course the proposition was never presented in that form. The promoters' demand forecasting methodology had in fact been outlined earlier in the proceedings by counsel: 'To a great extent the water engineers are guided by increases in the past, but they must also take into account, and do take into account, factors arising in the future.' (*Minutes*, 29 April, p. 8).

A gloss was added to the effect that 'the Water Resources Board itself has carried out considerable research on this topic', but that was (in 1969) a complete fabrication. The Engineer and Manager of MNWB claimed to have undertaken metering experiments regarding domestic use, but the only examples he could cite were: (i) four meters in his own house, where, he admitted, his own family had been trained to be cautious with water, and (ii) a barracks, where the resident population could be easily measured but which turned out to consume 106 ghd! Questioning by the Committee revealed that this figure included an indeterminate amount of water used for washing down army vehicles.

Amending the demand forecasts

My own summing-up of the discussion so far, aided generously by back-of-the-envelope calculations of the type most at home when parented by engineers, is that a more careful consideration of, first, the local implications of changing 'official' population forecasts in the late 1960s, and, second, the water demand trends of the 1960s, would have produced a demand-supply gap of 50 million gallons per day – the projected yield of Empingham – not in 1991, as the promoters claimed, but between five and ten years later.

It is most unlikely that this scale of lengthening of the period to full yield would have in itself defeated the reservoir proposal in the national economic and financial climate of the late 1960s. One possibility, however, is that the Committee might have reacted by recommending a smaller-scale development at Empingham. The four M.P.s could not, in any case, have been greatly impressed by part of the evidence of the promoters' leading engineering witness, Mr. T. E. Hawksley, who probably gave more away than he intended when stating that 'In an area where reservoir sites are scanty and demand is high and foreseeable future demands are going to grow, one wants to put in a reservoir as large as the country will permit, within reason, and we put the gross capacity of this one at this figure, as we consider that is as large as this area of country can accommodate.' (*Minutes,* 30 April, p. 41). An obvious case of the supply-fix mind at work!

Secondly, it is just possible that with a less dramatic deficiency emerging in the 'relevant area' before the earliest time that the Wash barrage could have been completed (in the early 1980s, it was then thought) the Parliamentary Committees might have been tempted to consider favourably solutions other than the supply-expansion variety, at least for the short term. This leads on naturally to a consideration of the Commons Committee's treatment of demand-management.

Demand-management possibilities

As might be expected, the *Minutes* reveal little serious examination of demand-side measures. In part this follows automatically from the lack of any information about the components of unmetered (and especially domestic) water use; without a detailed breakdown of consumption it is difficult to estimate the savings that might be made through, for example, the installation of economy variants of domestic water-using technology. Limited information was in fact available in the water supply industry in 1968 concerning two demand-side measures: domestic metering and dual-flush W.C.s.

Domestic metering

The small Commons Select Committee included one M. P., Mr. Stainton, who revealed much enthusiasm for the pricing mechanism on a number of occasions. His first interjection, when the MNWB Engineer and Manager (Mr. Brown) was in the witness box, was refreshingly direct: 'We have heard nothing so far about the pricing policy of water. It strikes me that this is fundamental to demand. Is it your intention to develop this point?' (*Minutes,* 6 May, p. 15). The record of the Proceedings shows that this contribution was completely ignored. Counsel cross-examining Mr. Brown (Mr. Ryan) immediately raised with his witness the fact that a previous answer had not been heard by the Committee reporter, and there followed an exchange on the matter. It was another three pages of *Minutes* before Mr. Stainton raised the matter again, asking for estimates of the 'elasticity of demand for water in different pricing structures'. The witness gave the conventional engineering answer, that the method of charging would not affect the demand for water (thus contradicting every investigation of the impact effect of metering ever reported; see Herrington & Tate 1971), and the same response was offered later by another witness, Mr. S. W. Hill, a financial adviser to MNWB.

Mr. Hill's evidence on metering was more thorough, and he drew the Committee's attention to a 1963 government report on water charges (Central Advisory Water Committee 1963), which quoted annual additional domestic metering costs of £2 but gave no source. The figure apparently applied to the universal metering of an area over a short period of time. The report was ambivalent in its attitude to metering; thus para. 19 included the statement that 'Almost all the evidence brought to us has suggested that economy is unlikely to be achieved by a reasonable system of charges whilst the cost of water remains at anything like its present low level', while in para. 26 metering was considered to be the 'ideal solution' for all supplies. Two years later Bird and Jackson, following their investigations in Malvern (where domestic metering has been in force since 1871), published their cost estimates for metering new homes: total annual additional costs, including capital, maintenance and extra reading, billing and collection, of £1.36 (Bird & Jackson 1965). The Commons Committee was not told about this additional evidence.

It must be accepted, however, that the lack of any firm estimates of the likely consumption savings arising from domestic metering in the United Kingdom meant that no quantification of the net benefits of even selective metering (e.g. of new dwellings only) was or would have been possible.

Dual-flush water closets

Definitive information about water savings following the installation of dual-flush W.C.s. became known in the industry in 1964, although they had been first manufactured in this country in the early 1930s. Two researchers had reported the results of dual-flush experiments undertaken in 36 Birmingham Corporation maisonettes and flats in the early 1960s (Sobolev & Lloyd 1964). Savings of 26% and 27% in W.C. water use were calculated in separate trials, equal to about 1½ ghd (7 lhd). In 1965 Leicester Water Department had become the first undertaking in the country to introduce a bye-law requiring all new and replacement installations of the washdown type to be dual-flush. Both these developments were ignored by the Committee and witnesses, although the Engineer and Manager of Leicester gave evidence at the proceedings. In the context of the expected growth of demand in the 'relevant area' the savings from even a total switch to dual-flush would not have been expected to defer Empingham's full yield date for more than a couple of years on the evidence then available.

Waste control

Waste of water – meaning water that leaks from the distribution system and therefore reaches no consumer – did not get much more attention. Mr. Brown estimated unaccounted-for water (the more precise but less controversial term used within the industry) in MNWB as 10% out of 51 ghd, claiming a 'fairly good' waste prevention service and implying there was little room for savings. The Committee did not pursue the question.

Scope for demand-management

The inevitable conclusion is that in 1969 there were few hard facts concerning water economy measures. Despite the pronounced bias of all witnesses to supply-augmentation solutions, it is doubtful whether, even if the Committee had been in full possession of those facts, it could have been expected to do any more than recommend the introduction of a Leicester-type bye-law throughout the 'relevant area'.

The 1970 decision: conclusion

The evidence has suggested that in 1969 the Commons Committee might have been persuaded that the demand forecasts were too high and there was a relatively small potential role for demand-management. Appropriate 'corrections' would have so reduced the future excess demand for water that the full yield of Empingham might not have been taken up until towards the year 2001, some ten years after the likely 'full yield' date accepted by Commons and Lords Committees. My judgement is that the uncertainties then surrounding the Wash barrage were such that these Committees would in these revised circumstances have arrived at a similar decision: they would have recommended that a reservoir at Empingham be authorised. It might, however, have been on a smaller scale.

How different is 1981 to 1970?

Having examined in detail the 1970 decision, it is a natural consequence to ask: knowing what we know now, a decade later, should the decision on Rutland Water have been a different one? In an important sense such a question is (appropriately, some would say) wholly academic, since if we take on board all the supply and demand developments up to 1981, including the establishment of a grid system in Anglian W.A. for pushing water supplies around the region, then it is of course impossible to compare like with like. For a latter-day Rutland Water would be proposed today in the context of the 'needs' of the region as a whole rather than those of the limited 'relevant area' discussed hitherto. Nevertheless I propose to ask the academic question, since a framework is thereby retained for assessing further changes in demand forecasts and the scope for demand measures.

Population forecasts

The decline in population forecasts continued up to and including the 1977-based OPCS set (Office of Poulation Censuses and Surveys 1980), although national projections are now being revised slowly upwards as the return to the long-term fertility rate occurs more rapidly than had been assumed a few years ago. Whereas the Commons Committee had been advised to expect a population increase over 1968 to 2001 of 98% for the combined MNWB and SLWB areas, the latest forecasts from Anglian W.A. suggest a 1968–2001 increase of only 52% for the combined Northampton and Peterborough divisions (see Table 1).

Per capita demand forecasts

In the second half of the last decade a number of water authorities - including Anglian - mounted important domestic water use studies. From a wide variety of survey and measurement techniques, there emerged a remarkable consistency of current per capita consumption estimates in the 100–120 lhd range, with a best single estimate of 115 lhd (Herrington 1979). Given what had already been established about unmetered commercial and miscellaneous use (fire-fighting, hydrant-testing, mains-scraping, etc.), the immediate implication was to confirm the controversial conclusion of the Technical Working Group on Waste of Water (DoE/NWC 1976) that waste was approaching an average of 25% of the public water supply in England and Wales. A detailed picture also emerged from the studies of the components of domestic water use, especially in Malvern and Mansfield (Thackray *et al.* 1978). Late in 1980 the National Water Council estimated average household use in England and Wales as shown in Table 3.

Now that the industry has begun to learn about the use of its own product, it is able to engage in more sophisticated forecasting. By making forecasts of the ownership of household water-related appliances and habits, the frequency of use and the water consumption associated with a single 'use' of each appliance or habit, it is possible to construct a statistical picture, often embodying a range of figures, of what future domestic demand may look like. This has three advantages over the traditional method of forecasting presented to the Commons Committee in 1969: first, at the very least it provides a check on extrapolation techniques so that if wide differences emerge between the two the forecaster is forced to face the question why (and that may turn out to be a productive exercise); second, the components method is, quite simply, more 'rational' (there is no reason why per capita unmeasured consumption should continue to increase in the future as it has done in the past, which is the essence of extrapolative methods); and third, attention is easily directed at the scope for demand-management measures.

Virtually all water authorities now grant a role to the components method in their long-term forecasting, and most have consequently revised their end-of-century unmeasured forecasts downwards, sometimes dramatically. The Anglian W. A. now uses a mixture of extrapolative and components methodology, and the resulting forecasts for the Northampton and Peterborough divisions are included in Table 2, above. They reveal enormous decreases on the corresponding projec-

Table 3. Average household water use in England and Wales, 1980.

Type of use	Amount (lhd)	(%)
W.C. flushing	40	(33)
Personal bathing/washing	30	(25)
Laundry	14	(12)
Washing up	12	(10)
Outside use	4	(3)
Drinking, cooking, cleaning etc.	20	(17)
	120	(100)

Source: (National Water Council 1980a).

tions for MNWB and SLWB in 1969, but three factors should be borne in mind: first, the experience of the 1970s, when metered consumption has not grown at all in many authorities and divisions, is a new factor; second, Anglian has already built into these 'demand' forecasts some significant demand-management in the form of expected waste control (waste 'demands' figure as part of unmetered consumption); and third, it is believed that planning margins have not been allowed for in the Table 2 1980 projections. The expectation is, nevertheless, clearly of a very much slower increase in per capita water use[4], for even a planning margin of 10% for 2001 would produce an overall per capita figure of only 72½ ghd for that year. This would represent a 40% increase on the 1968 figure, instead of the 85% increase implicit in the 1969 forecasts.

Amending the demand forecasts

The combined effect of these revisions to the long-term population and water demand forecasts is obviously very large. From the figures in Tables 1 and 2, it may be calculated that the demands for the public water supply would 'at present' be expected to increase in the old MNWB and SLWB areas by about 30 mgd over 1968/2001 (or 35 mgd, if a 10% planning margin is associated with the 2001 forecasts), as compared with the 80 mgd increase expected in 1969. The *overall* demand for 2001 is thus now forecast as a *little over half* what it was forecast in 1969; the *increase* from 1968 to 2001, only slightly more than the excess demands forecast for 2001, is now forecast as at most 44% of the original expectation.

Demand-management possibilities

It is not only in the area of demand forecasting that additional knowledge has begun to have an impact. The known scope for demand-management has also broadened considerably in the last decade, as

(i) the upheavals of the 1974 reorganisation brought into the water authorities specialists interested in economic concepts and analysis;

(ii) water use studies have pinpointed the important components of demand;

(iii) more technical advances have been researched, tested and publicised;

and (iv) financial constraints originating at the level of the national economy have begun to bear strongly upon the industry.

It would be impossible to cover adequately in this paper all the measures which might be placed under the heading of demand-management. A full description of these possibilities, including low-water-use washing machines and dishwashers, atomiser sprays for showers, and spray taps (in public buildings and office/factory cloakrooms) will be found in Rump (1978) and is summarised in Herrington (1979). Attention will be concentrated here on three ways of influencing demands: waste control, water closet technology and domestic metering.

Waste control

Waste control has in the past retained a Cinderella role in the water industry. In an environment so obviously dedicated to supply-fix, there have been few plaudits for people and policies dedicated to reducing the throughput in the public water supply system. It was, in any case, still widely believed in the industry up to the mid-1970s that wastage was very low and that therefore even intensive leakage detection and control could not provide much of a return. For example, the Director of the prestigious Water Research Centre was claiming as late as May 1976 that leakage of more than 5% to 10% of the public water supply was found 'only in exceptional circumstances' (letter to The Guardian, 19 May 1976). The DoE/NWC Technical Working Group report and the demand studies already referred to changed all that, however, so that by the end of the last decade there could be no serious doubts as to the physical (i.e. water use) savings that might result from sophisticated (and therefore perhaps costly) 'detection and cure' exercises. Thus Yorkshire Water Authority was, in its 1978 series of 'component' demand forecasts, predicting that per capita unmeasured consumption would increase from 192 lhd in 1977 to 246 lhd in 2001 if unaccounted-for water remained at the estimated figure of 27%, but only to 224 lhd if the unaccounted-for element could be reduced to 18%: a 40% reduction in the forecast increase[5] (Yorkshire Water Authority 1978).

The important question that remains is the extent to which it is economically desirable to reduce waste. It is easy to state the principle that waste detection and alleviation should be pursued in an area up to the point where the difference between the total benefits of leakage detection and cure and the total costs is maximised. The real problem is to identify and evaluate correctly these costs and benefits. An important step forward in this respect has now been taken with the publication of the final report of the Technical Group on Waste of Water (DoE/NWC 1980), most of which comprises a manual for leakage control practice and determining leakage control policy. The policy section sets out for the first time a detailed methodology for the correct calculation, using discounting techniques, of the costs of different types (and intensities) of leakage control and the costs of leakage itself, expressed as operating resource savings and the benefits from deferring capital expenditures.

Presentation of data in this way leads to each possible 'level' of leakage control ('passive', regular sounding, district metering, waste metering, and combined district and waste metering) being associated with an overall total cost of leakage and leakage control denominated in £/property/year. Minimisation of total cost is formally equivalent to maximisation of net benefit (since the costs of leakage are the benefits of leakage control), and thus the optimal control 'level' may be identified for a given area. It is difficult to generalise, from the case studies in the report, concerning the scope for further economical leakage control in England and Wales. The Welsh Water Authority, however, was so impressed with the draft report that in March 1980 it decided not to bother to carry out financial calculations before embarking on further leakage control. It was satisfied from the examples quoted that district metering would be economical in virtually the whole of the authority and began to alter its policies accordingly (MacRae 1980).

Water closet technology

Rump has estimated that about 20% of the total public water supply in England and Wales is used in flushing W.C.s (Rump 1978). As long as waste nationally is more than 20%, then W.C.s must consume more than a quarter of all potable water which reaches a consumer, metered or not. The W.C. is thus the single most important user of water; it is an obvious target for water savings. Following the Birmingham Corporation experiments, at least four further studies of dual-flush savings were reported over 1971–1979 (Herrington 1979). Wessex Water Authority tested kits for converting existing W.C.s. in 100 properties in Melksham, Wiltshire, in 1977 and 1978, and found that dual-flush and variable-flush conversions showed average W.C. water use savings of 33% and 39% respectively. The Building Research Establishment (BRE) has also carried out tests on low-volume flush toilets and believes that user-control could be added, giving rise to even larger savings (50% and 70% respectively). Table 4 summarises

Table 4. Economics of alternative W.C. options as compared with standard 9 litre flush (late 1976 prices).

Option	Lump-sum cost (£)	Equivalent annual cost[a] (£)	Annual savings[b] % = lhd	Value of net social benefits[c] (£/year)
Dual-flush conversion[d]	5	0.48	30% = 10.8	1.98
User-controlled flush conversion	4	0.41	40% = 14.4	2.88
Low-flush W.C. (4½ l.) as replacement[e]	50	3.92	50% = 18.0	0.19
Low-flush W.C., with user-control as replacement[f]	52	4.07	70% = 25.2	1.68
Low-flush W.C. in new installation[e]	5	0.48	50% = 18.0	3.63
Low-flush W.C., with user-control in new installation[f]	7	0.63	70% = 25.2	5.12

[a] 5% discount rate and 20 year life assumed; also administrative costs of £2 per installation, equal to £0.10 p.a.
[b] Traditional mono-flush W.C. use of 36 lhd and 2.5 people per W.C. assumed.
[c] Follows *Paying for Water* (National Water Council 1976) in assuming long-run marginal cost of water supplies and sewage treatment to be 21 p/m³ at 1975 prices, = 25 p/m³ at 1976 prices.
[d] Conversion or extra cost of dual-flush replacement.
[e] Estimates from BRE experiments.
[f] Theoretical possibility; no trials or experiments yet reported.

these results and the economics of the alternatives.

It is concluded that (a) the savings are significant and (with two exceptions) proven, and (b) the economics of the various options make them extremely attractive. Of the fully developed options, the user-controlled flush conversion offers the largest net social benefits. The annual rate of return to a water authority from investment in this kit is estimated to be in excess of 50%.

High net benefits also occur as a result of the installation of time switches to restrict the automatic flushing of urinals. If a four-gallon cistern is flushing every 20 minutes for 24 hours of the day and 365 days of the year and then shut off for (i) one day a week, or (ii) 16 hours a day with complete shutdown at weekends, this could give rise to gross annual social benefits equal to (i) 50%, and (ii) three times, the capital and installation cost of the timing device (Phillips 1981). Since additional operating costs would be minimal, the economic return would be extremely high. The new Water Byelaw 58A requires all new automatic flushing cisterns for urinals to be fitted with such a device. Even though existing cisterns in the offices and factories of metered consumers will be adapted if the financial savings to be realised are properly identified, the economic case for compulsory adaptation is strong.

It may be objected that the mandatory conversion of existing W.C.s and automatic cisterns would be resisted by consumers. Certainly that is possible, but it is not at issue here. The question is rather whether the community can afford not to make use of such advances in technology, granted the size of the economic benefits which would accrue from their adoption.

Domestic metering

Under this heading progress in understanding is less marked. Contrary to the recommendations of the Third Report of the Jukes Committee, in only two water authorities – Wessex and Severn-Trent – are officers known to have attempted to undertake 'research and, where practicable, field experiments to establish a firmer factual basis for the assessment of the advantages and disadvantages of introducing metering in selected areas' (Department of the Environment 1974a, para. 5.16). In both cases the attempt was vetoed, by central government and authority members respectively. The two United Kingdom studies providing evidence as to the size of the impact effect of domestic metering have been in Fylde, where the estimated cutback was 14½% for 291 properties switched from rateable value to quantity charging in 1971–72[6] (Jenking 1973), and Mansfield/Malvern, where the difference in consumption due to metering at Malvern was estimated to be of the order of 10% in 1976 (Thackray *et al.* 1978). The former is probably the more reliable guide since it is an attempt to estimate the impact effect directly. Even that may provide an underestimate since it would not have included any savings arising from changes in household appliances (towards water-economy varieties) which would be expected as a consequence of a permanent move to unit quantity charging.[7]

No consensus obtains on the costs of domestic metering, the differences between a number of estimates covering the years 1962 to 1973 (Herrington 1974b), the 1976 NWC estimates (National Water Council 1976) and the 1980 NWC estimates (National Water Council 1980b) being considerably at variance with price level changes. Both 1976 and 1980 NWC estimates (the latter working out at £8 to £11 per annum at 1980 prices[8]) appear not to include estimates for the costs of metering new dwellings, when installation would be cheaper.

Indeed, it is a regrettable feature of the debate on domestic metering in Britain that the arguments have normally been centred upon either universal metering, which could probably be justified on economic efficiency grounds alone in only a few high-marginal-water-cost areas of the country, or optional metering, which may soon be adopted by every water authority in England and Wales in the name of equity but which does little for efficiency save in an extremely long-term perspective. The idea of selective metering, e.g. of new housing developments, has received little attention.

One of the major driving forces for optional metering has been section 30(5) of the 1973 Water Act, requiring water authority charges to be 'such as not to show undue preference to, or discriminate unduly against, any class of perons'. It is the risk of action in the courts that has probably hastened recent moves in this direction. The substantial NWC report on domestic charging, which strongly recommended optional metering, should be commended as contributing further to the slow breakdown of anti-metering prejudices in the industry

(National Water Council 1980b); its brief treatment of selective metering is less satisfactory, however, since that is the way which on efficiency grounds is clearly the best way forward, generating valuable savings and cost information on the way. The 'public' view may, of course, turn out to be that selective metering of new houses as they are built, not freely chosen by the individual owning or renting householder, is unacceptable. The economist cannot object qua economist to such a 'decision', so long as the extra resource costs of its adoption are known and accepted as the price that has to be paid.

Where there does seem to have been a significant change in the last decade is towards almost total acceptance of the idea that in time the metering of all domestic (and other) supplies must come. The sheer illogicality of trying to cater for all demands with a marginal value to the consumer of greater than zero, no matter what the social costs of provision, is bound to appeal as a more powerful argument as

(i) water demands become more 'trivial' as they are associated with luxury goods (sprinklers, hoses, swimming pools, etc.),

(ii) thus the idea of the public water supply as a social service becomes more and more distant from the whole truth,

and (iii) the real cost of water increases as water authorities turn to increasingly expensive supply sources.

Changing trends in the metered and umetered sectors

Table 5 presents a similar message in a different way, revealing that over 1960 to 1979 the part of the public water supply not subject to any price mechanism (unmeasured) rose at a percentage rate about 3½ times that of the part subject to pricing (measured). Less than half of the change in trend of the metered component since 1970 can be explained by the slower industrial growth of the 1970s; while metered consumption and the Index of Industrial Production both rose about 30% in the 1960s, the 1970–79 *increase* in the Index of 12½% was accompanied by a *fall* in metered use of 7%. The strongest candidates for explaining the remaining part of the trend change are the price of water to industry as an input and the volumetric price of effluent disposal, both of which are zero in the unmeasured sector.

Table 5. Potable public water supply in England and Wales, 1960–79 (thousands of megalitres/day).

	Unmetered	(%)	Metered	(%)	Total	(%)
1960	6.27	(63)	3.66	(37)	9.93	(100)
1970	8.88	(65)	4.80	(35)	13.68	(100)
1979	10.97	(71)	4.48	(29)	15.45	(100)
Change, 1960–79:	+75%		+22%		+56%	

Sources: 1960 figures derived from W1 returns of local water undertakings to the Ministry of Housing and Local Government. 1970 and 1979 figures from WDU7 returns to the Water Data Unit.

The '1981 decision': conclusion

This discussion of demand-management factors suggests that in 1981 the known scope for economically sensible measures is much greater than it was a decade ago. When this is added to the conclusion reached above, that the 1968–2001 demand increases would now be forecast at less than 50% of the 1969 forecast, then it seems that a 'similar' decision taken today on Rutland Water, with the extra knowledge available but still constrained to the demands of the original Empingham reservoir 'relevant area', should on economic grounds be a negative one.

Although I find it impossible to offer even a 'sound judgement' on how the wider regional case for a Rutland Water integrated into the supply-system would fare, I would speculate that a development originally costed at £25m in 1968, and therefore between £75m and £100m in current prices, would be unlikely to gain ministerial or Parliamentary approval in 1981. The demand-side measures listed and analysed above would contribute to such an outcome, but the single most important factor would of course be the continuing pressures on public sector capital spending.

Concluding remarks

A framework for decisions

Save in the case of 'essential' social services, dispassionate examination of the possible ways in which future demands and supplies of a public sector output may be reconciled requires that attention be directed at demand-management as well as supply-expansion solutions. Do-nothing will only be a viable long-term solution if consumers can be persuaded that the present standard of service is far higher than they, or the community as a whole, can afford. The public interest is better served if demand-management options are taken seriously.

Limitations of economic criteria

It is worth reiterating that economics cannot *decide* on the wisdom or otherwise of most inland and estuary water supply proposals. Such proposed solutions to alleged supply deficiencies are bound to be controversial in a tightly-packed island which values highly its unspoilt countryside. The ultimate decisions will thus often be political and for the larger schemes are properly made by a minister or Parliament. Feasible demand-based and supply-based alternatives of all kinds should, however, be known and analysed. If a solution is adopted which is not the most economically efficient, i.e. does not maximise the difference between identifiable social benefits and costs, then the economist cannot object in his professional role so long as the decision-taker and his/her advisers are fully aware of the extra resource-use which has to be suffered.

Institutional methods of taking decisions

Whether major water supply proposals are publicly debated in the context of a Private Bill presented to Parliament (as in the case of Empingham) or of a Public Inquiry (as in the case of Carsington, Roadford and Broad Oak reservoirs in recent years), there is a need to balance demands, resource development, economic efficiency and any effects on the environment. The implications of different solutions should be examined and assessed in the inquiry procedures. Select Committees of Parliament and inspectors at Public Inquiries need engineering, environmental and economic assistance to assist them to sort out claims and counter-claims. The Parliamentary Bill method is grossly undemocratic since opposition to the bill has to proceed through Parliamentary Agents (expensive) and counsel (even more so). So Public Inquiries, at which ordinary people may object without fee, are preferable.

The tradition is now established that inspectors at major water supply Public Inquiries are assisted by engineering and environmental (or sometimes chemical) assessors. No economic assessors have yet sat, however, and thus no objective guidance has been available to inspectors on sometimes complex economic issues, concerning which promoters and opposers may disagree. The engineering assessor at the Carsington inquiry revealed limited understanding of the economics of domestic metering, since he (a) confused total with marginal costs, (b) understated the savings shown in the Fylde experiment, (c) claimed quite incomprehensibly that metering could hardly be 'an economic operation if no new source works is provided in the process' and (d) took no account of the marginal resource costs of new supplies. The inspector at Carsington accepted all his engineer told him about domestic metering and therefore based his negative conclusion on the issue on mistaken analysis. (Department of the Environment 1974b; *Report,* para. 88, and *Assessment,* para. 13–20).

On the other hand, the inspector at the Broad Oak Inquiry did not follow the water authority line; he was 'not convinced that the cost [of metering for all domestic premises] would be prohibitive', and seemingly could have used the services of an impartial economic assessor on this and other demand issues (Department of the Environment 1980, para. 8.10).

National or regional decisions on demand-management?

It has been objected (e.g. by Kent County Council in the context of the Broad Oak Public Enquiry) that demand-management measures of the type discussed above are most appropriately examined and assesed at the national level, policy being made centrally and then passed down to the water authorities. For some measures, this is sensible; obviously it would be confusing to have different water regulations in adjoining authorities with, for example, very different upper limits on automatic washing machine water use. For other techniques, however, such as domestic metering and waste control, strict 'central' policies might themselves have severe economic costs, since what is economically efficient in a high-water-cost area may be economically costly where water is cheap and abundant. Thus it will often be wise for authorities to proceed at different speeds on such matters, with regional discretion important.

This implies that the case for demand-management will probably have to be argued out at each major inquiry in future, just as it was at Broad Oak in 1979. If demand-management had been put forward in 1969 in the Empingham proceedings, there would have been little difference in the outcome. Today, however, there is more awareness of both the costs of water supply schemes and of the various ways by which some of these costs may be avoided, so that a Rutland Water Proposal in 1981 would probably have a very rough ride indeed.

Epilogue

Following this tour of the developments of the last decade, it is reasonable to conclude that in this nation's approach to water resources planning an escape from prison has begun. We are certainly out of the darkened cell, but it appears that we are not yet clear of the prison yard.

Footnotes

[1] The 'Rutland Water pipeline *versus* Carsington' saga reads as follows. Using Rutland Water to provide a piped supply from Wing to Leicester would have allowed Leicester's present direct supply from the River Derwent to be redeployed to the future 'excess demand' areas of the north Midlands, thus obviating the need to build the controversial Carsington reservoir. If the Anglian Water Authority needed to claw back early next century the Rutland water sent to Leicester, the north Midlands could, it has been argued, be satisfied by a conjunctive use development combining Derwent and Dove waters and underground storage in the Trassic sandstones. Debate was originally joined at the re-opened Carsington Reservoir Enquiry in May 1976, when the Central Water Planning Unit (CWPU; now, alas, a victim of the present government's paranoia about quangos) set out the economic and financial implications of constructing such a pipeline and using it (and the redirection of Derwent water) as a medium-term solution to the north Midlands problem. Assuming the Wing-Leicester transfer would reduce to zero over 2001-2006 and allowing no credit for the pipeline to Leicester remaining in existence after 2006, it was calculated using discounted costs that less resources would be used for the pipeline solution than for the Carsington reservoir. If Severn-Trent Water Authority had, however, in addition been forced to pay a proportionate share of the loan charges on the Empingham reservoir capital, Carsington would have been *financially* the cheaper solution. Because (a) the difference between economic and financial costs was not fully appreciated (loan charges represent a financial but not a resource cost), and/or (b) the two Water Authorities could not agree on the price to be paid for transfers to Leicester, and/or (c) regional autonomy was considered more important than the resource savings, the economically efficient solution was not adopted. It was also strongly rejected by Severn-Trent and the National Water Council in 1980 when the issue surfaced again following the announcement that Anglian planned to stop pumping into Rutland Water to save electricity costs. The CWPU statement to the re-opened Carsington Inquiry is reproduced as Appendix 1 in (Central Water Planning Unit 1977).

[2] In 1970 the estimated official 1971-81 population increase in the East Midlands standard region was 344 000 on a base of 3.4 m. Evidence to the House of Lords Select Committee was that the population of the reservoir's 'relevant area' would increase from 660 000 to 990 000 over 1968-81, which implies an increase for that part of the area in the East Midlands of about 200 000 on a base of some 450 000.

[3] The Parliamentary Record Office in the House of Lords, which is where the *Minutes* are kept and may be consulted, unfortunately does not maintain a collection of the graphs and other documents handed round during the course of Select Committee proceedings. These would have provided much extra information on per capita demand projections.

[4] Experience has shown that it is usually more sensible to analyse and predict metered consumption on an aggregate rather than a per capita basis. In the case of substantial urban developments, however, where the population is expected to increase but the type of employment cannot be forseen, it may make more sense to think of (and thus forecast) metered consumption on a per capita basis.

[5] The extrapolative forecast, which Yorkshire W. A. still preferred in 1978, suggested 288 lhd as the 2001 figure.

[6] Because households were guaranteed that they would not have to pay more in 1971-72 than would have been the case with a rateable value charge, 77 records out of the total sample of 368 properties could not be used. The remaining 291 properties revealed an overall reduction in consumption of 11%. Since the year-on-year growth in the whole of the Fylde Water Board area was 3½% per capita (and therefore assumedly per property) over 1970-71 to 1971-72, the total impact effect is estimated as 14½%.

[7] Note, however, the dangers of double-counting if we were to attempt to add together the effect of bye-law changes requiring water-economy varieties of some domestic appliances and estimates of the effect of permanent domestic metering.

[8] The 1980 NWC estimates add together continuing costs, including meter provision and replacement, of £7-£8 per annum and an installation cost of £20 to £60. The latter range is equivalent to an annual cost of £1-£3, using the government's recommended 5% discount rate. Hence the overall cost of £8-£11. See pp. 23-26 of National Water Council (1980b).

Note added in proof: The figures for water volume used in this paper are given in the units of the original studies to which reference is made. Before 1974 this was in gallons; after 1974 in litres (1 gallon = 4.5 litres).

References

Bird, P. A. & Jackson, C. I., 1965. Water meters: why not? New Society, 17 June, 11–13.

Cambridgeshire County Council, 1980. Cambridgeshire County War Plan.

Central Advisory Water Committee, 1963. Report of Sub-Committee on Water Charges. London: HMSO.

Central Water Planning Unit, 1977. Annual Report 1976/77. Reading: CWPU.

Department of the Environment, 1974a. The Water Services: Economic and Financial Policies, Third Report to the Secretary of State for the Environment. London: HMSO.

Department of the Environment, 1974b. Inspector's Report on Public Local Inquiry into Carsington Reservoir and Assessment of Engineering Evidence, 1973. Unpublished DoE papers.

Department of the Environment, 1980. Inspector's Report on Public Local Inquiry into Broad Oak Reservoir and Ancillary Works, 1979. Unpublished DoE paper.

DoE/NWC, 1976. Technical Working Group on Waste of Water, First Report. London: National Water Council.

DoE/NWC, 1980. Technical Working Group on Waste of Water, Leakage Control Policy and Practice. London: National Water Council.

Herrington, P. R. & Tate, J. C., 1971. The Metering of Residential Water Supplies: Some Empirical Evidence. Department of Economics. Leicester: University of Leicester.

Herrington, P. R., 1974a. Demand, a better basis for the water industry? Surveyor 144: 14–15.

Herrington, P. R., 1974b. The costs of domestic metering. Water Services 78: 306–310.

Herrington, P. R., 1979. Broad Oak (Canterbury) reservoir proposal. Proof of evidence. Appendices. London: Council for the Protection of Rural England.

House of Commons, 1969. Minutes of Proceedings taken before the Committee on the Welland and Nene (Empingham Reservoir) and Mid-Northamptonshire Water Bill. Mimeographed and lodged in the Parliamentary Record Office, London.

House of Lords, 1970. Special Report of the Select Committee on the Welland and Nene (Empingham Reservoir) and Mid-Northamptonshire Water Bill. London: HMSO.

Jenking, R. C., 1973. Fylde Metering. Blackpool: Fylde Water Board.

MacRae, J., 1980. Is your new reservoir really necessary? Municipal Engineering, 17 June, p. 407.

National Water Council, 1976. Paying for water. London: National Water Council.

National Water Council, 1980a. The water industry in figures. London: National Water Council.

National Water Council, 1980b. Charging households for water. London: National Water council.

Office of Population Censuses and Surveys, 1980. Population Projections, Series PP3, No. 3. London: OPCS.

Phillips, A., 1981. The economics of water supply. Unpublished undergraduate dissertation, Department of Economics, University of Leicester.

Rump, M. E., 1978. Potential water economy measures in dwellings: their feasibility and economics. Watford: Building Research Establishment.

Sobolev, A. & Lloyd, C. J., 1964. Trials of dual-flush cisterns. J. Instn. Water Engrs, 18: 53–58.

Thackray, J., 1978. Metering water demand, Water, No. 19, 7–9.

Thackray, J., Cocker, V. & Archibald, G., 1978. The Malvern and Mansfield studies of domestic water usage. Proc. Inst. Civil Engrs, Part 1, vol. 64, pp. 37–61, and ibid.: Discussion, Part 1, vol. 64, pp. 483–502.

Walker, D. L., 1977. Water: a national and regional resource. In: Drudy, P. J., (ed.) Water Planning and the Regions. Discussion Paper 9. London: Regional Studies Association.

Warford, J. J., 1966. Water requirements: the investment decision in the water supply industry. Manchester School, vol. 34 (1).

Water Resources Board, 1966. Water Supplies in South-East England. London: HMSO.

Yorkshire Water Authority, 1978. 1978 Series Forecast of the Demand for Potable Piped Water. Unpublished YWA Directorate of Resource Planning paper.

Discussion (Herrington)

Mr Herrington introduced his paper by emphasising that its arguments operated on two levels. At the general level, propositions are made about the framework within which water planning has been and is undertaken in the United Kingdom. At the particular level of Rutland Water, the arguments attempt to answer the question of whether in an economic sense Rutland Water was or could now be said to constitute a sensible investment decision.

At the general level, the thrust of the argument is that up to the early part of the 1970s the water industry was preoccupied almost entirely with the supply of water and other services with virtually no attention being paid to demands. The cost of the product was low, accentuated by subsidies from central government and by increasingly outdated accounting procedures which were based upon the recovery through depreciation of only historical capital costs, increasingly insufficient to pay for the replacement of plant and equipment.

In the last decade, this picture has begun to change. Reorganisation of the water industry in 1974 with the creation of the new Regional Water Authorities has meant new directorates, new jobs and, inevitably, new ideas. Central government subsidies have been cut out almost completely.

There has been a long-overdue move to current cost accounting, reflecting the true costs of replacing capital, and this has put up charges and rate demands. In this new era concepts like the price, demand and reliability of supplies have begun to be more carefully examined. In the area of waste detection for example, progress promises to be rapid. Various studies have been made since 1975 which have indicated that waste in supply is about a quarter of total water production. Waste detection and alleviation is now a matter of serious concern among regional water authorities and at higher levels in government, with recognition of the enormous scope for sensible economy and possibly large pay-offs in terms of economics, finance and environmental criteria.

In the area of domestic metering, progress is much slower. Few experiments are being made, and huge opportunities to engage in selective metering – in the construction of new housing estates – have been missed.

At the particular level of Rutland Water, the first exercise had been to reexamine the 1969 arguments for the reservoir in the context of the information available in that year. This suggested that the population forecasts were too high. Also, the water demand forecasts were very probably too high, although the lack of available information about how water was used (how much and by whom) meant that extrapolative demand forecasting, with all its inaccuracies, was then the order of the day. Finally the only proven demand-management instrument available at that time was the installation of dual-flush cisterns.

The result of a re-examination of these three points is that the scheme as it was proposed in 1969 would probably not have reached full yield until 2001 rather than 1991 as the promoters had claimed. This would be unlikely to have altered the decision to go ahead, however.

Changes in the last decade have been dramatic. Population forecasts for the remainder of this century show about half the growth predicted in 1969. Water demand forecasts for the region, based upon the Anglian Water Authority's projections are likewise down by about half. Finally, a great deal has been learned about the methods and technology of demand management – for example, waste detection, domestic metering and water closet technology. These developments all suggest that a project the size of this one would be most unlikely to gain approval in 1981.

Opening the discussion, K. Guiver of the Southern Water Authority, sought the author's opinions on future economic developments in water supply and charging policies. Firstly, in view of the high proportion of our potable water supply which is used for non-potable purposes such as toilet flushing, did the author foresee the introduction of dual-supply systems? Secondly, would the author enlarge upon his views of universal metering compared with the existing system of rateable value costs of water supply? Mr. Guiver expressed the view that there could be social costs of metering domestic water supplies – such as the problems for low income families with several children in housing of low rateable value – and he felt that the present cost of water was nowhere near high enough to make a large proportion of the public aware of its bills if supplies were metered. Thirdly, Mr. Guiver wondered what, if any, were the adverse effects of building a larger reservoir at Rutland than might have been authorised had the arguments put forward in the paper been accepted in 1969?

Replying, Mr. Herrington felt that there would be very little scope for dual supply systems except for a non-potable supply to high concentrations of industrial users who could accept an inferior quality water. He agreed that the universal metering of domestic consumers was not a panacea, and would no doubt provide problems. He emphasised, however that it was something which should be considered at a local level in relation to new supply schemes, because there was no sign of the national authorities giving a lead. All the evidence available from studies in the U.K. suggested that the savings possible from metering – 10–15% of consumption – did not disappear after the installation but remained; even if demand subsequently rose it was consistently lower than the unmetered demand would have been. Mr. Herrington felt that, as far as reservoir size was concerned, the major adverse effect of a reservoir larger than was necessary was in irrevocable land-take and in the social effects of this attitude to reservoir size repeated throughout the country.

Several speakers from the floor, drawing upon personal experience of metering, suggested that one of its problems was the inaccuracy of individual meters, and expressed their scepticism of permanent savings in water consumption when domestic meters were installed. Mr. K. Stroud suggested that a cheaper alternative to metering was no charge households on a rateable value basis with weightings for the number of water-using domestic appliances the household owned. Mr. Herrington re-emphasised the consistent results of all the domestic metering experiments reported in the U.K., savings of 10–15% in water use. He did not think that rateable values with weightings would be an alternative; it was far too cumbersome and difficult to administer fairly.

P. R. Langford suggested that as Rutland Water was now supplying 50% of its design capacity a mere three years after its commissioning, this was clear evidence that a reservoir of this size was necessary. The author rejected this method of justification of the reservoir, on the basis that once a reservoir has been constructed and integrated into a flexible water grid system such as that used by the Anglian Water Authority, the use of the reservoir depends upon so many different factors that its actual yield can in no way be used as justification for the original investment decision.

The landscape of Rutland Water

Dame Sylvia Crowe
Landscape Architect, 182 Gloucester Place, London NW1, England

Keywords: Rutland Water, landscape, amenity planning, tree planting, reservoir, draw-down

Abstract

The landscape problems encountered at Rutland Water were: 1) the reconciliation of the reservoir and its works with the rural surroundings; 2) siting and treatment of all ancillary works; 3) mitigation of exposure of shore by draw-down; 4) the provision of opportunities for recreation without damage to the environment; and 5) conservation of nature. The means adopted were: 1) forward planning; 2) consultation and co-operation with all interests concerned; 3) extensive planting of native species and of ground shaping; and 4) arrangements for appropriate maintenance.

Introduction

The proposal to create a reservoir in the agricultural countryside of Rutland aroused natural and inevitable opposition.

The objections voiced at the public enquiry were on two counts: loss of valuable agricultural land and damage to the landscape. The first objection was valid and undeniable, but I did not agree with the second. Provided all the necessary steps were taken to ensure that the reservoir and its related works were designed in full sympathy with the surrounding landscape, I believed that the water would even prove an enhancement to its surroundings.

The impact of the reservoir

In common with most reservoirs, Rutland Water's impact on the landscape could be foreseen as two-fold. First was the impact of the reservoir and all its related works, and secondly the effect of the car-borne crowds who would be attracted by the various opportunities for recreation, sailing, fishing, bird-watching, walking, picnicking and just looking. To see how these two impacts could be met, it was first necessary to assess the character of the surrounding landscape.

Rutland was the quintessence of a quiet agricultural countryside of low, gently rolling hills and shallow valleys, hedged fields for mixed farming and small woodlands rich in flowers and wild life.

The buildings, for the most part, were traditional villages and farms with the occasional mansion set in parkland. The whole character was of human scale and tranquillity rather than drama.

One fear expressed by opponents to the scheme was that the size of the water would be out of scale. This danger was mitigated by the fortunate shape of the reservoir, with the Hambledon Peninsula dividing the greater part of the area into two arms, diversified by indentations.

The elements essential to the scheme, which required special care to avoid incongruity, were:

1) the perimeter, where draw-down would expose a varying width of shore; 2) the necessarily long dam; 3) the pumphouses; and 4) the canalized stream below the dam.

1) The problem of draw-down can never be

Hydrobiologia 88, 43–46 (1982). 0018-8158/82/0881-0043/$00.80.

completely solved, particularly if the difference between high and low water is great, and if, as in the case of Rutland Water, the shore gradient is gentle. But its effects can be mitigated.

Many years ago, when dealing with my first reservoir, I explored the possibility of using plants which would survive partial submersion. Visiting many old reservoirs, I found *Salix caprea,* and more plentifully *Salix cinerea,* had established themselves well below top-water level. These, and alder (*Alnus glutinosa*), have been planted at salient points to break up the expanse of exposed shore. At certain key points, a grass, *Agrostis stolonifera subacqua,* which was recommended by the Nature Conservancy Council, has been sown to extend green sward below the top-water level.

The whole of the shallow end, where the problem might have been most severe, is a nature reserve, whose mud flats are enjoyed by the waders and where the landscape is made interesting by a series of lagoons, bunds and islands. These were formed during construction work, as a result of superb co-operation between engineering, conservation and landscape.

2) The long dam is fortunately also shallow and on a gentle slope. It is graded into the surrounding ground and the grass swept up and over the crest road, right up to the wave wall. The appearance from below is that of a large field. While, for safety, there can be no planting on the dam itself, groups of trees have been planted just clear of the construction to relate it to the scale and pattern of the surrounding landscape.

3) Three pumphouses are included in the project, one below the dam and two off-site, at Wansford and Tinwell. The architect has most skilfully designed them to provide the necessary size, without appearing out of scale with the countryside buildings.

In each case, land-shaping and planting have been used to tie them into their surroundings. In the case of the one below the dam (a larger complex than the others) a wooded knoll has been formed to break the scale and conserve the old landscape pattern.

4) The stream below the dam has been taken on a meandering course between groups of alder. The stabilization of the banks necessary for the passage of flood water has been carried out in a material which will green over with time.

Care for the landscape has extended beyond the perimeter of the reservoir and has included planting in relation to road diversions and to the sites of the tunnel shafts.

Recreation

Recreational impact is often more difficult to deal with than operational requirements. It is less predictable, harder to control and involves far more people and their cars.

The types of recreation on Rutland Water are: sailing, fishing, bird-watching, walking, picnicking and viewing. Car parks are needed for all these pursuits, and they are potentially the most intrusive elements in the landscape, far more so than any of the reservoir constructions.

The site of each facility and its car park was carefully considered in relation to traffic access as well as to the best location for each pursuit. The two access routes for the general public were the A606 north of the reservoir, and the road on the south side passing through Edith Weston.

At Sykes Lane, accessible from the A606, a considerable quantity of material was excavated for the dam construction. Advantage was taken of this structural necessity to form a large car park, well below road level. This gives visitors a good view over the water and access to the water-side paths. With the addition of grassed mounds and planting, it also allows the cars to be screened from general view. This car park now holds the main information centre and the great bronze sculpture.

The Whitwell peninsula, with its access road from the village, was selected to provide fishing facilities, on one side, and a picnic area on the other. Here again, ground shaping and planting contain the cars from view while providing walks and picnic areas beside the water.

Perhaps the most ambitious picnic area is Barnsdale. Access is given by the road which was cut by the reservoir. From this, a road was taken through a woodland to an open, whale-backed hill. A fine engineering feat has cut roads and parking places, swinging in hairpin bends, down the landward side of the hill. The cars are thereby completely hidden from the surrounding countryside and from the Hambledon Peninsula; while their occupants by dismounting from their cars get a superb view from the hill top and sea-ward slopes.

A waterside and woodland path runs all the way from Sykes Lane to Barnsdale. A similar path skirts the south shore from the nature reserve to the dam, via the Normanton car park and picnic area. Here the principles of screening cars and revealing views have been followed.

An attraction of the Normanton area is the church, jutting out into the water. The question as to whether this church could or should be saved was hotly argued. In the event a compromise was reached, which, like all compromises, is not wholly satisfactory, giving, as it does, a truncated proportion to the building. Yet, I think, it was worth saving, even in this form, and one would be sorry to lose it. A solid group of willows at the junction of the causeway has been planted to anchor it visually to the shore.

There have been five main elements in the development of Rutland Water landscape: 1) planning; 2) co-operation between engineers, architect and landscape architect; 3) land-shaping; 4) planting; and 5) maintenance.

The planning was a joint exercise with the water Authority and the Leicestershire County Council. All the recreational interests were also represented. In addition to operational needs, the process took note of traffic, visitors' desire, landscape capacity and the interests of residents. A firm stand was taken against caravan camps and such noisy pursuits as water ski-ing.

Co-operation has been exceptionally good throughout. In particular, there has been ceaseless care for the landscape throughout the operations. This co-operation has made it possible to consider landscape requirements from the earliest stages of the works, instead of trying to adapt or remedy unsympathetic elements. For instance, the working perimeter track was formed as a pleasant, curving path, passing sympathetically through the woodlands and inclining to the contours. Such co-operation not only gets the best results, it also saves money.

Land contouring has been used extensively in the landscape work. The nature of the natural landscape, with its small scale undulations, made this particularly appropriate. It also gives immediate effect while planting requires time to develop.

Often the boundary fence of a project can be its worst feature. But fortunately there was an early decision to use agricultural post and rail, accompanied by thorn field hedges. This ensures that there is no visual break between the Authority's land and the surrounding farm land. On the same principle planting is almost entirely of native species, the only exceptions being a few naturalized British citizens, such as Norway maple, balsam poplar and horse-chestnut.

The character of existing woodlands is being maintained, with the addition of water-tolerant species near the water's edge.

The needs of wild life have been considered throughout, as well as being specifically catered for in the Nature Reserve.

At an early stage it was decided to establish a nursery. This has proved very successful. Besides the economic advantage, it makes the plants available as and when required.

Maintenance must always be a major landscape concern. It must be efficient and economic. The great majority of the perimeter land is either grazed by sheep or maintained as woodland. This is desirable from every point of view. It not only reduces upkeep costs, but it also carries on the tradition of countryside which it has been our endeavour to conserve, despite the needs of water supply and the insatiable appetite of the public for water-based recreation.

Acknowledgements

Particular credit for success in the landscape planning is due to: the late Mr. Kennedy (County Planning Officer, Leicestershire County Council), and the Anglian Water Authority and its officers, with special regard to Mr. Knights (Reservoir Engineer), Mr. Ford (Forester) and Mr. Dickens (Architect).

Discussion

Introducing her paper, Dame Sylvia emphasised the two impacts of the reservoir upon the environment. The first is the impact of the water itself and the ancilliary structures, which was relatively easy to cope with through careful planning. A key element in the success of this strategy was early and persistent cooperation with engineers and contractors during the construction of the reservoir, which gave better results and saved a great deal of money when compared with later landscaping.

The second impact is that of the visitors themselves – particularly their cars. It has fortunately been possible to site the car parks down into the landscape rather than upon it, screened and tucked away using tree-plantings, embankments and natural landscape features. Ready availability of plant was a great help here, and starting early on in the life of the project means that the car parks now have a mature look about them and are well hidden in the landscape.

The discussion was opened by Mr. D. Wightman, Leicestershire County Council, who confirmed the success of Dame Sylvia's landscaping and building designs by referring to the favourable opinions expressed in visitor surveys conducted by the Council. This has vindicated the policy of including recreational planning from the beginning of the scheme, unlike other reservoirs where it has only been considered after construction. He wondered however, whether Dame Sylvia felt that any similar natural resource development in the 1980s could ever afford such concern for the environment? He further wondered what her feelings were about the proposed 8 m drawdown which had been under consideration in 1980 as a means of saving pumping costs?

Dame Sylvia agreed that, in today's economic climate, the standard of landscaping would not be as high but a great deal could be done with careful planning from the outset and inexpensive methods such as establishment of a nursery, use of mechanical plant already on site, and use of cheaper materials, e.g. wood instead of stone for buildings. She felt that it would be a great shame if an extensive summer drawdown was allowed to take place in the early years of the reservoir because this would destroy most of the water's edge planting which had focussed upon the zone within 2 m of top water level. If there was to be a regular drawdown in excess of 2 m when the reservoir was in full use then she hoped the Authority would consider extending the water's edge planting.

Mr. Clarke, Anglian Water Authority, felt that, although maintenance of top water level for as much of the time as possible would be necessary in future years to ensure the security of supplies, extensive drawdown such as experienced in the 1976 drought at other reservoirs was bound to occur. Mr. Warren, Wessex Water Authority, felt that drawdown in most reservoirs would increase in the future as the economics of operation became tighter.

Mr. Langford and Mr. Knights, Anglian Water Authority, emphasised that Dame Sylvia's approach had been an overall 'concept' rather than a specific set of rules, which as well as being very cost effective during construction, had survived the end of her contract period and continued to be the Authority's policy for the reservoir. A keystone in its success had been the Authority's purchase of perimeter land – around 600 ha in addition to the 1 300 ha of water surface. Much of this land was re-let to the original owners for grazing but the important point was that the Authority controlled its use.

Perimeter forestry and landscape works at Rutland Water

A. Ford
Forester, Anglian Water Authority, Rutland Water, Whitwell, Oakham, Leicestershire, England

Keywords: Landscape, tree planting, water-tolerant species

Abstract

The forestry management programme at Rutland Water is described. It incorporated management of existing old woodlands as well as establishment of new woodland and hedgerow areas together with amenity planting at car park and picnic sites. The focus of the new planting programme has been the water's edge zone within 2 m of top water level.

Introduction

The landscape plans prepared for the reservoir provided for extensive tree planting (Crowe 1982). A forester to coordinate these plans was appointed by the Authority in July 1972 and was provided with a work force which has averaged five men since that time.

Tree nursery

An early requirement at the site was for a nursery to grow-on cuttings and transplants. An area extending to over a hectare was established close to the north side of the dam; this site had fertile, deep, neutral soil in a sheltered aspect with easy access.

Initially cuttings of poplar (*Populus* spp.) and willow (*Salix* spp.) were obtained from within the reservoir site clearance area and transplanted to the nursery. *S. cinerea* was not common at the site and so cuttings of this species were obtained from nearby Fineshade forest for eventual water's edge planting. Most of the cuttings established in the nursery grew to around 2 m in 3 years.

Approximately 20 000 small transplants of the species recommended in the landscape plans (Crowe 1982) were purchased for growing-on in the nursery. In addition the future requirements of the Nature Reserve (Appleton 1982) were catered for in a separate section of the nursery established at this time.

Site clearance

The requirements for the construction of the reservoir were that the clearance of trees and hedgerows should take place only below the top water line and careful check was kept on the contractor's site clearance work. Water tolerant species such as poplar (*Populus* spp.) and willow (*Salix* spp.) found growing within 1 m of top water level were marked for retention and a further 150 individuals were moved to this zone from areas to be deeply flooded. Several hundred oak (*Quercus* sp.) and birch (*Betula* sp.) of 2–3 m in height were moved from the area of Hambledon wood to be cleared and were planted in the new open space areas. Approximately 500 hawthorns (*Crataegus* spp.) of 2–3 m in height were moved from roadside hedges within the site area and, after heavy pruning, successfully transplanted into new scrub areas.

Hydrobiologia 88, 47–50 (1982). 0018-8158/82/0881-0047/$00.80.

Access tracks

The construction of access tracks approximately 3 m wide around the perimeter of the reservoir was an early priority. Where these passed through woodland they were positioned to follow contours close to top water level but sufficiently within the woodland for traffic to be screened by it. Minimal felling of trees was used during construction of the track and advantage was taken of natural breaks in the canopy. The track often changed direction to avoid the felling of mature trees. The result, stone surfaced, soon blended into the landscape and now provides attractive woodland walks in addition to its operational function.

Water's edge planting

An early priority in the programme was the planting of water-tolerant species in order for them to become well established prior to inundation and its associated wave action. Initially this planting had to be confined to woodland edges because existing agricultural land was still in use. Nursery grown specimens of *Salix* and *Populus* were deep planted (between 0.5 and 1 m) within 1 m of top water level. In addition, stakes of *Salix* have been deep planted using a tractor-mounted soil auger. This has proved a successful and inexpensive method of establishment.

In addition to the tree planting in the water's edge zone, a water-tolerant grass, *Agrostis stolonifera palustris,* was seeded in a 15 m band below top water level in high amenity areas associated with car parks and picnic areas as well as on Nature Reserve bunds to contribute to their stability. This has now become successfully established and is spreading naturally. Some steep areas have suffered from wave erosion exposing the roots.

During the period of three months when the reservoir was kept up to top water level (necessary for the commissioning of the dam), this deep flooding caused the trees to put out adventitious roots into the water. Eventual draw-down left these roots exposed; they dried up and in some cases where anaerobic conditions had been created below ground level, tree death occurred.

During the winter following this deep flooding, ice formation on the reservoir (up to 7 cm thick) caused physical damage to the trees' thin bark and when the ice eventually broke up it was washed in on waves and again caused abrasive damage.

Wave action loosened, and in some cases overturned, trees. These quite often put out roots where branches touched the mud and stabilised where the water receded.

Car park and picnic areas

Construction of the first of the five car parks was at Whitwell in 1973 and as the nursery plantings had not grown to adequate size stock was purchased and trees and shrubs planted with an auger.

Whitwell car park and the Rutland Sailing Club were the only areas to be grass seeded by contractors; subsequent amenity areas were seeded and maintained using the Authority's own labour force.

Most of the grassed picnic areas are sheep grazed to reduce maintenance and the many trees planted here have protective fences. Sheep are excluded from the immediate vicinity of the car parks and this has resulted in 37 ha which require a mowing regime. Many plantings are left unmown to encourage undergrowth and to 'break up' large grassed areas. Vandalism of planting has been minimal with only a few trees broken in the early days. No severe litter problem has developed, helped by the policy of regular collections and encouragement to visitors to 'Take your litter home'.

Older woodland

The existing woodland on the reservoir site (around 48 ha) had poor access with wet rides before the Authority's purchase. The advent of the access tracks made management and timber extraction easier.

There are occasional good timber trees, e.g. of *Q. robur* and *F. excelsior*, but the majority are of poor quality. They present a very 'natural' landscape, however, together with the understory of *Corylus avellana* and *Crataegus* spp. and ground layer of such species as *Primula veris* and *Endymion non-scriptus.* This is much appreciated by naturalists and the general public.

Early management consisted of clearing Dutch-elm-diseased trees to make the woods safe for public use. About 6.5 ha have been cleared so far and at the same time the understory of over-mature *C. avellana* and *Crataegus* spp. has been cut back to encourage re-growth. The clearings are being planted with *F. excelsior* and *Q. robor* and *Fagus sylvatica* where shade is dense. Woodland margins have occasional *Prunus avium* and *Acer campestre.*

Approximately 4.5 ha of the woodlands consisted 15-year-old mixed plantations of conifer and *F. sylvatica.* These are growing well and prior to thinning are being cleared of invasive *Ulmus* spp. and *A. pseudoplatanus.*

Wind throw

Site clearance in the older woodlands created newly exposed woodland edges in these areas, where trees approached 18 m in height, and were growing on heavy textured surface water gley soils. Windthrow was expected in such areas, but fortunately was confined to small pockets not more than 0. I ha associated with extra wet spring areas and in these it was large *Populus* which were the main casualties.

New woodland

New plantings (not including car parks and picnic areas) amount to 70 ha established since 1972. Most of the sites are on heavy clay grassland. The severe drought of 1976 caused many losses amongst trees planted the previous season and vole populations in the grassland caused some damage in the form of stem ring barking at ground level. Open areas have been left associated with the woodland to give it a more 'natural' appearance and to encourage the colonisation by species such as bramble, elder and hawthorn.

Hew hedgerows

The outer boundary of the Authority land purchases was initially established as a five-rail oak fence, which has an expected life of 15–20 years. A programme of quickthorn hedge planting is being carried out to produce hedges which will replace the fences at the end of this period. Approximately 13.6 km of hedgerow has been planted to date, which is just over half the programme. The new hedges have to be protected from sheep and will eventually have to be laid to make an effective stock-proof boundary.

Acknowledgements

The author wishes to thank the Anglian Water Authority for permission to publish this paper and to his many colleagues for assistance in the management programme.

References

Appleton, T. P., 1982. Rutland Water Nature Reserve: concept, design and management. In: Harper, D. M. & Bullock, J. A. (Eds) Rutland Water - Decade of Change, pp. 211–224. Junk, The Hague.

Crowe, S., 1982. The landscape of Rutland Water. In: Harper, D. M. & Bullock, J. A. (Eds) Rutland Water - Decade of Change, pp. 43–46. Junk, The Hague.

Appendix: Planting by the Water Authority at Rutland Water from 1972 to 1980

Quickthorn (*Crataegus* spp.) hedging	93000
Ash (*Fraxinus excelsior*)	15000
Oak (*Quercus robur*)	12500
Willow (*Salix* spp.)	7355
Beech (*Fagus sylvatica*)	5700
Hawthorn (*Crataegus* spp.)	5650
Field maple (*Acer campestre*)	5600
Wild cherry (*Prunus avium*)	4900
Hazel (*Corylus avellana*)	4820
Alder (*Alnus glutinosa*)	3230
Scots pine (*Pinus sylvestris*)	2600
Dogrose (*Rosa canina*)	1500
Poplar (*Populus* spp.)	1205
Birch (*Betula pubescens*)	880
Sycamore (*Acer pseudoplatanus*)	500
Blackthorn (*Prunus spinosa*)	400
Norway maple (*Acer platanoides*)	400
Lime (*Tilia* spp.)	200
Dogwood (*Cornus sanguinea*)	200
Guelder rose (*Viburnum opulus*)	200
Broom (*Cytisus scoparius*)	200
Gorse (*Ulex europaeus*)	150
Horse chestnut (*Aesculus hippocastanum*)	120
Whitebeam (*Sorbus aria*)	30
	Total 166340

Willow species:
Salix cinerea
atrocinerea
viminalis
viminalis × *cinerea*
vitellina
daphnoides
alba
fragilis
caprea
triandra
britzensis
colodendron
fragilis latifolia
purpurea

Poplar species:
Populus alba
balsamifera
robusta
gelrica
tremula

Tree pre-planting at Rutland Water

C. J. Gill
Department of Applied Biology, University of Cambridge[1]
[1] *Present address: The College, Old Wish Road, Eastbourne, England*

Keywords: tree-planting, reservoir, drawdown

Abstract

The aesthetic problems posed by reservoir margins have been recognised for many years. They stem from a complex pattern of action and interaction of several environmental variables – water manipulation patterns, wave action, substrate characteristics, grazing and other factors.

These factors usually result in reservoir margins being not only visually unattractive and physically unstable, but also of low ecological diversity.

The Natural Environment Research Council and Water Research Centre have, since 1968, together funded a research programme to investigate the use of plant species to combat these problems.

In isolated instances in the UK, woody species in particular have succeeded in colonising margins naturally, to very good visual effect (e.g. at Lake Vyrnwy). 'Biotechnical' treatments – using artificially planted shrubs and trees – have been successfully used in large Central European reservoirs to combat soil erosion by waves.

Following trials of various woody species on the margins of established reservoirs, 10 species/cultivars were 'pre-planted' above and below the future top water level (TWL) of Rutland Water in 1973, 5 years before reservoir filling was completed. Pre-planting theoretically allows better survival by trees of subsequent partial inundation. The precise siting of pre-planted trees in the vertical dimension should normally be done using modelled predictions of reservoir summer median water level (a variable important to tree survival). The complex factors affecting the operation of Rutland Water made precise prediction impossible; the experimental plot eventually extended from 500 mm above TWL, to 2 000 mm below TWL, in 500 mm steps.

252 trees were involved in the plot and their growth has been monitored for the past 7 years. The species being predominantly hydrophytic, drought was a problem in the years prior to reservoir filling. Mulching with grass-cuttings proved to be helpful in this respect. Subsequently, prolonged periods of high water level have been deleterious to tree survival in the lower areas of the plot. Once abstraction/recharging patterns become more cyclic, it is anticipated that tree growth will stabilise.

Hydrobiologia 88, 51 (1982). 0018-8158/82/0881-0051/$00.20.

Area structure plans of Leicestershire County and Rutland District Councils

D. Wightman[1] & P. Burford[2]
[1] *Planning and Transportation Department, Leicestershire County Council, County Hall, Glenfield, Leicester, England*
[2] *Planning Department, Rutland District Council, Oakham, Leicestershire, England*

Keywords: reservoir, planning policies

Abstract

The poster summarised Leicestershire County and Rutland District Councils Planning policies operative in the area surrounding Rutland Water.

The policies are contained in the Structure Plan for Rutland, approved in December 1979, Rutland District Council's Settlement Policy and the Empingham District Plan preliminary discussion document. The policies of more specific application to the area surrounding Rutland Water are those of the Empingham District Plan which relate to land use, recreation and traffic management in the Parishes of Burley, Edith Weston, Egleton, Empingham, Exton, Gunthorpe, Hambleton, Horn, Lyndon, Manton, Normanton and Whitwell.

The theme reflected in all policies covering this area is one of conservation. It is expressed in one of the Structure Plan's basic policies as follows: 'Rutland's Planning Role should be one of positive conservation particularly of natural resources with an emphasis on minimising the loss of, and where practicable improving the natural assets of farmland, woodland, minerals and water.'

Hydrobiologia 88, 52 (1982). 0018-8158/82/0881-0052/$00.20.

Traffic and visitor surveys at Rutland Water

D. Wightman
Planning and Transportation Department, Leicestershire County Council, County Hall, Glenfield, Leicester, England

Keywords: reservoir, visitors, traffic, amenity

Abstract

The creation of England's largest lowland lake has resulted in major changes to existing patterns of recreation demand in the East Midlands. In 1977 and 1978, Leicestershire County Council carried out a number of surveys at Rutland Water as part of a regular monitoring of all large countryside recreation sites in the county. In addition to establishing visitor levels and characteristics, the surveys sought to establish the extent to which Rutland Water has attracted visitors from other beauty spots in the region or indeed has generated an additional demand. Further studies were undertaken to assess the impact of the large number of visitors on the surrounding roads and villages. Lessons learnt in planning for recreation at Rutland Water may well have a general application to similar new developments elsewhere in the country.

Over 600 interviews were carried out (involving about 2 000 visitors, approximately one third of all visitors during survey times). Parking surveys were undertaken in all the official car parks on selected days during 1977 and 1978, and records were also kept of cars parked on the roadside verges around the reservoir. Traffic counts were undertaken on all approach roads to Rutland Water, and at major road junctions, and compared with counts undertaken before the reservoir was opened.

Main survey findings

About 400 000 people per year visit the Rutland Water picnic areas. Over half come on Sundays – on the days when surveys were being undertaken, totals ranged from 5 000 to 12 000 visitors. Visitor numbers have risen steadily but not as rapidly as originally forecast. Over 80% of visitors come from the three counties of Leicestershire (49%), Lincolnshire (20%), and Northamptonshire (14%). The average distance travelled was about 30 miles, although some had come more than 100 miles. Practically all came by car, and a surprisingly high percentage did not venture away from the car parks during their visit. The average length of stay was just over one hour. A small percentage combined a visit to Rutland Water with a trip to a nearby stately home or a tour of the Rutland countryside, but most came straight to the reservoir and went straight home.

Most of those interviewed, on their first visit to Rutland Water, had been impressed by what they had seen, and although asking for some additional facilities, did not want the area commercialised. A number of local residents, although using the reservoir's facilities, still regretted its intrusion into the Rutland landscape.

About one-fifth of visitors did not use the car parks, but parked on the roadside verge. Some of the approach roads to the reservoir are unsuitable as recreational traffic routes, and whilst most visitors followed the recommended roads, some minor roads and village roads were taking a lot of day-trip traffic. Several road junctions can get congested at peak times, and a further traffic management scheme could be required if visitor numbers increase greatly.

As well as helping to relieve pressure on some overcrowded country parks elsewhere in Leicestershire, the

Hydrobiologia 88, 53–54 (1982). 0018-8158/82/0881-0053/$00.40.

opening of Rutland Water has stimulated additional recreation demand for countryside trips, and visits to the Leicestershire countryside have increased significantly faster over the last decade than the regional average. Pressures have arisen for hotel and other related developments in villages close to the reservoir, to cater for the tourist as well as the day-tripper, and careful planning policies will be required if the character of the surrounding area is not to be lost.

Rutland 2000: educational resources of Rutland Water

Bryan Waites
Secretary, Rutland 2000, 6, Chater Road, Oakham, Leicestershire, England

Keywords: reservoir, education

Abstract

Rutland, once England's smallest county, is in the middle of an environmental revolution. Since 1970 it has increasingly attracted commuters who can live in rural surroundings yet reach their work in Leicester, Nottingham, Peterborough and even Northampton quite easily. It is now the site of Britain's largest man-made lake which occupies about 3% of the old county and is introducing a completely new recreational pattern to the area. In 1974 Rutland vanished into Leicestershire and a new strategy plan, then district plans, threatened more change. Not far away another environmental revolution may be on the way in the Vale of Belvoir where one of the largest coalfields in western Europe has been discovered.

Such changes provided both a threat and an opportunity. Teachers in the locality were stimulated to form an environmental education group known as *Rutland 2000* in 1974 to develop resources for use in schools and to encourage field and classroom projects in connection with the environmental issues.

This group expanded to include librarians, rural officers, museum curators, planners, etc. thus providing the only forum for such representatives to meet together.

So far the group has produced a *Register of Rutland Resources* which catalogues manuscript and published sources, maps, aerial photographs, etc. A *Rutland Social Survey* has been produced by fieldwork of local schools in co-operation with Leicestershire Planning Department particularly examining rural transport and Rutland's special identity. A series of teaching units under the general title *Rutland 2000* is being prepared. So far five units have been published:

Unit 1: Introducing Rutland
Unit 2: The Voice of the People
Unit 3: Landscape and Land Use
Unit 4: People, homes and jobs
Unit 7: Rutland Heritage

Rutland 2000 has organised role playing activities and competitions for schools related to Rutland Water and planning issues. It has now produced slide sets, town trails and a *Rutland Water Guide* and offers its help to visiting parties to Rutland. It has the support of the Schools' Council, Leicester University School of Education, the County and District Planning Departments, many local firms and its publications have been in demand in Australia, New Zealand and elsewhere as well as being displayed in the recent UNESCO Environmental Education Conference in Tbilisi, USSR.

Hydrobiologia 88, 55 (1982). 0018-8158/82/0881-0055/$00.20.

Archaeological discoveries at Rutland Water

A. W. Adams[1], T. H. McK. Clough[2], M. S. Gorin[3], N. M. Reynolds[4] & M. Todd[5]
[1] *Rutland Field Research Group, c/o Rutland County Museum, Oakham, Leicestershire, England*
[2] *Rutland County Museum, Catmos Street, Oakham, Leicestershire, England*
[3] *Newark Technical College, Newark on Trent, Nottinghamshire, England*
[4] *Inspectorate of Ancient Monuments, Scottish Development Department, Atholl Crescent, Edinburgh, Scotland*
[5] *Department of History, University of Exeter, Exeter, Devon, England*

Keywords: Rutland Water, archaeological excavation, Iron Age, Romano British, Medieval, Anglo Saxon

Abstract

This paper summarises the results of archaeological excavations on the Rutland Water site from 1967 to date. They range from Iron Age to Medieval and include Romano-British farmsteads and two Anglo-Saxon cemeteries.

Introduction (T.H. McK. C.)

Archaeologically, Rutland Water came too soon. There was no permanent archaeological presence, amateur or professional, in Rutland when the enabling Act was passed by Parliament in 1970, and that Act made no provision for archaeological investigation of the Rutland Water area. Soon, however, the situation was to change completely. By the time water began to flow from the pumps, archaeological excavations had taken place on a number of sites, there was a well-established local museum, and a small but active archaeological research group had come into being.

Hitherto, our knowledge of the prehistory and history of the Gwash valley was scanty, to say the least. Now, one can offer a record of a substantial pre-Roman Iron Age site at Whitwell; a major Roman farmstead at Whitwell and others less well defined in Empingham parish; two pagan Anglo-Saxon cemeteries, whose burials were accompanied by grave goods of considerable importance, at Empingham, and medieval village settlements at Whitwell and Nether Hambleton (Fig. 1). Extensive field-walking on land acquired for reservoir purposes and along the routes of its associated long-distance pipelines resulted in a number of more minor discoveries, of which the details are recorded at the Rutland County Museum.

This paper summarises the main results of the work undertaken in the field by the various organisations involved. The full archaeological publication of the several excavations and the subsequent research will take place elsewhere (e.g. Todd 1981). The opportunity is taken here to demonstrate how great a contribution can be made to the advancement of our archaeological and historical knowledge as a result of public works on this scale. This can only be properly fruitful where there is willing co-operation between the archaeologist and the governing body. No matter how hard one may examine the ground on foot or from the air in advance of development, the evidence of the past may remain concealed until the earth is disturbed. Rutland Water proved to be no exception, with many of the major sites being found by chance while contractors' works were in progress.

The Iron Age and Roman settlements at Whitwell (M.T.)

Discovery

In July and August 1976, preparations were being made for the construction of the administrative headquarters of the Anglian Water Authority at the reservoir of Rutland Water. The site chosen lay on a flat-topped spur projecting into the flooded

Hydrobiologia 88, 57–66 (1982). 0018–8158/82/0881–0057/$02.00.

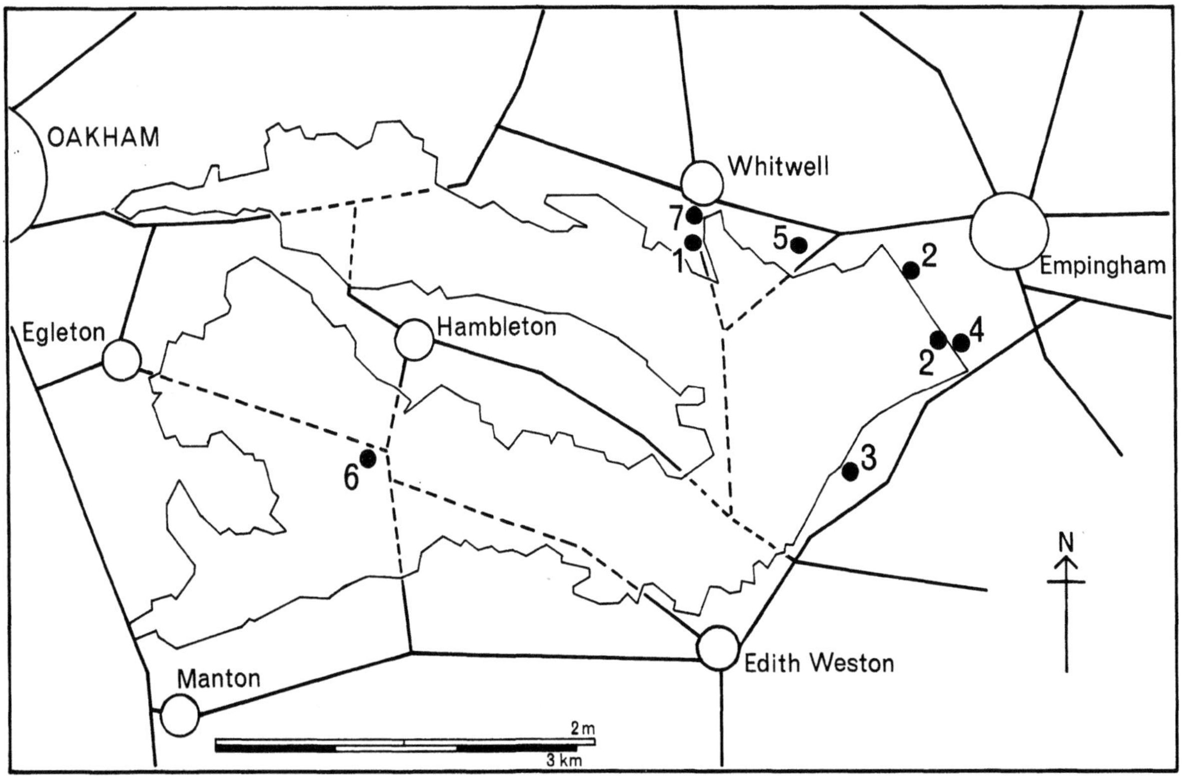

Fig. 1. Location diagram showing the principal archaeological sites excavated at Rutland Water: 1, Whitwell Iron Age and Roman; 2, Empingham Roman; 3, Normanton Roman; 4, Empingham Anglo-Saxon I; 5, Empingham Anglo-Saxon II; 6, Hambleton medieval; 7, Whitwell medieval.

valley of the river Gwash half a mile south of the little village of Whitwell and four miles east of Oakham (SK 925081). In addition to the administrative building, lodges for sailing and fishing had been planned and large areas had been reserved for car parks. The removal of topsoil in the preliminary stages of work on this project revealed over a considerable area the presence of Iron Age and Roman pottery, traces of stone buildings, pits, ditches and other indications of early settlement.

The discovery of this site, or group of sites, was reported by members of the Anglian Water Authority to Mr T. H. McK. Clough of the Rutland County Museum at Oakham. A preliminary examination of the areas already cleared, carried out by Mr Clough and members of the Rutland Field Research Group, defined with greater clarity the size and nature of the surviving remains. It was evident that concerted excavation was urgently needed. The Ancient Monuments Division of the Department of the Environment agreed to finance the work and later generously increased the measure of its support when the importance of the site was made manifest. The writer was invited to organise and direct the excavation and an excavation team was assembled at very short notice. Work began in late August and continued until 1 October, and a return was made to the site in December 1976 and January 1977, albeit in very wet and difficult conditions, to complete the work of salvage on those parts of the area still available for examination.

The discovery of this complex of Iron Age and Roman sites came as something entirely unexpected. No ancient material or other finds had been recorded here previously, a circumstance to be explained by the long use of the fields on the spur as pasture. Traces of ridge and furrow are still evident running down the slope on the eastern side of the spur and it is clear that there has been no ploughing of the ground for an appreciable time. Nor has the area immediately around the village of Whitwell been productive of sites of prehistoric and Roman date. A small quantity of Roman pottery from a site to the north of the village is the only well attested find. The total area over which the remains recorded here extended amounted to some seven acres (about 2.8 ha) and of this about two acres (0.8 ha) had been stripped of topsoil. There is no doubt that if further removal of the turf had been feasible much more of both the Iron Age and the Roman sites would have been laid bare. For the foreseeable future, however, the areas undisturbed in 1976 will remain under the protection of the covering of turf and humus. The site has now been fully published (Todd 1981).

The settlements (Figs. 2 and 3)

Although slight traces of earlier occupation were recovered, the first major settlement of the Whitwell site occurred in the later first millennium BC. Although no certain structural evidence was obtained, this settlement is likely to have been a single homestead consisting of a group of circular houses, perhaps with outbuildings and possibly set within an enclosure. The site was abandoned well before the Roman conquest, perhaps a century or more earlier, to be re-occupied shortly after AD 43 and certainly by AD 50. The early Roman site possessed a group of enclosures bounded by substantial ditches. The nucleus of the settlement at this date may have lain at the south end of the area examined, in Area 3. Iron-smelting appears to have been carried out in this phase. The artifacts associated with the first century settlement in particular were of notably high quality, especially the pottery. The massive bronze strap-junction (Fig. 3) also belonged to this phase.

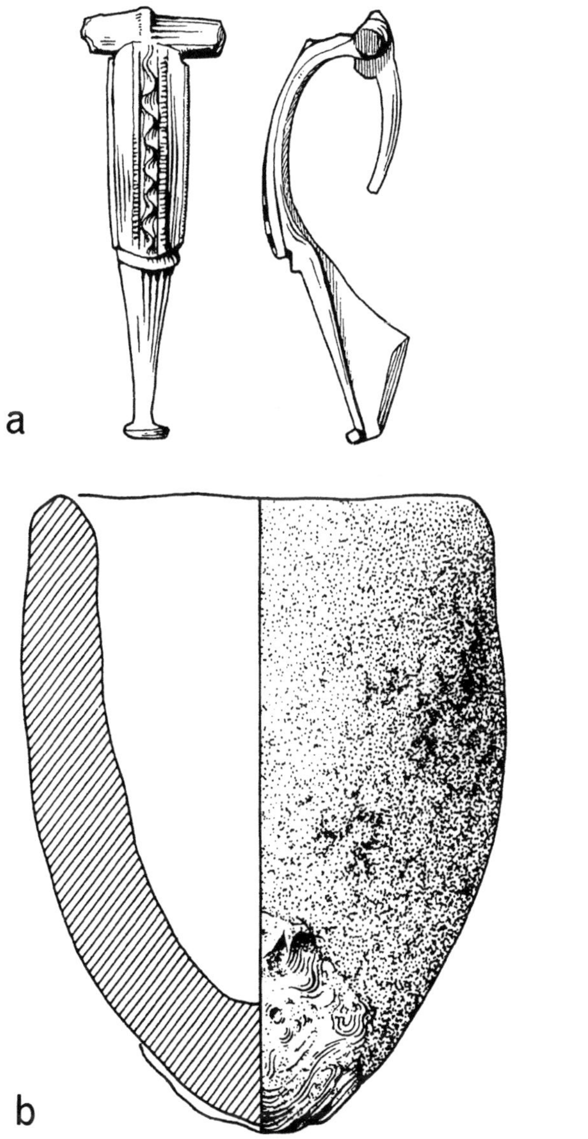

Fig. 2. (a) Roman bronze brooch of Hod Hill type, mid-1st century AD; (b) Roman clay crucible, possibly used in the smelting of bronze.

Fig. 3. Iron Age bronze strap-junction with enamelled decoration, from Whitwell, mid-1st century AD (photograph copyright Leicestershire Museums).

Occupation appears to have continued without break throughout the first and second centuries. Towards the end of the second century or very early in the third, a timber aisled building was constructed in Area 1, which now became the focus of settlement and remained so until the close. Before the end of the third century this building had been replaced by a rectangular structure with stone foundations. This, in turn, was abandoned about the middle of the fourth century. A rectilinear field-system situated to the north was associated with these successive buildings. To the south of the farmstead lay a fenced enclosure and beyond this an area containing corn-drying furnaces and metal-working installations. After its abandonment in the mid-fourth century, the site lay desolate for a long interval. A few small sherds of medieval pottery are probably to be linked with later agricultural activity, centred on the village of Whitwell, half a mile to the north.

The site recorded here is only one of a number of Iron Age and Romano-British settlements excavated in this stretch of the Gwash valley, most of them before the construction of the Rutland Water reservoir. They include two Romano-British farmsteads sharing much in common with that at Whitwell.

Romano-British and Anglo-Saxon sites at Empingham, 1967–1974 (M.S.G.)

The first archaeological work at Empingham was arranged through Miss Christine Mahany at Stamford and directed by the late Malcolm J. Dean. Mr Dean's attention was drawn to surface finds of Anglo-Saxon material, including a cruciform brooch, a bronze-bound wooden bucket, sherds of Anglian pottery and fragments of a bronze bowl (SK 94470776). Financed by the then Ministry of Public Building & Works, excavations took place in 1967 and again, briefly, in 1969. The major discoveries were a Saxon hut and two groups of burials.

The post-holes of the hut suggest it was a rectangular building, measuring 3.65 m by 3.40 m, using part of an earlier Romano-British trackway as a forecourt. A number of burials were later inserted into the floor of the hut, suggesting a date in the early 5th century for its occupation.

The first group of burials, six in all, was separated by some 12 m from the second group of four. In his manuscript notes, Mr Dean states that the former are clearly pagan, but expostulates that the latter represent Christian interments. His study of the grave goods reveals an apparent time gap of some 70–80 years between the latest pagan burial

Fig. 4. A fine silver-gilt square-headed brooch from the first Anglo-Saxon cemetery at Empingham (photograph copyright Leicestershire Museums).

and the earliest possible date for Christianity. Material from Group 1 (e.g. Fig. 4) is datable to AD 425–550 while Group 2 burials, lacking the rich grave goods of Group 1, were supine in neat rectangular graves, deeper than those of the pagan burials.

The line of the trackway, referred to above, leads into the field adjacent to the west, passing through an area where a scatter of Romano-British pottery and building materials had been discovered. Early features on this site (Empingham I, excavated 1969–1971) revealed evidence for occupation in the second century, but the main phase of building occurred in the third century, with the site comprising an aisled barn, a well-shed, a one-roomed building and a farm-yard area. The east walls of each of the main structures were on one building line, implying their contemporaneity. Two largely featureless strips across the site immediately south of the aisled barn and south of the one-roomed building suggested that the site was originally terraced into three building platforms.

The aisled barn, measuring 23 m from east to west and 12 m from north to south, had three clear and massive post bases supporting the northern aisle. It appears to have been used for light industrial activity as evidenced by the number of hearths. Some internal partition walls were observed, and a stone spread outside to the east may have been a foundation for an outside staircase.

A well, just south of the aisled barn, supplied the site until about AD 270 when it was filled deliberately with rubbish. Thereafter water was presumably carried from a nearby spring – no other well was discovered when the area was scraped by machines. A number of wooden objects, including an unusual shoe, were recovered from the fill of the well.

Access to the one-roomed building was by a door from the inner yard or by a larger opening from the open yard to the east. It may have served for storage – a number of large and pitched slabs formed two bases – or, perhaps, for dairy-like activities.

A keyhole-shaped corn-drier had occupied the corner of a building further south until alterations took place and it was filled in. A pair of rather flimsy parallel walls some 15 m long and just over 1 m apart formed the southernmost extension of the site but their purpose is undecided. A large area of the site, mostly east of the 'wall-line', was cobbled with ironstone and limestone blocks to form a farm-yard.

Another Romano-British farmstead (Empingham II, excavated 1971) was situated some 400 m from Empingham I on the north side of the River Gwash. This also took the form of an aisled farmhouse, equipped with a heating system with hypocausts in one aisle. It is not as firmly dated as Empingham I, but it was probably contemporary with it.

The sites are further linked by the discovery during construction work of two limestone coffins, one on each side of the river. They were aligned on one another, with the feet of both burials pointing to the river. Some relationship between the male (south side) and female (north side) burials must be argued from their locations.

During the construction work of 1972 and 1973 a watching brief was kept on the reservoir site. A number of archaeological sites were revealed and some were simply recorded while others were excavated more fully. A summary of the major results can be given as follows:

North of River Gwash

(i) Two Anglo-Saxon *grübenhauser* (sunken huts) and adjacent features (SK 936079).

(ii) An Iron Age settlement of which one circular house was completely excavated. Dating revolves around the distinctive scored pottery found here, at Whitwell and at other Iron Age sites in the East Midlands. The site is unlikely to date after the mid-first century BC (SK 937079).

South of River Gwash – near to Normanton Church

Excavations by the contractors for the inlet pipes revealed sufficient scattered pottery and limestone foundations for another Romano-British farmstead of third/four century date to be identified. Two features were excavated – a well and an H-shaped corn-drier (SK 936067).

All in all the evidence revealed during the years 1967–1974 points to almost continuous occupation of the Gwash valley west of Empingham since about 250 BC. Similar small river valley situations, it may be implied, would provide like evidence, if the opportunity arose to investigate them.

The second Anglo-Saxon cemetery at Empingham (N.M.R.)

In the course of constructing the Sykes Lane car park some 1.4 km W of Empingham village (SK 936082), the graves of a second Anglo-Saxon cemetery were disturbed. The site was excavated in 1974 by Mr M. S. Gorin, and in 1975 by the writer, on behalf of the Anglian Water Authority and the Department of the Environment, with considerable help from the Rutland County Museum, and from local volunteers. The cemetery has proved to be an important addition to the series of pagan Anglo-Saxon burial grounds in the East Midlands, and will hopefully form a key to earlier, more haphazard, collections of finds from the region (summarised, for example, in Clough *et al.* 1975).

A total of 132 inhumation graves, and one cremation, were found along the line of a hollow-way, probably in use in the Roman period, and in amongst a complex of silted-up Iron Age ditches. The layout of the graves appeared to be largely random, and most were comparatively shallow. No grave cut into another, perhaps indicating that their positions were originally marked in some way. One grave was cut into the centre of the hollow-way. Despite the way in which the site was discovered, and despite the extensive machine-stripping of the top soil, it seems likely that a majority of the graves of the cemetery were recovered. A preliminary comment on the grave structure has already been published (Reynolds 1976).

If the grave goods with which the dead were buried are an indication of wealth, the community was a rich one: 76% of the bodies represented were buried with grave goods of one sort or another. A large and varied collection of fine objects had been buried with the dead. Many of the men had been interred with weapons: spears, knives, the iron bosses and fittings surviving from wooden and leather shields, and buckles and belt-fittings. Many of the women had been festooned with beads, apparently both as necklaces and costume-fittings, made of both coloured glass and amber. Some wore long bronze pins, some had their sleeves fastened with bronze (and in one case silver) clasps, some wore bronze rings on their fingers; one young girl had a fine sheet-bronze ring round her neck, and a thin bronze wire ring round her ankle. Two of the burials had bone combs, and one each a cauldron, a bronze bowl and a bucket. A variety of objects hung from the waists of the women – some apparently toilet instruments, others the so-called 'chatelaines', hung with iron objects resembling keys.

A wide variety of bronze brooches adorned the women's costumes, including annular, penannular and disc brooches, open-work swastika brooches, and cruciform brooches, which have a roughly cross-shaped head and an animal-head foot on either side of the central bow. A particularly fine collection of these cruciform brooches from seven graves on the site had been elaborately moulded with human masks and birds' heads; the bronze had been gilded, and some had panels of silver foil. They belong to the development of a florid art-style in late-pagan Anglo-Saxon metalwork, flourishing in the East Midlands and East Anglia at the end of the 6th and early 7th centuries AD. The Empingham examples appear to form a distinct stylistic group, perhaps indicating a single workshop. Although examples of earlier metalwork occur in the cemetery, most would seem to agree with the broad date suggested by the florid cruciform brooches for the main period of burial on the site.

The 132 graves contained a total of some 150 bodies. About a third of these were children. Among the adults, about half had died in the 17–25 age range, but a fair number lived well into their 30s and 40s. In height, the males ranged between 5′3″ and 6′1″ (1.60–1.85 m), the females between 4′10″ and 5′11″ (1.47–1.80 m), with the majority between 5′1″ and 5′7″ (1.55–1.70 m). There is no indication of the reason for the presence of a single cremation, nor is there any suggestion of its being an amateurish one-off: the cremation had been efficiently carried out, the bone well-collected. The body was that of a child about 12 months old. In general, the population would seem to have been rather sturdy and masculine-looking. The skeletons displayed a variety of interesting genetic features, which are forming the subject of further study: there is some possibility that familial links may be discernible.

Once all the work has been completed in analysing the metalwork and collating the material from the graves with the skeletal information, a most important body of information on an early Anglo-Saxon population will have been recovered. All too few of the cemeteries of the region excavated in earlier years paid any attention to the actual people who wore the jewellery so eagerly collected into museums. Empingham II may give us a first insight into a settled rural population in the decades just before the conversion of the pagan Anglo-Saxons to Christianity.

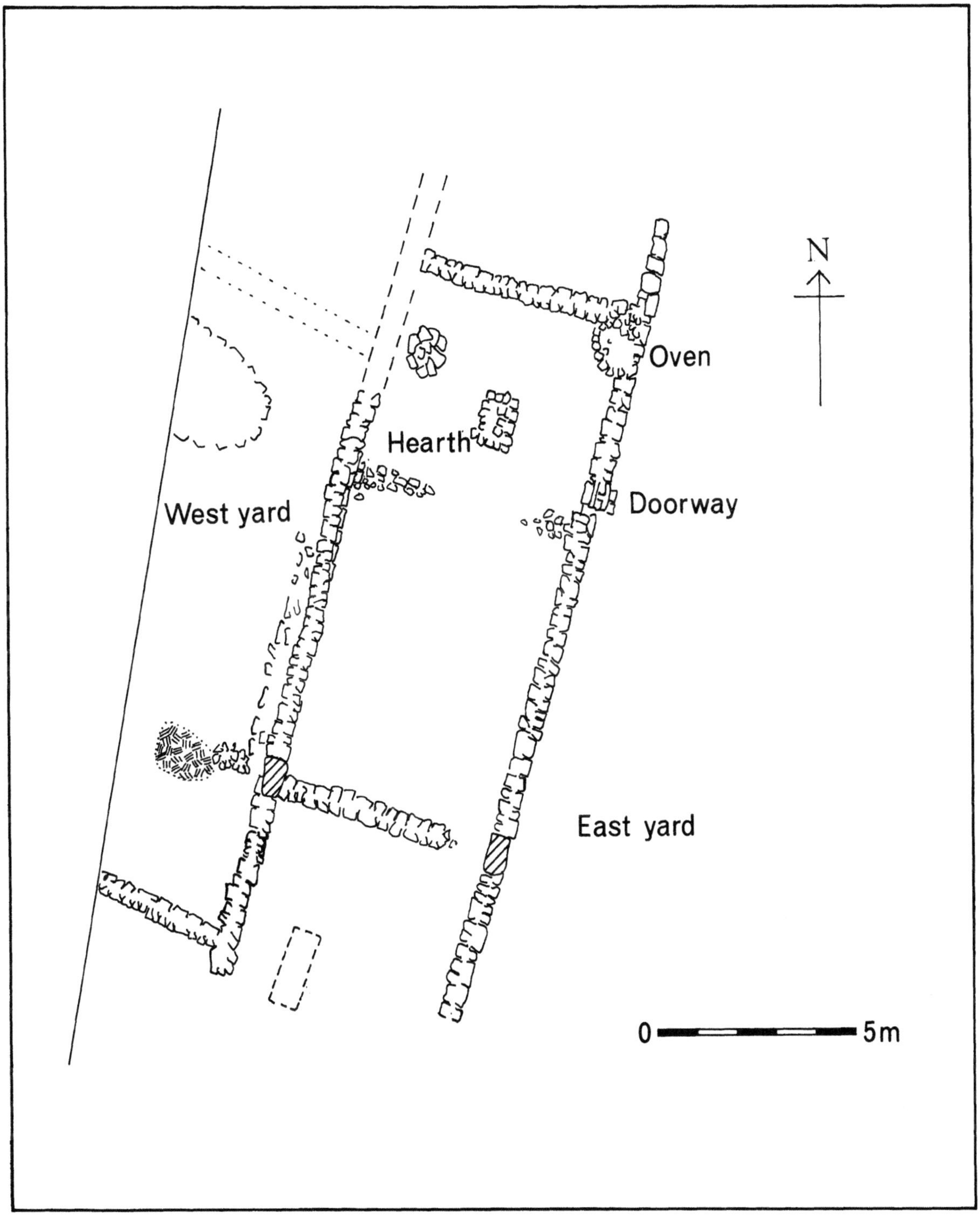

Fig. 5. Outline plan of the medieval house excavated at Nether Hambleton.

A medieval building at Nether Hambleton, 1973–76 (A.W.A.)

Field examination and survey work carried out by members of the Rutland Field Research Group in 1971 and 1972 confirmed the existence of a number of early buildings under earthworks adjacent to surviving old properties at Nether Hambleton, on land acquired for Rutland Water (SK 894067). In view of the anticipated filling of the reservoir, permission to carry out a small excavation in the area was obtained from the Anglian Water Authority and the tenant farmer. Following survey and ground examination a promising area was selected and work commenced in April 1973. The initial excavation was limited to an area 50 ft (about 15 m) square which was later expanded to 17 m by 20 m to include part of the medieval village street, indicated by a green way running the full length of the field.

Artefacts were found immediately following the initial removal of turf and soil, and the site continued to produce an abundance of pottery sherds and small finds throughout the three year period of activity on the site. As the work progressed, the outline of a substantial building was laid bare (Fig. 5), and the pottery recovered suggested a date range between the 11th and 15th centuries. The main rectangular structure appeared to contain two main rooms, separated by a lightweight wall. The smaller northern room contained a well-built paved hearth and a later wall-oven in the corner. Large deposits of ash and charcoal suggested continuous use over a long period. Coins found near the hearth confirmed a 13th century date, and most of the pottery sherds found within the room floors were of the same period. Part-glazed (in yellow and green) pottery was later assembled to re-form a large jug approximately 35 cm in height and 25 cm in diameter. Pottery mainly from the adjoining yard areas ranged from early Stamford cream ware, through developed Stamford wares, to Nottingham and Midland Purple. A large number of medieval nails spread across the site suggested the locations of collapsed roof timbers. Investigation at the northern end of the site revealed earlier foundations of large weatherworn wall slabs on the same orientation as the main building. Extension of the excavation into the village street revealed additions to the walls suggestive of lean-to rooms, and a trial trench cut from the southern room into the street produced a hoard of long-cross pennies and halfpennies of mid-13th century date. The road showed three layers of stone and pebble metalling. Further investigation of the site was precluded by the filling of the reservoir in 1976.

In all, the site produced over 4 500 sherds of pottery, some 250 small finds, including the remains of iron tools in poor condition, bronze items, and a selection of hones and whetstones. Bones and teeth recovered were mainly from cattle, ox, horse, pig, sheep and possibly deer. The hones and whetstones have been examined at the British Museum (Natural History), and the pottery and small finds are in course of examination by the Leicestershire Museums Service in co-operation with the excavators.

Documentary research carried out at the Leicester University Library, the Leicestershire Record Office, and the Lincoln Archives Office, has produced a wide range of references to the Hambleton area. The village of 'Hameldune' appears to have pre-Conquest origins, and it is possible that Nether Hambleton was included in the seven Berewics referred to in Domesday Book. Although the older buildings, still in existence before completion of the reservoir, can be traced back to the early 18th century, further research into older documents may confirm that the area in which the excavated building lies was allowed to become derelict after a plague period in the late 14th or early 15th century.

The Anglian Water Authority's deposit of all the recovered material at the Rutland County Museum has enabled the Group to re-assemble some of the pottery and provide representative displays of interest to visitors to the Museum. The site has produced the most varied and comprehensive collection of medieval material so far recovered from Rutland. The site will be fully published when post-excavation work and documentary research are complete.

A medieval complex of Whitwell (A.W.A.)

During the latter part of 1976, machinery being used for site clearance adjacent to Bull Brig Lane, Whitwell, uncovered the stonework of a Romano-British farmstead near the site of the present Fishing Lodge and Rescue Centre (Todd 1981).

Further work near the Old Hall, Whitwell, also exposed large quantities of worked stone. The examination of a very old mill dam, the collection of several hundred pottery sherds, and the location of earthworks indicated that several medieval building platforms existed under and around the scraped area. The excavation of the Romano-British site had been entrusted to the Trent Valley Archaeological Research Group and it was therefore suggested that the Rutland Field Research Group should investigate the medieval area (SK 924086). Although there was some Roman pottery amongst the unstratified material from the surface, the majority of sherds were medieval, ranging from early Stamford ware to Midland Purple and Cistercian wares.

The clearance of a machine-scraped area adjacent to the Water Authority's northern boundary fence was followed by excavation work up to the fence. Over the period 1977–79 an area some 25 m from east to west by a maximum of 15 m southwards from the fence was excavated down to a blue lias clay level. A complex of heavy limestone wall footings was revealed, mainly of rectangular outline but joined at the eastern end by part of a semicircular stone drain on which three courses of stone slabs had been laid. The main rectangular feature has well-built walls some 60 cm or more in thickness and measures some 13 m by 5 m. The wall between the rectangular area and the heavy drained wall also stood on a 'V' drain, and this dividing wall and the western wall ran under the present fence into the adjoining property. The area within the walls was paved with worn stone rubble over blue clay and no earlier features were found. It was evident that the western wall had been rebuilt on earlier larger foundations. Adjacent to it were the remains of a partition wall and an area paved with large purple pebbles. Artefacts recovered from the higher levels included a good collection of Cistercian pottery sherds, a variety of metal objects of iron and bronze, some flints and a range of smelted iron and charcoal samples. Pottery sherds found adjacent to and in the wall surfaces were of 13th century type. Towards the end of 1979 permission was obtained to expand the excavation into the adjacent grass paddock. This was considered essential in order that any northern boundary wall could be located as well as the continuation of the west wall and the dividing wall.

The end of the 1979 season and all the available working periods of 1980 concentrated on the removal of as much turf and subsoil as possible from the paddock area of the site. Removal of rubble from the western part of the area has revealed the continuation of the earlier west wall, a paved area delineated by large pebbles and, proceeding into the next higher house platform, a large circular area of iron ore, smelt and slag and, in the central area, a well constructed house wall of four courses of limestone. This wall appears to have been well robbed, but the clearance of rubble adjacent to it produced the footings of another north to south wall, and also a slab-paved floor approximately 4 m square. Removal of deeper layers of rubble near the boundary fence uncovered large slabs used as drain covers, and their position indicates the continuation of the dividing wall and drain from the southern area of the yard or building.

The pottery from the northern part of the site is mainly middle and late medieval, and the finds are mainly ferrous items including six table knives and a variety of nails. A few small bronze items were found in the form of short headed pins. A long-cross penny dating from about 1260 was found alongside the west wall, and another medieval coin came from the paved floor area. A fine pointed Neolithic arrowhead, evidently a stray, was recovered from the rubble in the same area.

It would appear that the buildings were occupied over a period ranging from the 12th to the 17th century, but the later part of this may have been concerned with quarrying for other building construction and also for ironstone burning. The seasons ahead should produce further building features and sufficient artefacts to confirm the present dating evidence. The material so far recovered is mainly held at the Rutland County Museum and awaits further study.

Acknowledgements

The authors express their gratitude to the Anglian Water Authority, their staff, consultants and contractors, for their very real assistance. The finds from the various excavations are deposited at the Rutland County Museum by courtesy of the Authority, as and when post-excavation research is complete.

The main archaeological participants in the excavations are mentioned in the text above, and the many individuals who have contributed to them will be acknowledged in the appropriate archaeological publications.

This paper has been compiled by the Rutland Field Research group in conjunction with the Rutland County Museum.

References

Clough, T. H. McK., Dornier, A. & Rutland, R. A., 1975. Anglo-Saxon and Viking Leicestershire, including Rutland. Leicestershire Museums.

Reynolds, N. M., 1976. The structure of Anglo-Saxon graves. Antiquity L: 140–144.

Todd, M., 1981. The Iron Age and Roman Settlement at Whitwell, Leicestershire. Leicestershire Museums Archaeological Report 1.

Circulation patterns; their investigation and prediction

E. J. Myers[1], R. S. Oldham[1] & A. J. D. Ferguson[2]
[1] *School of Life Sciences, Leicester Polytechnic, Leicester, England*
[2] *Welland and Nene Division, Anglian Water Authority, Oundle, Northamptonshire, England*

Keywords: Rutland Water, wind, water currents, drogues, algal distribution, physical model

Abstract

The effect of wind on the water currents in Rutland Water is being studied with a view to predicting algal distribution patterns so that, by exercising different abstraction options, water treatment problems can be avoided. Investigations using a scale, physical model of the reservoir in a wind tunnel suggest that under most wind directions the circulations tend to separate the water body into three distinct, topographic regions. Water currents in the reservoir itself have been plotted by tracking the paths of drogues using theodolites. This has indicated that in the central basin the circulation may be of the 'conveyor belt' type, rather than the solely horizontal type of the model. Surface chlorophyll readings are shown to indicate wind-driven water currents. Algal distribution has been predicted from wind data; however, the analysis requires modification in order to apply to all wind conditions. The effect of wind on the vertical distribution of algae and its implication for abstraction are discussed.

Introduction

Reservoirs in Britain with water treatment plants similar to that at Rutland Water commonly experience serious treatment problems one year in every six or seven due to high algal biomass (WRA 1971). The algae tend to clog the filtration system. It is hoped that these problems can be minimised in Rutland Water.

After considering the effects of the Earth's rotation on water circulation the present positions of the river inlet jets and two draw-off towers were chosen as optimal for algal management (Harper 1978). The input jets would cause the nutrient-rich river water to circulate westward along the surface to the southern arm and in turn this would cause a deep return drift eastwards to the central basin (Fig. 1). Coriolis force would then influence the water to continue into the northern arm. This predicted circulation was estimated to take approximately two months. Phytoplankton growth would be vigorous in the incoming nutrient-rich water, so that when the water had circulated as far as the draw-off towers the nutrient levels, and thus the algal biomass, would have fallen considerably.

Normally water is abstracted from the primary draw-off tower in the central basin but in times of exceptionally high algal biomass, water could be abstracted from the secondary draw-off tower in the north arm. It is impractical to wait until water containing unacceptably high algal biomass has reached the treatment works before deciding to switch to the secondary draw-off tower, especially since water trapped in the aqueduct between the two draw-off towers must be pumped through before normal abstraction can continue. In order to maximise the use of the facility it is important that a decision can be made before the algae reach the primary draw-off tower. It is therefore desirable to find a predictive relationship between the wind speed and direction, the consequent water circulation, and the movement and distribution of the plankton.

Hydrobiologia 88, 67–77 (1982). 0018-8158/82/0881-0067/$02.20.

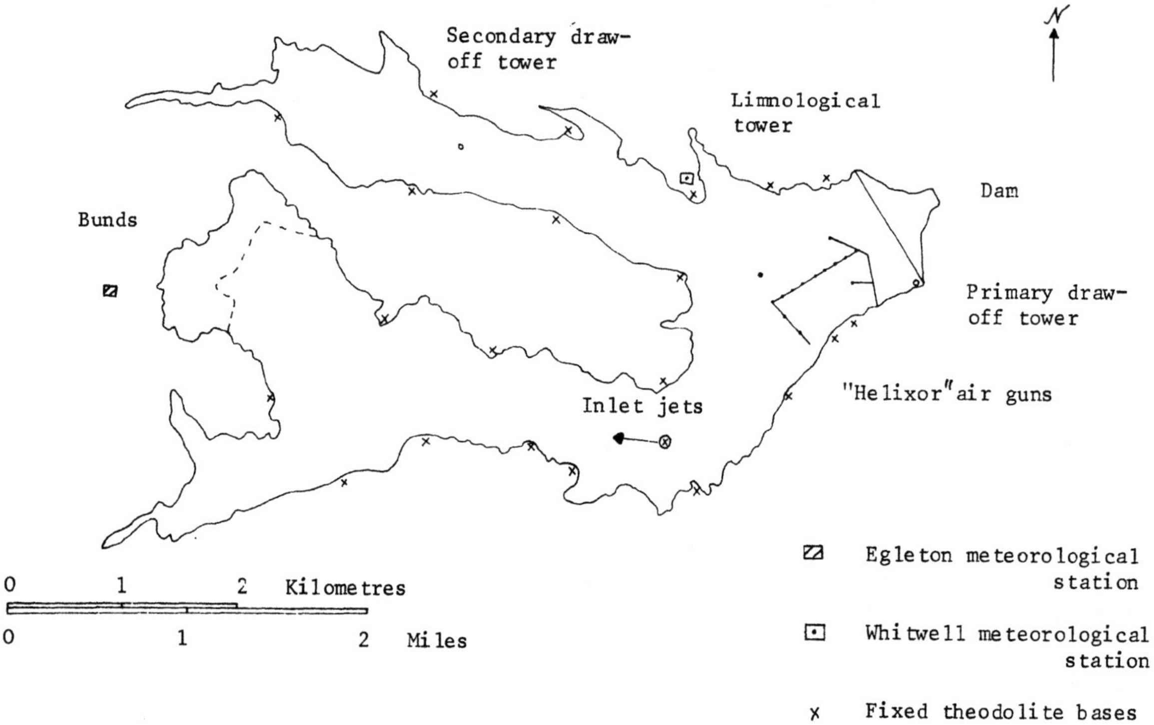

Fig. 1. Rutland Water showing positioning of water management features, the meteorological stations at Egleton and Whitwell and the 21 permanent theodolite bases.

At present the effect of wind on water circulation is being studied in two ways:

(a) by use of a physical model of the reservoir, and

(b) by *in situ* measurements at Rutland Water.

Physical model

The fibre-glass model is housed in a wind tunnel at the University of Nottingham. It measures approximately 3 m by 2 m, having a horizontal scale of 1 : 2 500 and a vertical scale of 1 : 625. The 4 : 1 vertical distortion is necessary; with a vertical scale of 1 : 2 500 the maximum water depth would be only 13 mm. The model incorporates the topography immediately surrounding the reservoir.

To date only surface water currents have been studied but if these are seen to reflect the true circulation patterns then deeper currents will also be investigated.

Patches of graphite powder were shaken on to the water surface, the wind propellor turned on and kept at a constant speed. After 10 min, which allows the circulation patterns to reach a steady state, the graphite patches were recorded photographically from above. Photographs were taken every 5 min, and each experiment ran for 90 min. Initially several experiments were conducted using wind of different speeds from the same direction. However, the same pattern developed each time, differing only in current speeds, and a single wind speed was used subsequently. This measured 45 cm s^{-1} at 16 mm above the water surface. Between each experiment the model was rotated through 45° to give results from eight points of the compass.

Results

Figures 2a-h show the graphite paths for each wind direction. The surface circulations are perhaps best explained as adjacent discs rotating clockwise and anticlockwise at different speeds with water able to flow from one to the next at the edges. Under every wind direction, except east, a

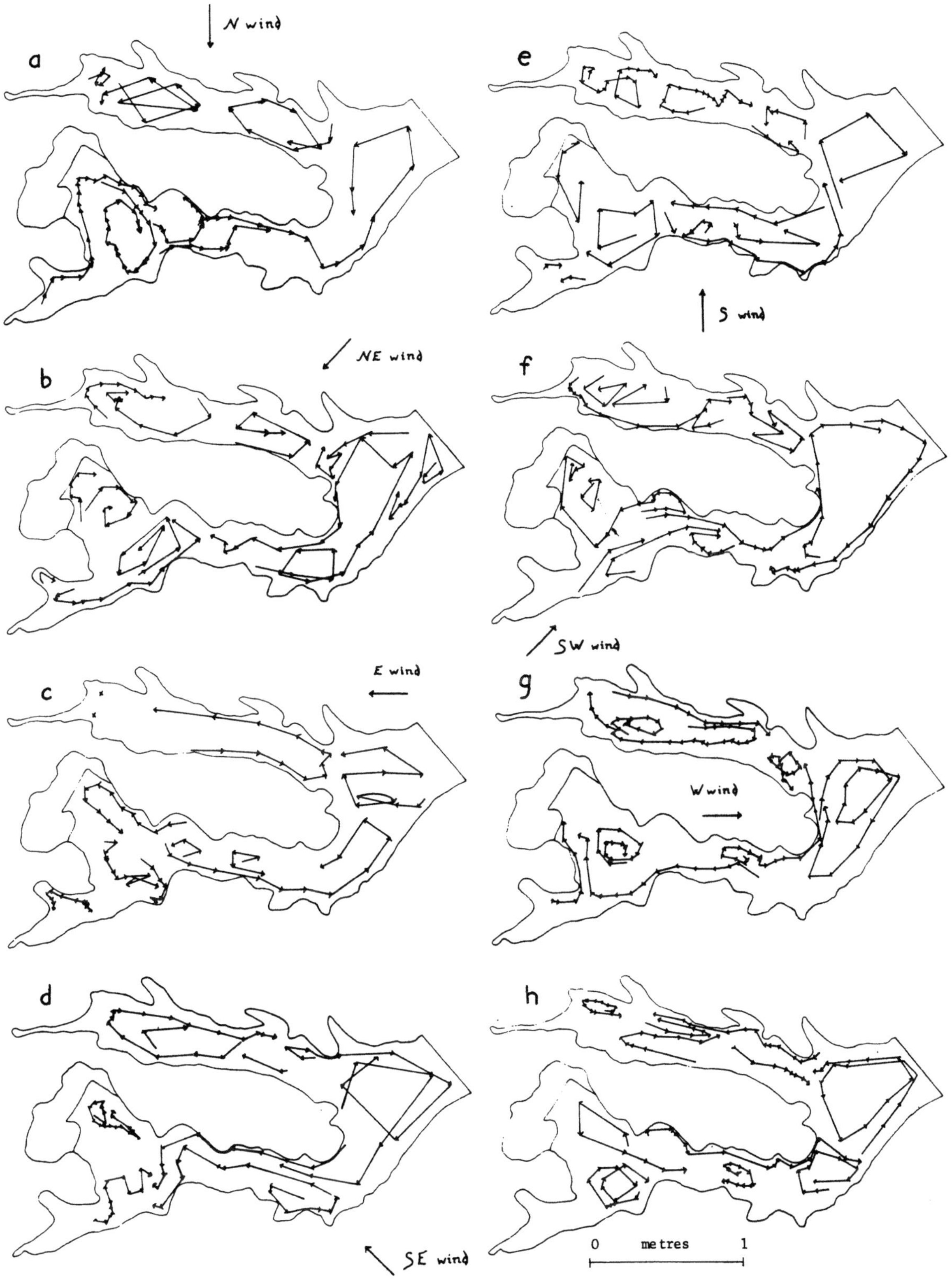

Fig. 2a–h. Surface water currents produced on a physical scale model of Rutland Water in a wind tunnel. The wind direction is given by a single arrow. Consecutive arrows show the movement of surface graphite patches at 5 min intervals.

single, fast circulation develops in the east of the reservoir. The graphite patches gave speeds between 3 and 7 cm min^{-1}. For all directions from southeast to west the rotation was clockwise and for all directions from northwest to northeast it was anticlockwise. These large circulations in the east tend to separate this region of the model from the two arms, only allowing exchange of water along the north or south bank, and around the end of the peninsula from one arm to the other. This hydrological division of the reservoir into three topographical components, i.e. the central basin, north arm and south arm, is supported by significant variations in algal biomass in the actual reservoir (Harper 1978). Consequently unacceptable levels of algal biomass may not pertain in the central basin and in the north arm simultaneously. If water were unacceptable in only one of them, then a switch to abstracting from the alternative tower could be made until either algal production had decreased or the wind had carried the algae away from the central basin.

When the wind is from the east, the single circulation in the central basin is broken down into at least two smaller ones, and with the rapid circulation in the north arm and eastern region of the south arm, the mixing of water between the three topographic regions may be greater than under other wind conditions.

Unless a water body is particularly shallow, water currents are thought to involve upwelling and downwelling at the enclosing walls (Smith 1979). In a morphometrically simple lake under conditions of steady wind there is a build up of water at the down wind end resulting in a slope on the water surface in addition to that due to wind stress. The slope can force a gradient current to flow back to the upwind end of the lake. This is a uniform flow throughout the depth of the lake, except at the bottom where it is reduced by friction. As the water level rises at the downwind end an equilibrium condition is eventually reached when the downwind drift transport is equal to that of the gradient return flow. A 'conveyor belt' type of circulation results from superimposing gradient and drift currents. The only clear indication of this in the model occurred with an east wind. Two graphite patches towards the west end of the north arm failed to move further west, suggesting that the water flow was downwards, and thereafter presumably in a reverse, easterly direction along the bottom. This is the only example of the graphite halting, suggesting that the model usually displays only circulations in a horizontal plane (i.e. the graphite paths may indicate the tops of rotating vertical cells of water).

There is evidence from the reservoir itself that there are conveyor belt circulations. This comes mostly from an analysis of temperature profiles on transects across the reservoir (Maddocks, pers. comm.). When the water is thermally stratified, downwelling involves the transport of warm water to depth and vice versa for upwelling. However, the positioning of these vertical movements is not always easily related to the wind direction and it may be that what appears as a distinct horizontal circulation in the model occurs as a distinct conveyor belt circulation in the reservoir. This has yet to be confirmed.

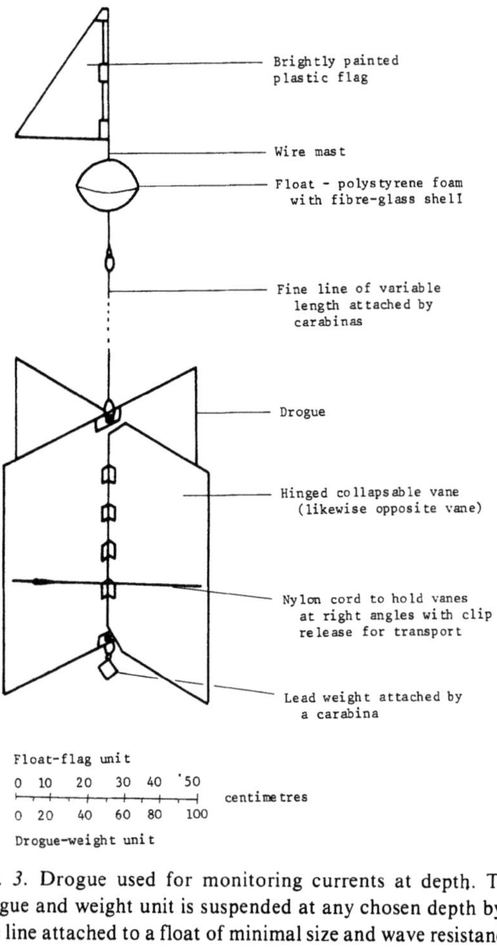

Fig. 3. Drogue used for monitoring currents at depth. The drogue and weight unit is suspended at any chosen depth by a fine line attached to a float of minimal size and wave resistance. Progress is tracked by simultaneous theodolite fixes on a marker flag carried by the float.

Fig. 4a–e. Paths of drogues at 1 m depth in the central basin. The positions of the theodolites are shown (+). The wind speed and direction on the day of tracking and four previous days are shown in progressive vector diagrams.

In situ current measurements

The second method of determining the circulation patterns is by *in situ* current measurements at Rutland Water.

Since Lagrangian flow, rather than Eulerian flow, is ultimately of importance, drogues and not current meters are being used. The drogues, each attached by a fine line to a surface float and marker flag, have been designed for use at any depth in the reservoir. The drogue consists of four vertical polypropylene vanes each measuring 61 cm × 122 cm and forming a cross when viewed from above (Fig. 3). The surface float consists of a fibre-glass shell filled with polyurethane foam, and having stiff wire as a mast for a marker flag. The drogue and float are joined by a fine line of any chosen length. As the marker flag, rather than the drogue, is tracked it is necessary that the drogue hangs as nearly vertical below the float as possible. A 2.3 kg lead weight is attached to the centre bottom of the drogue to achieve this effect. During transport, all the components (float-flag unit, line, drogue, and weight) are detached and the drogue can be collapsed into a single plane.

The drogues are released by boat and left undisturbed whilst their paths are tracked by simultaneous fixes from two land-based theodolites. Theodolites were chosen as the best tracking method after consideration of range, accuracy and cost (Goransson & Svensson 1976). There are 21 permanent bases around the perimeter of the reservoir from which the theodolites can be operated (Fig. 1). The drogue paths are then plotted on a map of Rutland Water. Wind data for correlation with water currents are available from the Egleton meteorological station, from the Whitwell fishing lodge and also from readings taken *in situ*.

The following results are from drogues at 1 m and so are comparable with the results from the model. The wind speed and direction for the day on which the drogues were tracked, and for the 4 preceding days are shown as a series of vectors in each figure.

4 March 1981 (Fig. 4a)

On the previous day the wind was from the northeast, at a speed of 6 m s^{-1}, changing through 180° only on 4 March.

The actual directions of the drogues vary from east-southeast to northeast suggesting that the recent turn in the wind has not completely overruled the surface currents caused by the previous day's wind, although as all three drogues show a trend in a northwesterly direction the response to wind change has been relatively rapid in the surface waters.

Under the gentle winds of 4 March (1 m s^{-1}) the drogues travelled slowly, at only 1.5 cm s^{-1}. This speed agrees with the rule of thumb (Haines & Bryson 1961) that surface water currents are 1–2% of the wind speeds.

These results do not contradict those of the model (Fig. 2f), but a complete comparison is not possible as the drogue paths are so short and the wind history is not uniform.

12 March 1981 (Fig. 4b)

The wind history for 12 March is far more uniform, being a constant south-westerly on the two previous days, and remaining so on the 12th. It was also stronger, averaging 5 m s^{-1}.

The drogues followed a strong, almost unidirectional, flow. They travelled at a speed of approximately 10 cm s^{-1}, which is 2% of the wind speed.

Under southwesterly winds the model also showed very strong, straight currents along the edge of the central basin but they were in the opposite direction (Fig. 2f). This probably indicates that there is a conveyor belt circulation operating in the central basin and not the purely horizontal circulation of the model. This is supported by the fact that the leading drogue was swept on to the dam wall rather than northwards around the basin.

17 March 1981 (Fig. 4c)

The wind averaged 4 m s^{-1} from the north for the two preceding days and turned to the northwest and increased to 6 m s^{-1} on 17 March. The drogues clearly show predominantly southerly rather than southeasterly flowing currents. This apparent lack of response to the wind may be due to the deflection caused by Coriolis force (Ekman 1905; Smith 1979), but this possibility has not yet been investigated.

As in many of the model results, the water current around the end of the peninsula is very rapid, averaging 13 cm s^{-1}. The results indicate a conveyor belt circulation, with downwelling at the southern shore and upwelling at the northern shore, rather than the horizontal circulation of the model.

7 May 1981 (Fig. 4d)

The wind history was extremely variable, changing through 225° and ranging from 0 to 13 m s^{-1}. On 7 May the wind was from the east, increasing from 5 to 10 m s^{-1} during the day.

In spite of such a strong wind both drogues initially travelled northwest though gradually turning further west. Their speeds varied from 4 to 9 cm s^{-1} i.e. approximately 1% of the wind speed.

The deviation from the wind direction is probably due to complicated residual circulations caused by the previous days' wind combined with a deflection to the right caused by Coriolis force. The drogues show no evidence of the two oppositely rotating horizontal circulations found in the central basin of the model during an east wind (Fig. 2c). Instead they show that much of the surface water from the central basin flows towards the north arm.

28 May 1981 (Fig. 4e)

Four days prior to 28 May the wind was steady from the south at an average speed of 6.5 m s^{-1}. During the intervening three days it varied in direction and decreased in strength. On 28 May it was southwesterly increasing from 3 to 5 m s^{-1} during the day.

Two drogues travelled in the same direction as the wind, and one showed deviation to the right. The speeds were 1–2% of the wind speed.

The model shows a single horizontal clockwise circulation under a southwest wind whereas in the reservoir most of the surface water from the central basin flows towards the dam. Unless there was a much faster return current from the dam along the northern bank the water reaching the dam must be sinking and returning at depth as a conveyor belt circulation. The latter seems more likely, though tests using more drogues at a range of depths will be necessary.

Phytoplankton

The aim of this study is to reduce filtration problems in water treatment and so it is necessary to determine to what extent water circulation is reflected by the pattern of algal distribution. In these initial studies only the total quantity of algae have been considered but, because the varying nature, shape and size of different species control the extent to which they hamper filtration, it is ultimately necessary to know both the biomass of algae, and the species composition.

Continuous fluorimetry readings

Figure 5 shows the results of continuous fluorimetry readings taken in zig-zag transects from the surface of the reservoir, on 10 August 1979. The wind was from the north.

It is clear that the chlorophyll distribution is markedly affected by the wind, resulting in an increase in algae towards the southern bank. However, the increase is not uniform, and there is evidence of a more complicated circulation pattern. There are indications of a fast current around the end of the peninsula from north to south. This agrees with results from the model.

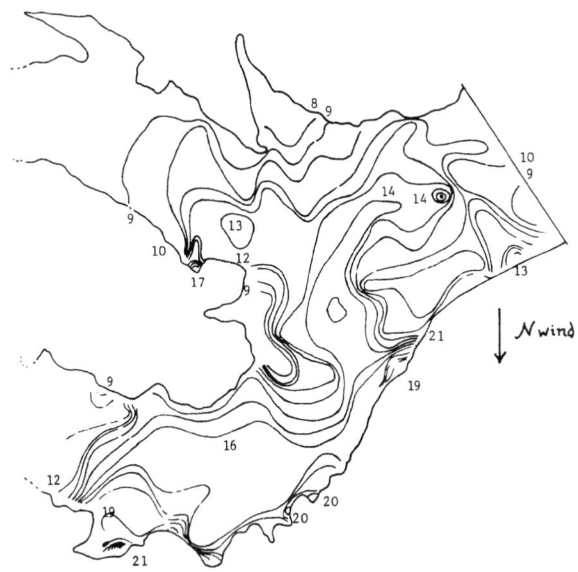

Fig. 5. Results of continuous fluorimetry readings taken in zig-zag transects from the surface of the reservoir on 10 August 1979, showing lines of equal fluorescence in g l^{-1}.

Table 1. Effective displacement index (E.D.I.) and displacement vector (D.V.) values, wind characteristics and the algal distributions they predict. From Small (1963).

E.D.I.	Wind speed	D.V.	Wind direction	Algal distribution
Small	Light	Small	Fluctuating	Patchy, disorganised distribution, with values at each station seemingly unrelated
Small	Light	Large	Uni-directional	Pile up on the leeward side of the lake with relatively little vertical mixing
Large	High	Small	Fluctuating	Homogeneous
Large	High	Large	Uni-directional	Pile up on leeward side of lake with much vertical mixing

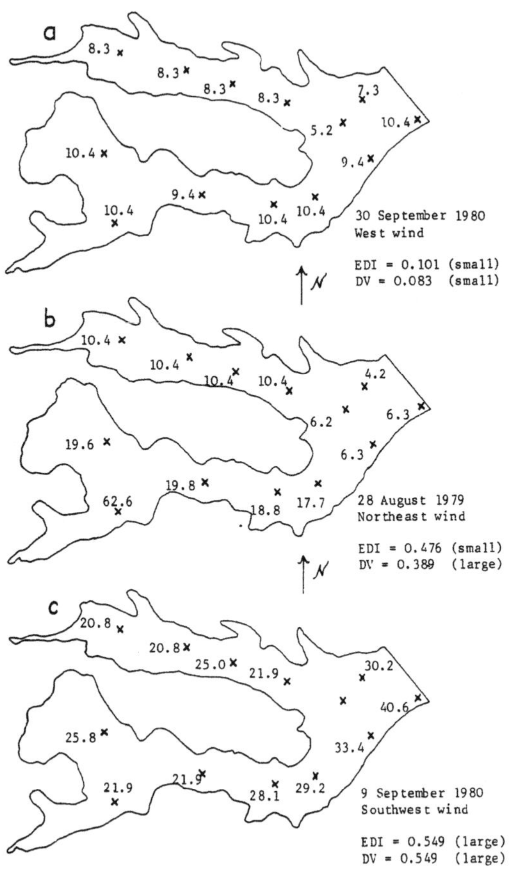

Fig. 6a, b and c. Three examples of the correct prediction of chlorophyll (g l^{-1}) distribution from Small's (1963) Effective Chlorophyll Displacement Index (E.D.I.) and Displacement Vector (D.V.). The E.D.I. and D.V. are calculated from wind data, provided by the meteorological station at Egleton, for the preceding ten days.

Usually the highest chlorophyll levels were found in bays along the south bank, the exception being an increase to 16 μg l^{-1} chlorophyll *a* in the middle of the central basin. This might be explained by a circular flow around the maximum, in which case the model may be nearer to reflecting the reservoir flow patterns than the drogue results imply.

Application of Small's (1963) wind analysis to Rutland Water

Small (1963) developed a method of analysing wind data in order to predict the chlorophyll distribution expected on Clear Lake, Iowa. By weighting the wind speeds on the day of sampling, and on the ten days preceding this such that the sampling date received the greatest weight, Small arrived at an Effective Chlorophyll Displacement Index (E.D.I.). This is a measure of the accumulated wind stress on the algae. By combining the E.D.I. with the most predominant direction in which it has acted, a Displacement Vector (D.V.) can be calculated. Small classified the EDI and DV as 'large' or 'small' and predicted the type of chlorophyll distribution expected under each combination of the two factors (Table 1).

Figure 6a, b, and c show selected examples, taken from Rutland Water, where the pattern of chlorophyll distribution can be predicted from the wind data in this manner. The chlorophyll levels are provided by the AWA weekly sampling programme in which water is collected from a 5 m tube at 13 sites around the reservoir. The wind data are taken from the meteorological station at Egleton and the Whitwell Fishing Lodge.

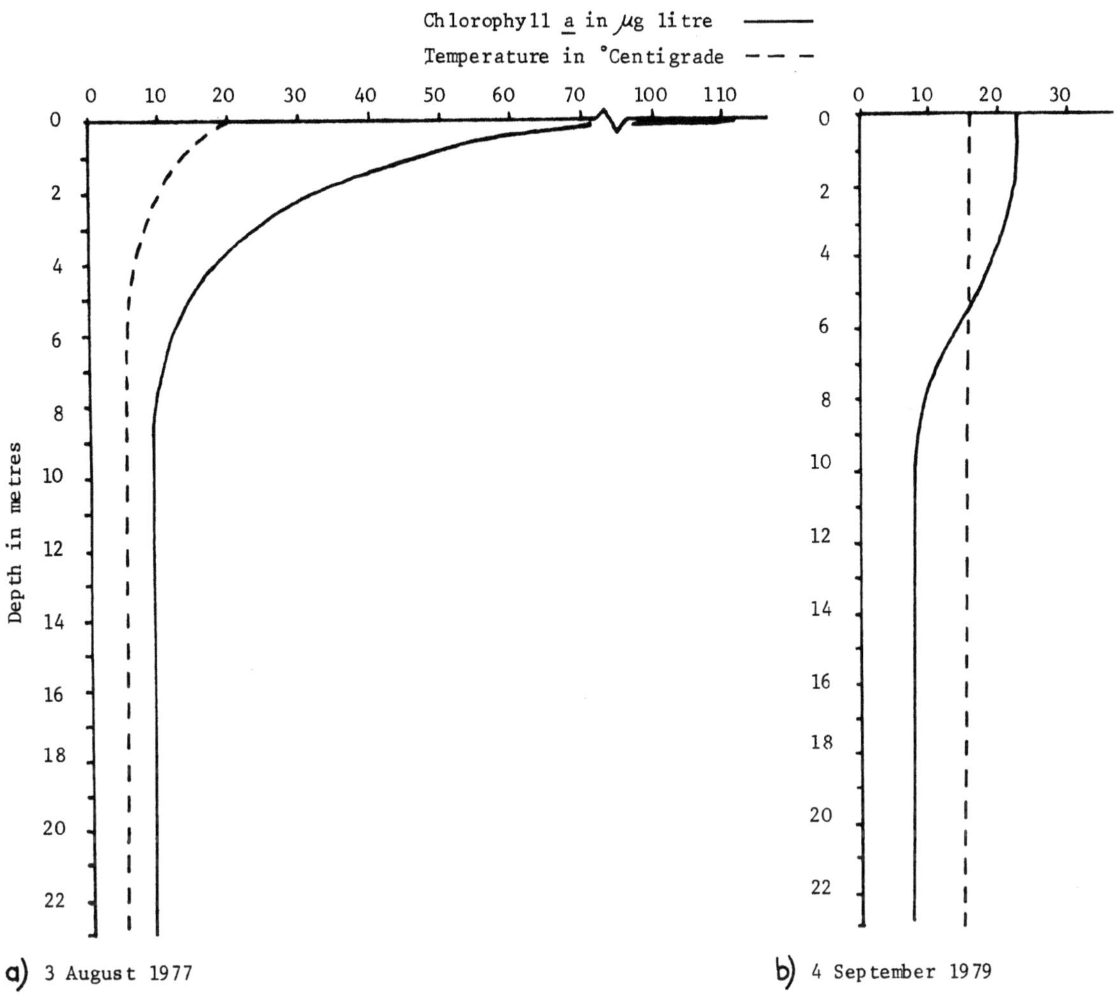

Fig. 7a and b. Vertical profiles of chlorophyll *a* (———) and temperature (---) measured at the Limnological Tower. a) Unidirectional wind with a speed of less then 3 m s^{-1}. b) Unidirectional wind with a speed of 6 m s^{-1}.

Although this analysis can, for certain days, predict algal distribution in Rutland Water from the appropriate wind data, it does not always do so. Clear Lake is smaller, shallower, and of a more regular shape than Rutland Water, and it is most likely that the complicated hydrology of the latter gives rise to the discrepancy. However, such predictive analysis would be invaluable and an attempt is being made to formulate a suitable analysis for use at Rutland Water.

Vertical profiles of chlorophyll a

Water can be abstracted from a range of depths at each draw-off tower. The primary draw-off tower has outlets at 79, 74, 69 and 55 m O.D., and the secondary at 79, 75 and 72 m O.D. Vertical profiles of chlorophyll *a* are regularly measured by AWA at the limnology tower, in the central basin. Figure 7 shows those taken when the surface chlorophyll readings were at or near the maximum for each year.

Results for 3 August 1977 show an enormous decrease in chlorophyll a and temperature within the top 2 m. The algae were mostly the buoyant blue-greens dominated by *Aphanizomenon*. At first such discrete concentrations may seem an advantage for abstraction purposes, as water could be drawn off at depth. However, the algae only form such dense surface blooms when the wind speed is below 3 m s^{-1}. Under continuous, gentle breezes from a constant direction the blooms move laterally and concentrate downwind (Small, *op cit.)*. For example on 3 August 1977 there were gentle (1.5 m s^{-1}) north-westerly winds and the algae accumulated in the southeast.

By contrast, on 4 September 1979 the chlorophyll profile shows no reduction until a depth of 4–5 m, and there is no thermal stratification. This greater uniformity with depth is due to the mixing effect of the much higher wind speeds on the sampling day and on preceding days (Smith 1979). The wind averaged 6 m s^{-1} and was steady from the south-west, causing the algae to accumulate in the north-east of the central basin.

Wind will mix surface blooms of algae into deeper water. Strong winds acting on water with weak or non-existent stratification can cause mixing throughout the whole depth of the reservoir so that abstracting from depth to avoid surface blooms is not feasible. However, provided there is strong stratification this mixing may only occur within the upper layer of water, in which case water abstracted from below the thermocline will have a greatly reduced algal content.

Discussion

There is a conflict between the type of circulations indicated by the model and the drogue work. The former suggests vertical, rotating cells of water, whereas the latter suggests conveyor belt circulation. At this stage in the study it is not possible to dismiss either, as the separation into three regions shown by the model is supported by observation in the reservoir itself. In order to resolve the conflict future work on the reservoir will include:

(1) Tracking a greater number of drogues on each occasion.
(2) Positioning of drogues at a range of depths.
(3) Continuous tracking of drogues over several consecutive days.

It is hoped that this will provide evidence of any deep return currents and the change of circulation under changing wind conditions.

Acknowledgements

We would like to thank M. Swan, M. Knight, T. Gibson and R. Hayes of Leicester Polytechnic for help in drogue tracking; M. H. Jones for advice and practical assistance; School of Building, Surveying and Land Economy, Leicester Polytechnic, for the loan of theodolites; Department of Architecture, University of Nottingham, for use of the wind tunnel; and the Anglian Water Authority for funding and assistance at all stages.

References

Ekman, V. W., 1905. On the influence of the earth's rotation on ocean currents. Ark. Met. Astr. Fys. 2(11): 1–52.

Goransson, C. G. & Svensson, T., 1976. Drogue measurements – computer program for evaluation incl. correction for deviating drag forces. Department of Hydraulics, Chalmers University of Technology, Gothenburg. Communication No. 85.

Haines, D. A. & Bryson, R. A., 1961. An empirical study of windfactor in Lake Mendota. Limnol. Oceanogr. 6: 356–364.

Harper, D., 1978. Limnology of Rutland Water. Verh. int. Verein. Limnol. 20: 1604–1611.

Small, L. F., 1963. Effect of wind on the distribution of chlorophyll a in Clear Lake, Iowa. Limnol. Oceanogr. 8: 462–432.

Smith, I. R., 1979. Hydraulic condition in isothermal lakes. Freshwat. Biol. 9: 119–145.

Water Research Association, 1971. Empingham Reservoir – biological considerations. W.R.A., Medmenham, U.K.

Discussion

In introducing her joint paper, Miss Myers explained that the prime aim of both field and wind tunnel studies was to provide information on the circulation in the reservoir under various wind regimes. From this information, the movement of algae could be predicted and essential decisions taken as to whether the primary or the secondary off-take point should be used. She then elaborated on the use of graphite applied to the water surface in wind tunnel experiments and the use of flag carrying drogues on the reservoir, and presented examples of the results.

Opening the discussion, Mr. Youngman said that he and his associates had found it difficult to track drogues even on small water bodies. He wondered if drogues were the best solution and enquired what alternatives had been considered, and why drogues tracked by theodolites had been used. He suggested that, if this was the most suitable approach, it might be better to concentrate on one part of the reservoir rather than hope to study the whole. He was doubtful if observations made in different areas under differing weather conditions could be satisfactorily related.

Turning to the use of a physical model, he was concerned that surface tracers such as graphite could not be used to study jetting and wondered how this aspect could be included in the model, particularly under 'wind-less' conditions. Finally, he enquired how the authors intended to study the effect of wind action on sub-surface currents.

In reply, Miss Myers said that, with experience, a drogue could be discerned with a theodolite at *ca* 2 km and successfully tracked, even when moving fast. She considered that, by droguing under similar weather conditions on different parts of the reservoir, an integrated picture would emerge. At present she was using four drogues but an increase to six or more was likely, particularly under more slow moving conditions.

Dr. Oldham added that they were well aware of the shortcomings of the physical model and that they envisaged it being used purely to generate hypotheses for field tests.

Dr. Bullock enquired what effect the 24 hr tide reported by Miss Maddocks might have on the circulation. Miss Myers replied that it would be certain to have an effect, particularly in the presence of a thermocline when the tide would be complemented by seiching. The effect would be much less than in a deep rectangular lake.

Mr. Stroud commented that he had used small floats which stayed a short distance below the surface in a physical model, so that it was not entirely a surface effect. He asked how this compared with the use of graphite. Miss Myers replied that graphite formed a very fine layer on the surface which she believed was most appropriate for modelling surface movement.

Vertical movements of the surface water and thermal stratification in Rutland Water

S. L. Maddocks[1]
Oceanography Department, University College of Swansea, Swansea SA2 8PP, Wales
[1] *Present address: Department of Zoology, University of Leicester, Leicester LE11 7RH, England*

Keywords: seiches, wind-stress, stratification, upwelling, reservoirs

Abstract

Vertical movements of the water surface in Rutland Water were examined using water level gauges. The timing and extent of the surface seiches and slopes due to wind-stress were not well correlated with wind speed and direction; this is thought to be due to the complex shape of the reservoir. Unexplainable diel oscillations of the water surface occurred, and an examination of the thermal stratification showed the presence of upwelling and downwelling.

Introduction

Water currents in reservoirs are controlled mainly by friction, Coriolis and externally applied forces, which are wind-stresses on the water surface. Water movements are made more complex by the presence of shores, thermal stratification and water oscillations (seiches) induced by a change in wind speed or direction.

Large basins, where Coriolis force is significant, have complex surface and internal oscillations which develop when the lake is subject to wind-stresses (Smith 1975). Several axes of oscillation can occur and may rotate around the basin.

Nakamura & Honda (1911) studied the annular caldera Lake Toyaka in Japan and suggested the possibility of two seiches at right angles, with periods of 10.3 and 11.2 min. Endrös (1903, 1906) recorded seventeen different seiches in the Chiemsee of Bavaria, excluding periods of less than 3 min. The shorter period seiches produce very complex patterns due to islands present across the nodal lines; some seiche periods are sensitive to water level changes (Hutchinson 1957, pp. 299–331).

Pressure and wind changes start a lake oscillating and depend greatly on the geophysical position and surroundings of the lake. Seiches can be generated by local rain showers (Chiemsee and Loch Earn) and a sudden inflow of flood water (Hutchinson 1957). Seiching may help to keep bays and dendritic parts of the reservoir clear of sediment build-up although it creates extra stress on the dam. Therefore, seiche prediction is important to engineers for dam construction and upkeep.

Rutland Water is a man-made eutrophic reservoir lying 30 km east of Leicester. It has a length of 8 km, a surface area of 1 260 ha and a capacity of 135 $\times$ 10^6 m^3. The maximum depth is 34 m (in the central basin) with a mean depth of 10.7 m.

The complexity of Rutland Water arises from its overall shape, size, general topography and the surrounding countryside. The effect of these complexities are illustrated by the results of this study on water circulation. No previous work on seiching, wind-stress, and upwelling and downwelling have been carried out on this reservoir. However, thermal stratification is routinely measured at the Limnological Tower by the Anglian Water Authority, but not throughout the reservoir.

Hydrobiologia 88, 79–87 (1982). 0018-8158/82/0881-0079/$01.80.

Methods

Water height and wind recording

Negretti Zambra water level gauges were placed as far apart as possible, at either ends of the southern arm (Fig. 1). The sensors were fixed under the water, and a mark made on the gauge paper, indicating the starting time, and the water height above the sensor. The gauges were accurately levelled by the A.W.A. to enable a comparison of the records. The sensors were lowered because of unforeseen water loss due to evaporation and draw-down. The paper was changed and the gauges rewound every week. The error in determining water height is stated by the manufacturer as being ±2 cm. This factor refers to the accuracy of absolute determination and presents problems in the diagnosis of water-sloping but has no apparent significance in the recognition of the occurrence of a seiche. Complete records were collected as shown in Table 1. However the gauge at Sounding Bridge was malfunctioning for a few weeks of the survey. The wind data was obtained, with kind permission of the RAF, from Wittering near Peterborough. These data are comparable to the winds found at Rutland Water but are slightly greater in strength. RAF Wittering is 12 km from Tail Bay and 18 km from Sounding Bridge.

Temperature and oxygen measurements

A Yellow Springs temperature and oxygen probe was used at 14 stations throughout the reservoir

Table 1. Dates of complete records (+), collected from water level gauges at Tail Bay and Sounding Bridge.

Date From	To	Tail Bay	Sounding Bridge
02.06.80	09.06.80	+	+
09.06.80	15.06.80	+	+
15.06.80	22.06.80	+	
22.06.80	29.06.80	+	+
06.07.80	13.07.80	+	
20.07.80	27.07.80	+	
27.07.80	03.08.80	+	
15.09.80	22.09.80	+	+
22.09.80	28.09.80	+	+
28.09.80	07.10.80	+	+
07.10.80	14.10.80	+	+
14.10.80	19.10.80	+	

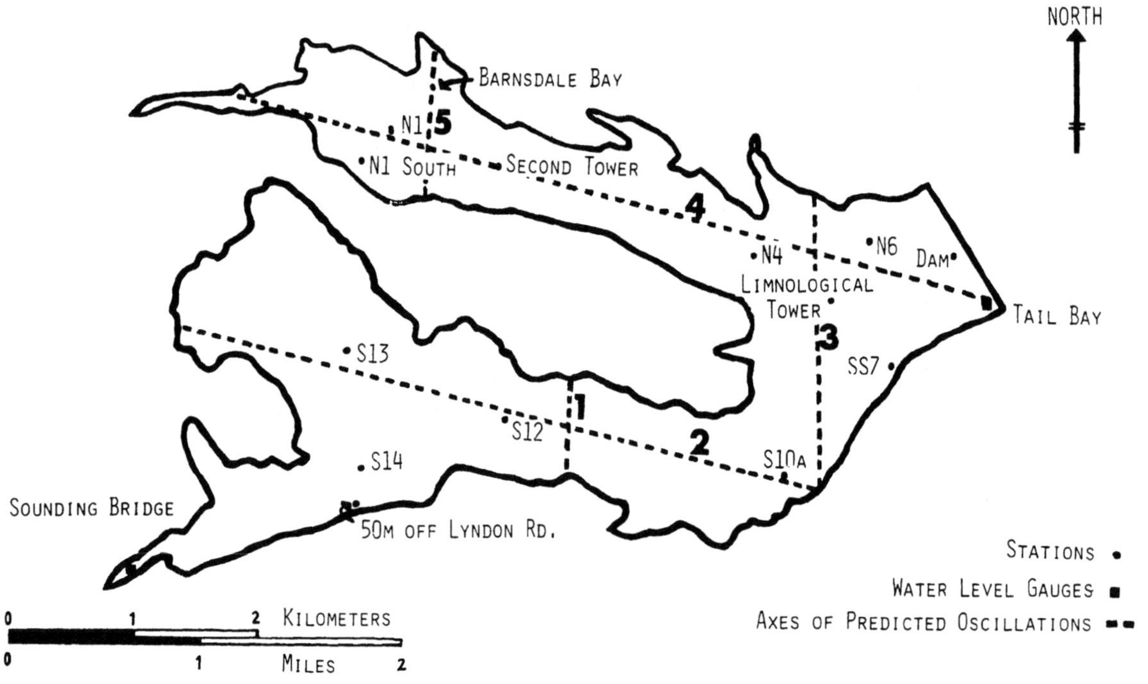

Fig. 1. Positions of stations and equipment used during the survey and predicted axes of oscillations (Table 4).

(Fig. 1). The readings were taken from the bottom to the surface at 1 m intervals, and the oxygen readings were converted to % oxygen saturation.

Results

Seiches

Seiches were recognised as a major deviation from the normal straight-line trace followed by a series of oscillations of decreasing amplitude. During the survey 18 such seiche records were obtained (Table 2). The amplitudes of oscillations ranged from 0.25 to 4.0 cm, and four characteristic records are dealt with in detail in this paper (Table 3).

On 5.6.80, a maximum oscillation of 0.5 cm from 1 930 to 2 000 h at Tail Bay (TB), and 1 900 to 2 300 h at Sounding Bridge (SB) occurred. From 1 800 to 2 400 h the water surface was 1.4 cm higher at SB than at TB. Before this the mean wind speed of 10.2 km h^{-1} gusting from 150° may have piled up water near SB, and reduced the amount at TB. When the wind changed direction and increased in speed, water may have been pushed further towards TB, thus releasing pressure at SB initiating the water movement away from the area. This would have started an oscillation of the water mass due to an overshoot of the equilibrium state. The TB oscillation was damped by wind-induced water movement towards TB.

Table 3. Wind speed, direction and gusts at times on certain days pertaining to seiches observed.

Date	Time (BST)	Wind		
		Direction	speed (km h^{-1})	Gust speed (km h^{-1})
05.06.80	1800	170	18.5	29.5
	1900	230	22	37
	2000	280	26	37
	2100	290	18.5	31.5
	2200	290	15	26
	2300	280	16.5	24
	2400	280	22	33.5
14.06.80	1000	40	28	39
	1100	50	24	39
	1200	60	18.5	31.5
	1300	60	15	31.5
	1400	180	22	35
	1500	170	29.5	52
	1600	170	29.5	52
30.09.80	0300	260	15	22
	0400	280	18.5	37
	0500	290	18.5	28
	0600	290	15	26
	0700	270	15	24
	0800	290	16.5	24
07.10.80	1200	230	44.5	66.5
	1300	230	41	70.5
	1400	240	37	65
	1500	250	44.5	76
	1600	250	41	59.5
	1700	240	37	57.4
	1800	230	39	59.5
	1900	230	37	55.5
	2000	230	37	61
	2100	230	35	55.5

Table 2. Amplitude (cm) and times of water oscillations recorded at Sounding Bridge and Tail Bay. The start = initial deviation from the normal line (heighest amp.). Finish = the time when there appeared to be no deviation from the normal, i.e. the vertical oscillations had stopped. Time in brackets = trace was broken, and the oscillations had not stopped. - = oscillations declined very slowly.

Date	Tail Bay			Sounding Bridge		
	Start	Finish	Amplitude	Start	Finish	Amplitude
05.06.80	1900	2300	0.5	1930	2000	0.5
10.06.80	0300	2100	0.75			
14.06.80	1200	2000	1.25			
16.09.80	0800	2400	0.75	1900	2400	0.75
17.09.80	1100	1300	0.25	1100	1300	0.25
18.09.80	1700	1800	0.25			
30.09.80	0500	(1400)	1.5			
02.10.80	0400	(1200)	2.5			
06.10.80	1800	-	2.0	1800	-	1.5
07.10.80	1400	-	4.0	1700	1900	0.75
08.10.80	1500	-	1.0			
11.10.80	1100	-	2.25			
15.10.80				1000	1200	1.0

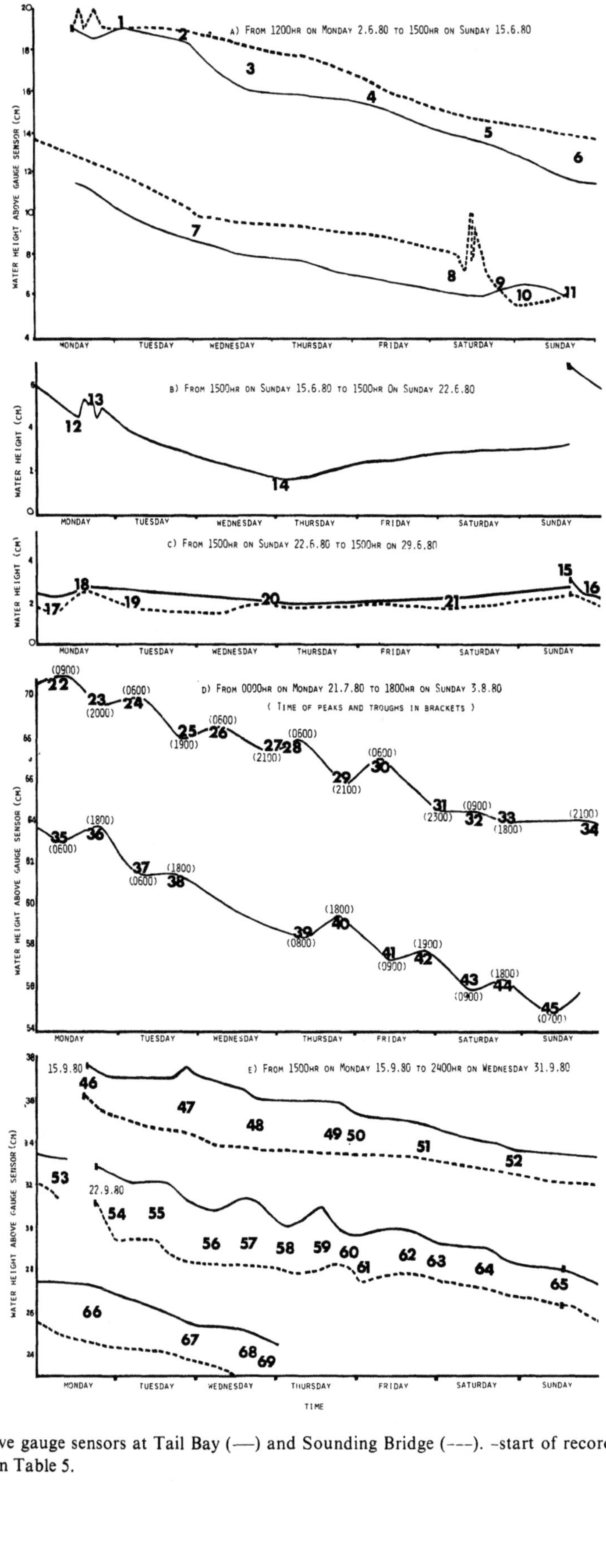

Fig. 2. Water heights above gauge sensors at Tail Bay (—) and Sounding Bridge (---). -start of record trace. The numbers 1–69 correspond to wind data on Table 5.

The seiche amplitude between 1 200 and 2 000 h on 14.6.80 at SB was 1.25 cm, and by 1 200 h, the SB water level was 4.4 cm higher than at TB. However, by 2000 h the water level was equal, with a wind direction of 190°, and a speed of 37 km h^{-1} gusting 61 km h^{-1}. From 1 400 h, the water level at TB began to increase due to the wind changing direction and relieving pressure at SB, resulting in water flowing back towards TB. At 0500 h on 30.9.80, the seiche amplitude was 1.5 cm maximum, and continued until 1 400 h when the trace was discontinued. Before 0500 h, the wind direction was 260° at 54 km h^{-1}. The wind increased up to 18.5 km h^{-1} gusting 37 at 0400 h, and then decreased to 28 km h^{-1} at 0500 h decreasing further until 0700 h. This decrease could account for the seiche traces at SB.

From 1 400 h at SB and 1 700 h at TB on 7.10.80, the seiche amplitudes were of 4 and 0.75 cm respectively. The TB trace lasted until 1 900 h, while the SB trace declined slowly into 8.10.80. The wind speed decreased from 48 to 37 km h^{-1} between 1 000 and 1 500 h, and then increased to 44.5 km h^{-1} at 1 500 h, while the direction changed from 230° (1 300 h) to 250° (1 500 h). The seiche could be due to either the sudden decrease and then increase in wind speed or gusting, or to the change of wind direction.

The records obtained were a summation of water movements, and therefore to pick up only one seiche between two points is very unlikely (Fig. 2 and Table 4). For example, a value recorded at position A on Fig. 2 may be due to an oscillation along axis 2 or axis 3, or a summation of the two

Table 4. Some predicted seiche periods (t), based upon Merian's formula:
$t = 2L/(g\bar{Z})^{1/2}$
where L = length of basin at surface, $\bar{Z}$ = mean depth of basin, and g = gravity = 980.6 cm^{-1} s^{-1}.

Number	Length (m) L	Mean depth (m) $\bar{Z}$	1st seiche period (min) t
1	1000	10.25	3.32
2	4000	10.25	13.30
3	2500	20.00	5.95
4	4000	8.75	14.40
5	1000	8.75	3.60
Whole reservoir	8000	10.70	26.03

oscillations. The direction of a seiche is mainly dependent on the wind characteristics, but many other factors have to be taken into account, e.g. shape of the reservoir. To obtain a more definite picture of water movements during seiching, more gauges would have to be placed around the reservoir.

The influence of any sudden movement of water could be greatly damped by the time the water fluctuations reach either end of the two arms. The areas of greatest influence are probably around the western portions of the two arms and within the central basin. There may not have been seiches occurring between SB and TB, but two separate seiches picked up by the two gauges. This would account for there being a seiche occurring at one station only for some of the time.

Wind-stress and sloping of the water surface

The water levels at TB and SB are displayed in Figs. 3a–e, and the relevant wind characteristics are shown in Table 5. It is generally accepted that water surface slopes are caused mainly by wind dragging the water; a change of wind speed or direction will cause a change in slope. If this is so, Figs. 3a–e and Table 5 show varying degrees of anomalous behaviour. For example, in No. 5 of Table 5, the water height at SB was 1.1 cm above that at TB with a wind speed of 18.5 km h^{-1} (Fig. 3a). However, by 2 000 h on 8.6.80, the water height difference increased to 2.4 cm at SB. The wind speed was only 5.5 km h^{-1} and from almost the same direction.

Diel oscillations occurred from 21.7.80 to 3.8.80 (Fig. 3d). From 21.7.80 to 26.7.80 the peaks were around 0600 h and troughs around 2 100 h. Later, from 27.7.90 to 3.8.80, the peaks were around 1 800 h and troughs at 0900 h. There is no definite explanation of this. It was not due to a sudden input or output of water, and there was no periodic wind speed or direction changes capable of producing the traces at TB. The explanation is complicated further by the peaks and troughs reversing through 12 hours after 26.7.80. The gauges do not appear to be sensitive to changes in air temperature or differences between air and water temperature. A possible explanation is barometric pressure, but this seems unlikely.

The TB trace was more variable than SB trace from 15.9.80 to 28.9.80 (Fig. 3e), decreasing in steps

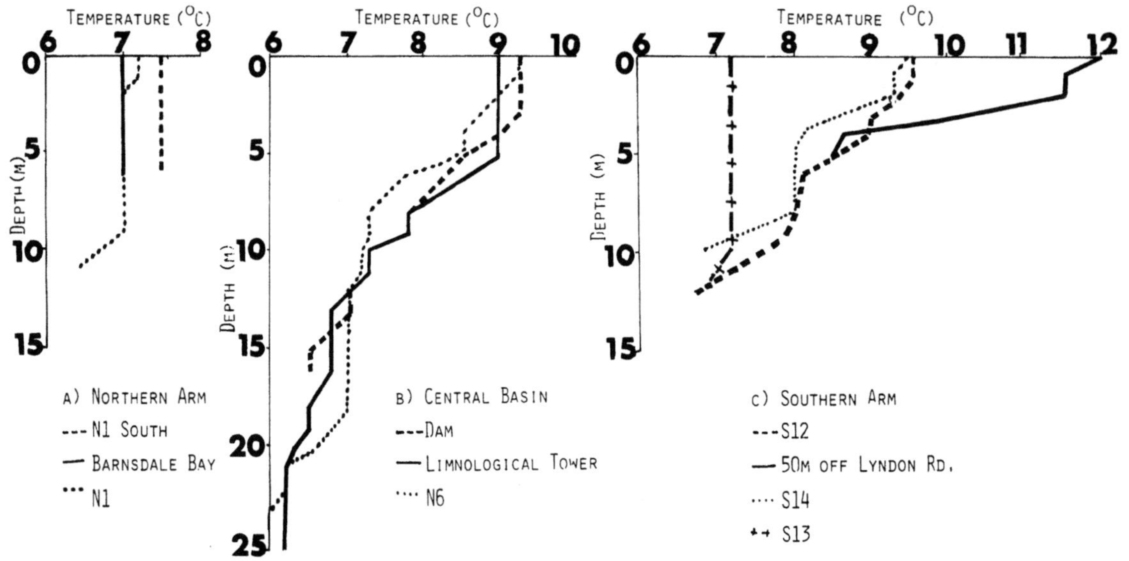

Fig. 3. Temperature profiles at 10 stations in Rutland Water (Fig. 1) on 17.4.80.

Table 5. Wind data from RAF Wittering, relating to Fig. 3. Time is in BST, speed in km h^{-1}, and direction in degrees.

No.	Date	Time	Wind		No.	Date	Time	Wind		No.	Date	Time	Wind	
			Dir.	Sp.				Dir.	Sp.				Dir.	Sp.
1	03.06.80	0400	250	24	24	22.07.80	0600	190	5.5	47	16.09.80	2100	180	26
2		2200	210	3.5	25		1930	180	15	48	17.09.80	1800	230	15
3	04.06.80	1800	130	15	26	23.07.80	0600	140	5.5	49	18.09.80	2100	150	20.5
4	06.06.80	0600	260	18.5	27		2100	240	11	50		2300	140	11
5	07.06.80	1800	260	18.5	28	24.07.80	0600	270	13	51	19.09.80	2100	180	5.5
6	08.06.80	2000	290	5.5	29		2100	80	13	52	20.09.80	2100	30	16.5
7	11.06.80	0300	10	18.5	30	25.07.80	0600	100	11	53	22.09.80	0600	110	5.5
8	14.06.80	0700	30	22	31		2300	110	5.5	54	23.09.80	0100	220	3.5
9		2000	190	37	32	26.07.80	0900	310	11	55		1500	220	22
10	15.06.80	0300	230	29.5	33		1800	230	9.5	56	24.09.80	0600	230	11
11		1500	190	18.5	34	27.07.80	2100	190	7.5	57		1500	270	20.5
12	16.06.80	1200	170	28	35	28.07.80	0600	180	3.5	58	25.09.80	0300	310	7.5
13		2000	230	31.5	36		1800	60	20.5	59		1200	150	5.5
14	19.06.80	0500	230	26	37	29.07.80	0600	30	20.5	60		2000	150	3.5
15	22.06.80	1500	210	22	38		1800	60	22	61	26.09.80	0100	190	5.5
16		1800	200	15	39	31.07.80	0800	210	15.5	62		1500	170	18.5
17	23.06.80	0600	220	7.5	40		1800	200	16.5	63	27.09.80	0100	200	9.5
18		1400	220	18.5	41	01.08.80	0900	100	11	64		1400	240	11
19	24.06.80	0600	230	13	42		1900	100	5.5	65	28.09.80	1200	10	16.5
20	25.06.80	1800	330	11	43	02.08.80	0900	80	7.5	66	29.09.80	1800	240	18.5
21	28.06.80	0500	180	16.5	44		1800	30	18.5	67	30.09.80	2300	250	9.5
22	21.07.80	0900	300	15	45	03.08.80	0700	190	3.5	68	01.10.80	1600	250	22
23		2000	250	5.5	46	15.09.80	1500	210	26	69		2100	270	15

during the early evening. The time of these changes is similar to the peaks and troughs found on Fig. 3d and therefore may have the same or similar causes.

Thermal stratification

On 17.4.80, the water temperature was higher on the windward (southern) shore. However, below 3 m the water temperature in the central basin was higher than in the southern arm. The northern arm had the lowest temperature on that day (Fig. 6). It is possible to segregate the profiles for the three different areas indicating separate circulation (Fig. 4).

Surface heating of the water accounted for the

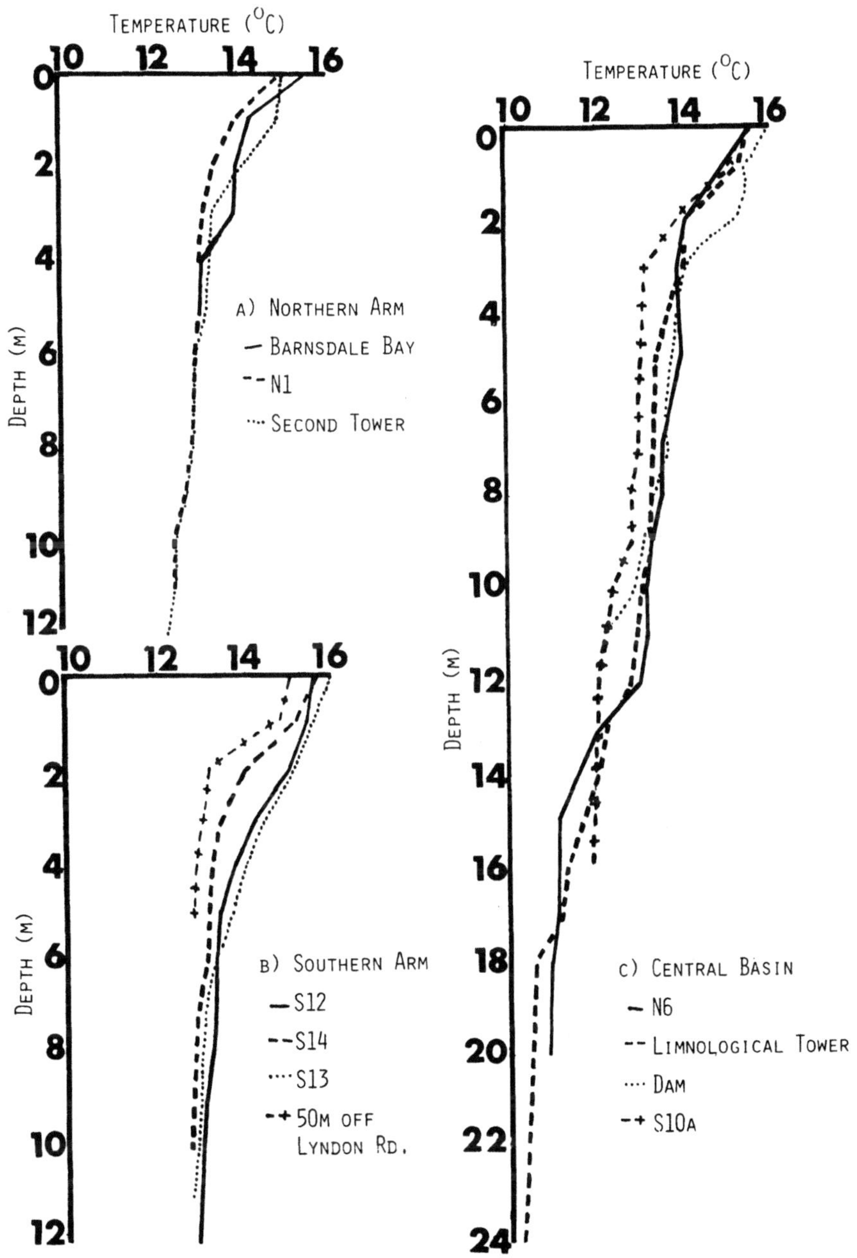

Fig. 4. Temperature profiles at the different stations on Rutland Water 5.6.80.

sharp drop in water temperature between 0 and 3 m which then mixed downwards (Fig. 5). A slight thermocline was present on 5.6.80 around 12 m at the Limnological Tower and N6, and no motion may have been present below 10 m in the central basin. Although there was a slight water movement, stagnation was slowly being reached at the Limnological Tower; the oxygen saturation was recorded as 58.9% at 24 m, and 123.5% at the surface. The presence of a thermocline and/or stagnation is normally reduced by a system of Helixor air-guns in the central basin which prevent deoxygenation of the hypolimnion.

In upwelling areas, cool, poorly oxygenated water rises, and in downwelling areas, warm and well oxygenated water sinks. The presence or absence of these phenomena is illustrated by looking at temperature and oxygen profiles as well as isoline contours along a transect of the reservoir. Where the contour lines come to the surface, upwelling is indicated, and downwelling is shown up as depressions of the iso-oxygen or isotherms. Figure 6 shows the areas where upwelling and downwelling were found to occur on 17.4.80 and 5.6.80. The majority of the findings were as expected: upwelling on the leeward shore and downwelling on the windward shore.

Discussion

Very complex seiche patterns occur depending on wind characteristics (the amount of change, final wind speed and direction), topography, and dimensions of the area. The records obtained were a summation of the surrounding vertical water movements consisting of one or more distinct seiches, i.e. with one or more different seiche axes. The timing or the height of the seiches are not predictable, and further work must be carried out to understand the complex water movements occurring in the different areas under varying conditions in the reservoir.

Wind-stress influences the water level at SB and TB. TB showed greater water height variations than SB and the amount of sloping depends on the wind characteristics, topography of the area and sheltering effect of the wind by the surrounding land. In most cases, when the wind changes direction or speed, the water height changes accordingly.

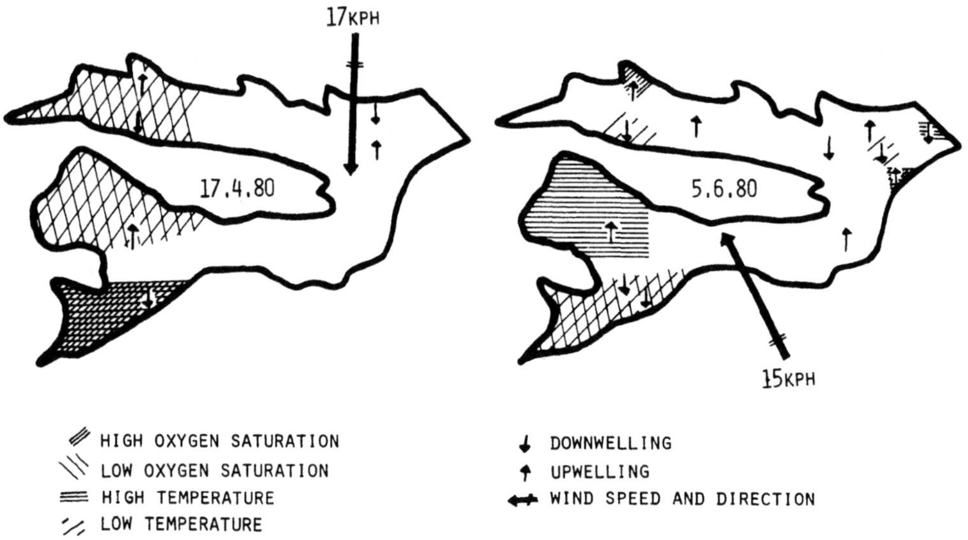

Fig. 5. Summary diagrams of oxygen and temperature patterns at the water surface (0–1 m). On 17.4.80, over half the area had an oxygen saturation >110%. The low oxygen saturated water was present in the Central Basin where the water was relatively horizontally homogeneous. On 5.6.80, most of the area had an oxygen saturation >110%, and highly oxygenated water was present in part of the Central Basin. The water there was not relatively homogeneous. On both days the Northern Arm was the least oxygen saturated and the Southern Arm was the most saturated. This may be due to biological factors. Separate circulation may exist in the three areas: Northern Arm, Central Basin and Southern Arm. No motion occurred below 10 m at the Second Tower, Limnological Tower, Dam and N4 on 5.6.80. (cf. Fig. 1).

Diel oscillations were found to occur at Tail Bay. A study of the literature has not revealed the existence of comparable phenomena, and so far there is no explanation of what caused these oscillations.

The temperature profiles tend to suggest that with cross-winds, a series of local cellular circulations occur which may remain separate (cf. Myers *et al.* 1982). No motion was found below 10 m in the central basin, and this may hold throughout the reservoir.

In most cases, upwelling and downwelling occurred on the leeward and windward shores respectively. The difference in the oxygen and temperature distribution, between the three areas on 17.4.80 and 5.6.80 may have been due to the presence or absence of flora and fauna which could influence the determination of upwelling and downwelling areas. Therefore the biota must be considered in this field of study. This may explain why there was downwelling at the Limnological Tower on 5.6.80; high concentration of blue/green algae (10 000 cells/ml) and a high number of zooplankton present. A further possibility is that the Limnological Tower acts as an obstruction and the water sinks.

Where the lake consists of many bays and dendritic parts, the oxygen concentration is very different and each unit often operates as essentially individual lakes. Therefore, the structure of oxygen distribution becomes highly variable horizontally, vertically and temporally.

Seiches, sloping of the water surface, upwelling and downwelling were found to occur as well as unexplained diel oscillations. However, further work must be done to enable us to predict the water circulation occurring in Rutland Water under varying wind conditions.

Acknowledgements

This project was carried out as part of a B.Sc. dissertation. I wish to thank all who helped, including Dr. M. B. Collins, the Anglian Water Authority for all their assistance throughout the period, and especially Dr. A. J. D. Ferguson.

References

Endrös, A., 1903. Seeschwankungen (Seiches) beobachtet am Chiemsee. Diss. K. Technischen Hochschule zu München, Traunstein. 117 pp.

Endrös, A., 1906. Die Seeschwankungen (Seiches) des Chiemsee. S. B. bayer. Akad. Wiss., Math.-Phys. Kl. 36: 297–350.

Hutchinson, G. E., 1957. A Treatise of Limnology, Vol. 1; Wiley, New York.

Myers, E. J., Oldham, R. S. & Ferguson, A. J. D., 1982. Circulation patterns; their investigation and prediction. In: Harper, D. M. & Bullock, J. A. (Eds.) Rutland Water – Decade of Change, pp. 67–77. Junk, The Hague.

Nakamura, S. & Honda, K., 1911. Seiches in some lakes of Japan. J. Coll. Sci. Tokyo 28, No. 5, 95 pp.

Smith, I. R., 1975. Turbulence in lakes and rivers. Freshwat. Biol. Ass. 29: 26–30.

The effect of Langmuir circulation on the distribution and settling of algae and suspended particles

Tavicha Buranathanitt[1], David J. Cockrell & Peter H. John
Engineering Department, Leicester University, Leicester, England
[1] *Prince of Songkla University, Thailand*

Keywords: reservoir, Langmuir circulation, algae, sedimentation

Abstract

The quantitative extent to which the large-scale organised water motion in the surface waters of lakes and reservoirs, known as Langmuir circulation, affects the distribution and settling of algae and other suspended particles is not known and has thus been ignored in conventionally used water quality models. Since the distribution and settling of these particles is important in determining water quality, this study set out to investigate these effects using a mathematical model based on the two-dimensional advection-diffusion mass transport equation describing the temporal and spatial distribution of suspended particles in a typical Langmuir cell. The Langmuir circulation flow field and turbulent diffusion coefficients are empirically modelled by relating them to environmental parameters (Buranathanitt *et al.*, in press).

The results show that Langmuir circulation does affect particle distribution and settling. For particles with small sinking speeds, such as the lighter algae, the circulation causes intense mixing, resulting in essentially uniform distribution of particles over the cell (as assumed in the 'well-mixed compartment model'). For particles with high sinking velocities, however, aggregation can occur, giving rise to significant reduction in sinking loss. For diatoms, effective sinking speeds are as much as 6% less than when cells are not operating while for coarse silt particles in a cell of large width to depth ratio a reduction of more than 60% is possible.

Two preliminary laboratory experiments have also been performed in order to obtain further information on both the physical characteristics of Langmuir cells and their effects on suspended particles. The results of a wind-wave tank experiment suggest that the Langmuir circulation scale of motion is related to the height of the surface waves, thus providing an empirical means of determining the size of Langmuir cells from environmental variables. A particle settling tank experiment confirmed, in a qualitative way, the results obtained in the analytical study, with both the uniformly mixed and aggregation phenomena being reproduced (Buranathanitt *et al.*, submitted).

Full details of the work described in this abstract will be published elsewhere (Buranathanitt *et al.*, in press; Buranathanitt *et al.*, submitted).)

References

Buranathanitt, T., Cockrell, D. J. & John, P. H., In press. Some effects of Langmuir circulation on the quality of water resource systems. Ecological Modelling.

Buranathanitt, T., Cockrell, D. J. & John, P. H. A laboratory experiment on the distribution and settling of non-buoyant particles in a Langmuir circulation. Submitted to J. Fluid Mechanics.

Hydrobiologia 88, 88 (1982). 0018-8158/82/0881-0088/$00.20.

Models of annual cycles in lentic water bodies

B. Henderson-Sellers & P. B. R. Archer
Department of Civil Engineering, Salford University, Salford, England

Keywords: mathematical models, stochastic, deterministic, validation, water quality

Abstract

A discussion of mathematical modelling of water quality, including a summary of the parameters considered, a comparison of the two major model types (stochastic and deterministic) and a review of the validation process, is presented. A water quality model currently being developed is discussed and a list of ecological models already developed is given.

Mathematical modelling

In a body of water, variations occur in all limnological parameters on many time and space scales. In order to predict how any of these parameters will respond as a result of e.g. a specific management strategy, a mathematical model may be used. Such a model represents a water body by a series of mathematical equations. The model may be specific to a given subset or for a specific water body, or may be designed to simulate the gross characteristics of e.g. current, temperature or algal biomass. The model must consider the factors involved in determining water quality and how they interact in order to predict a probable outcome. A summary of the factors involved appears as Table 1.

The manner in which models consider these factors categorises them into one of two groups: stochastic or deterministic. The principal features of each are summarised in Table 2.

Both deterministic and stochastic models may be used for one of two purposes: hypothesis testing and prediction (for more details see e.g. Smith and Henderson-Sellers 1981). In the former mode a suggestion about one of the biogeochemical feedbacks is implemented and the model results compared with observations. Agreement substantiates the hypothesis – often a quicker and cheaper method than full scale field or laboratory experiments. On the other hand, many questions may be asked about

Table 1. A summary of the factors involved in determining the water quality.

DEPTH	WIND SPEED
REFLECTIVITY	RADIATION PENETRATION
SPECIFIC HEAT	RAINFALL
SUNSHINE HOURS	SURFACE AREA
EDDY DIFFUSIVITY	VAPOUR PRESSURE
PHOTOSYNTHETIC RATE	TURBIDITY
RESPIRATION RATE	CONVECTIVE HEAT FLUX
TEMPERATURE	NUTRIENT CONCENTRATION
BIOMASS	ATMOSPHERIC PRESSURE

Hydrobiologia 88, 89–91 (1982). 0018-8158/82/0881-0089/$00.60.

Table 2. A summary of the features of stochastic and deterministic models.

STOCHASTIC	Elements of the model are allowed to vary in a random fashion. Inaccuracies are decreased because the input data are not condensed in any way. Results may be more realistic since they are associated with a degree of uncertainty. Complex and therefore time consuming and costly.
DETERMINISTIC	Functional representation of a system – The model always produces the same results for a given set of data. Mathematically simpler than stochastic models. Often gives the same result as that expected from a corresponding stochastic model. Easier to implement and thus less costly.

Table 3. A summary of the developmental stages in predictive model formation.

EQUATIONS
VERIFICATION
SENSITIVITY ANALYSIS
CALIBRATION
VALIDATION

the response of the system to a change in external stimuli (e.g. a new sewage inflow, a land use change or inclement weather). A fully validated mathematical model can be used to forecast such outcomes with reasonable accuracy. To do this, the model must first be tested against extensive data bases. The steps towards a full validation are depicted in Table 3 (see also Benarie 1980).

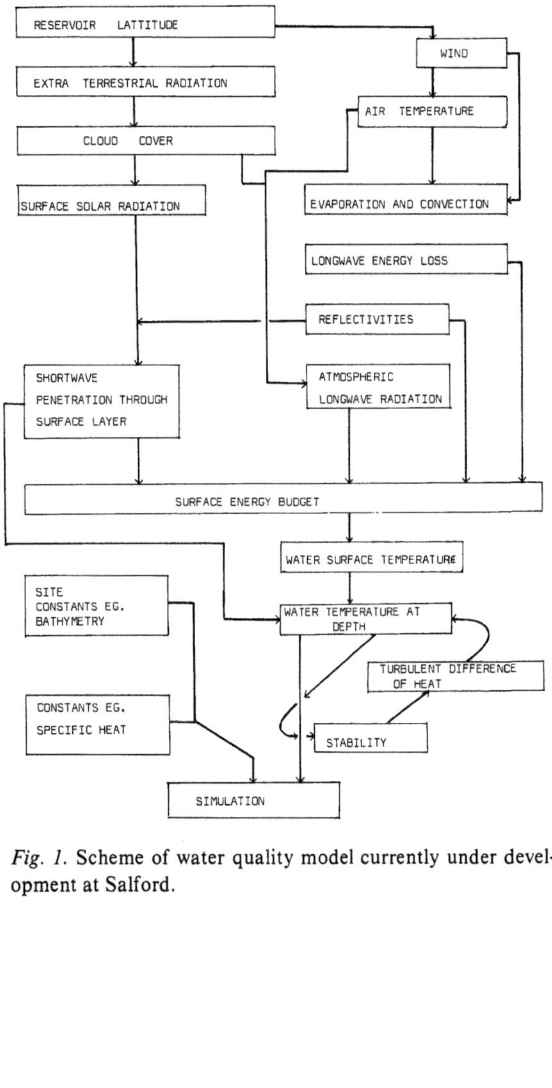

Fig. 1. Scheme of water quality model currently under development at Salford.

Quality model

One example of a deterministic model is given in Fig. 1. The scheme solves the heat transfer equation (i.e. energy absorbed at the surface is transported downwards by turbulence – thus increasing the heating at depth from solar radiation penetration, which is itself affected by phytoplankton mass and hence, turbidity), by using a finite difference grid. Here, the water body is divided into a number of horizontally homogeneous 'slices' of finite thickness and the values of limnological parameters such as temperature and phosphorus content are calculated as means throughout each slice volume (for details see e.g. Henderson-Sellers 1980). Results for the long-term mean thermal response (the 'hydroclimate') for two U.K. lakes are given in Fig. 2. The number of slices used has a direct bearing upon both accuracy and the length of computer time required.

Table 4 gives a summary of a wide range of ecological models (after Jorgensen 1980). Increasing complexity does not necessarily lead to better simulations and the importance of considering the possibility of restricting the number of the model variables has been recently discussed (e.g. Beck 1981).

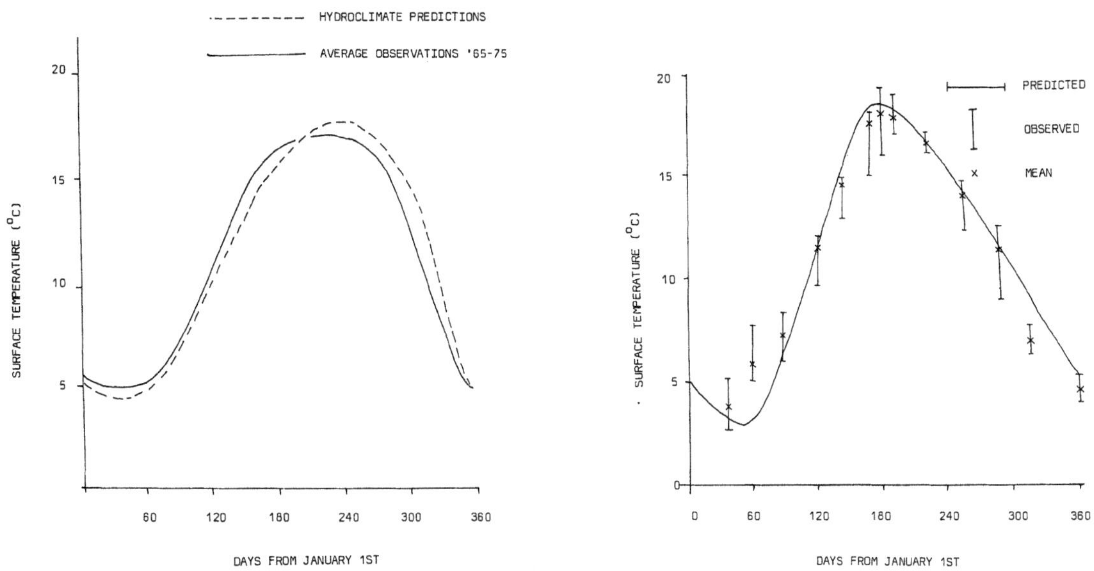

Fig. 2. Examples of results after modelling the long-term response for two lakes. Left: Windermere results compared with predictions (1965–1975 mean). Right: Graffham Water results compared with predictions (1974–1977 mean).

Conclusion

Mathematical modelling provides a quick and relatively inexpensive method of predicting the outcome of (e.g.) a specific management strategy. A full validation of the model by comparing its predictions with data from previous years (hindcasting) is essential before any confidence can be placed in it.

Acknowledgements

We wish to thank Mr. M. J. Pearson and the Anglian Water Authority for the data for Fig. 2 (Graffham Water).

References

Beck, M. B., 1981. Hard or soft environmental systems? Ecol. Modelling 11: 233–251.

Benarie, M. M., 1980. Urban Air Pollution Modelling. Macmillan, London, 405 pp.

Henderson-Sellers, B., 1980. Limnological modelling: simulation of vertical profiles of temperature, dissolved oxygen and nutrients. I. Theory. J. Limnol. Soc. S. Afr. 6: 49–54.

Jorgensen, S. E., 1980. Lake Management. Pergamon, Oxford.

Smith, I. R. & Henderson-Sellers, B., 1981. An introduction to mathematical modelling for ecologists and environmental scientists, Environ, Educ. Inform. 1: 143–151.

Table 4. Summary of some ecological models (after Jorgensen 1980).

Model	Number of state variables per layer or segment	Nutrient considered	Dimension (D) or Layers (L)	Calibrated (C) and/or Validated (V)	Number of case studies in literature
Vollenweider	1	P(N)	1L	C+V	many
Imboden	2	P	2L,1D		3
O'Melia	2	P	1D	C	1
Lorenzen	2	P	1L	C+V	1
Thomann (1)	8	P,N,C	3L	C+V	1
Thomann (2)	10	P,N,C	7L	C	1
Thomann (3)	15	P,N,C	7L	–	1
Chen & Orlob	15	P,N,C	7L	C	min 2
Patten	33	P,N,C	1L	C	1
Di Toro	7	O,N	1L	C+V	1
Bierman	14	P,N,Si	1L	C	1
Canale	25	O,N,Si	2L	C	1
Jorgensen	17	O,N,C	1-2L	C+V	3
CLEANER	40	P,N,C,Si	sev L	C	many
Nyholm	7	P,N	1-2L	C+V	13

Rutland Water raw water quality - problems and management

N. F. Low
Anglian Water Authority, Oundle, Peterborough, England

Keywords: reservoir, raw water, management, chemistry

Abstract

The options available for the management of Rutland Water are identified and their use discussed. The changes in the chemistry of the stored water with regard to sulphate, silica (molybdate-reactive), dissolved reactive phosphate and total oxidised nitrogen are shown.

Introduction

At the conception of Rutland Water it was recognised that water quality problems were likely, due to the eutrophic character of the input waters and the reservoir being sufficiently deep for stratification to occur. To control the extent of these problems the reservoir was designed incorporating a number of features and management options which have been described previously (Harper 1978).

Table 1. Options available for controlling water quality.

	Options		
INPUT WATERS			
(1) Natural catchment-Oakham S.T.W. effluent	Discharge to North arm	or	Discharge to land area and bunded section in the south arm
(2) R. Welland, Tinwell	Pump	or	Not pump
(3) R. Nene, Wansford	Pump	or	Not pump
RESERVOIR			
(1) Abstraction site	Central Basin	or	Secondary tower in north arm
(2) Draw-off level at central basin	79 or 74	or	69 or 55 m. O.D.
(3) Draw-off level at secondary tower	79 or 75	or	72 m O.D.
(4) Destratification air lift pumps (helixors)	On	or	Off

Options available for water quality control

Table 1 lists the options available for water quality control. These options are split into two groups, those that operate on the input waters, and those that affect the quality of the stored water.

Input water

Effective filling of the reservoir commenced in February 1975 (see Fig. 1) but only using water from the natural catchment. The option for modifying water quality from this source did not become operational until July 1977 when the Oakham Sewage Treatment Works effluent was diverted to a land treatment area prior to discharge into the southern arm. This discharge can be directed into (i) the main body of the southern arm, or (ii) into the bunded section of the nature reserve.

Commissioning of the pumps on the River Nene and River Welland was completed by February 1976. The main filling programme commenced in April 1976 when the drought of 1976 was being felt. Because of the drought, options on whether to pump or not were limited and water of a much poorer quality than expected had to be pumped. As Fig. 1 shows, the volume of water pumped from the two rivers was very small until September 1976 when the drought broke. Because at that time the extent of the rainfall for the autumn-winter period 1976/77 was uncertain (it was in fact one of the wettest winters of

Hydrobiologia 88, 93-101 (1982). 0018-8158/82/0881-0101/$01.80.

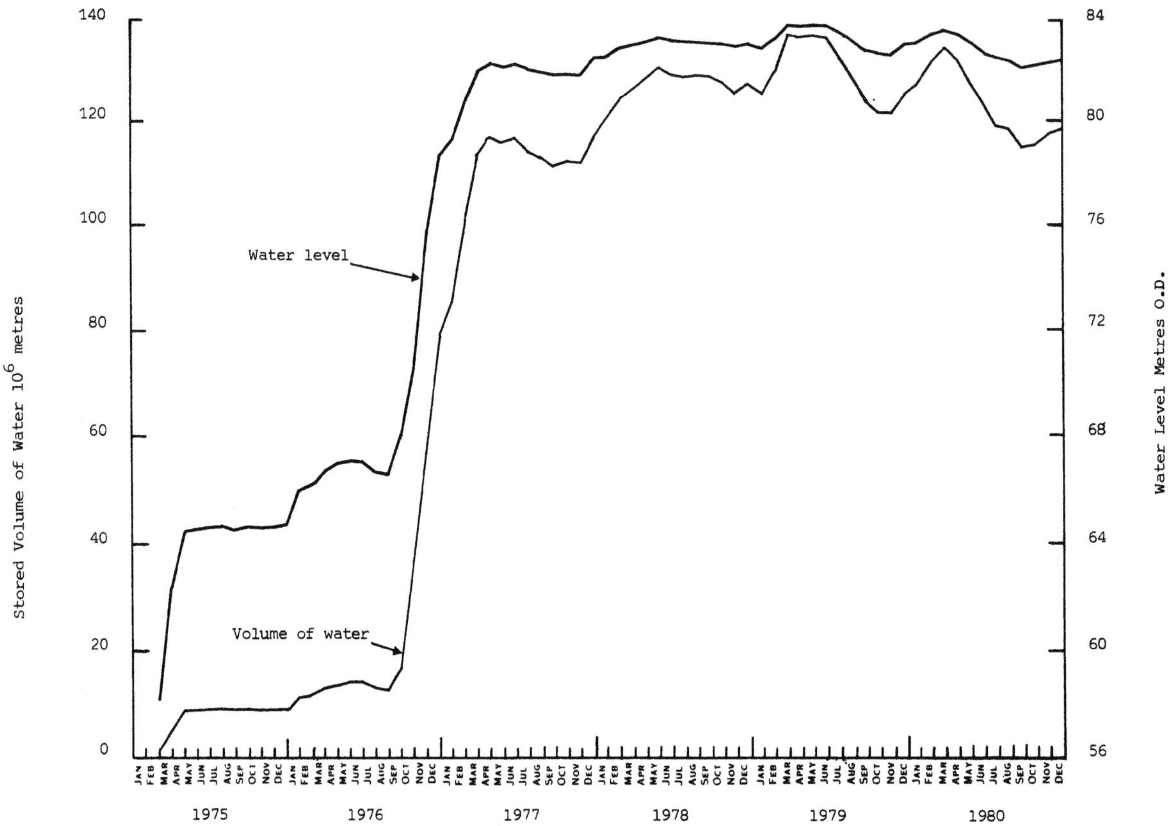

Fig. 1. Reservoir water level and volume of stored water 1975–1980.

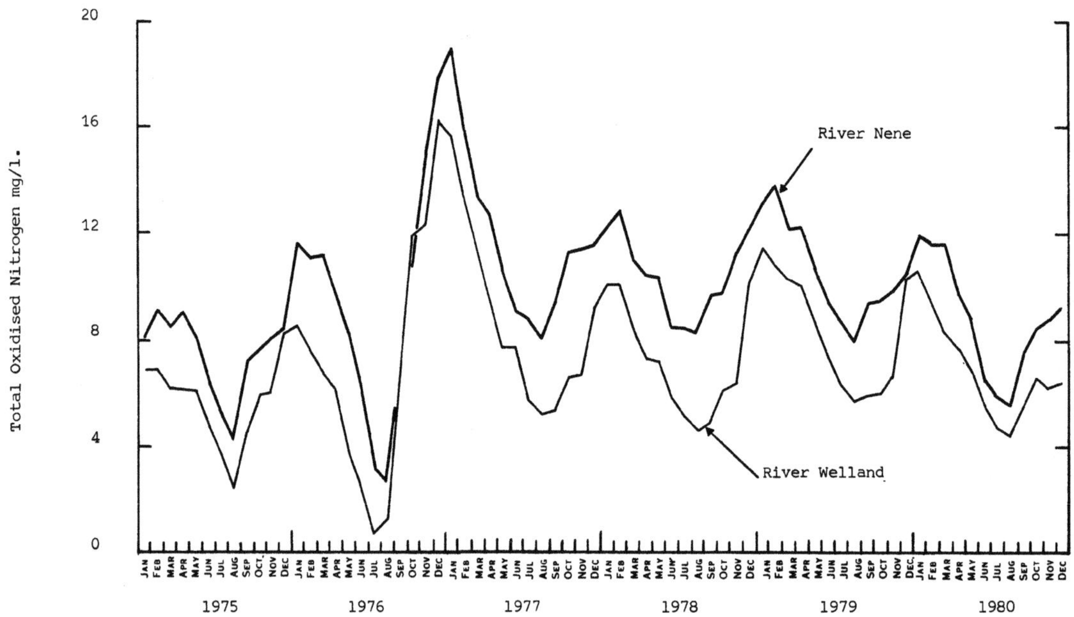

Fig. 2. Mean monthly total oxidised nitrogen concentration in River Nene and River Welland 1975–1980.

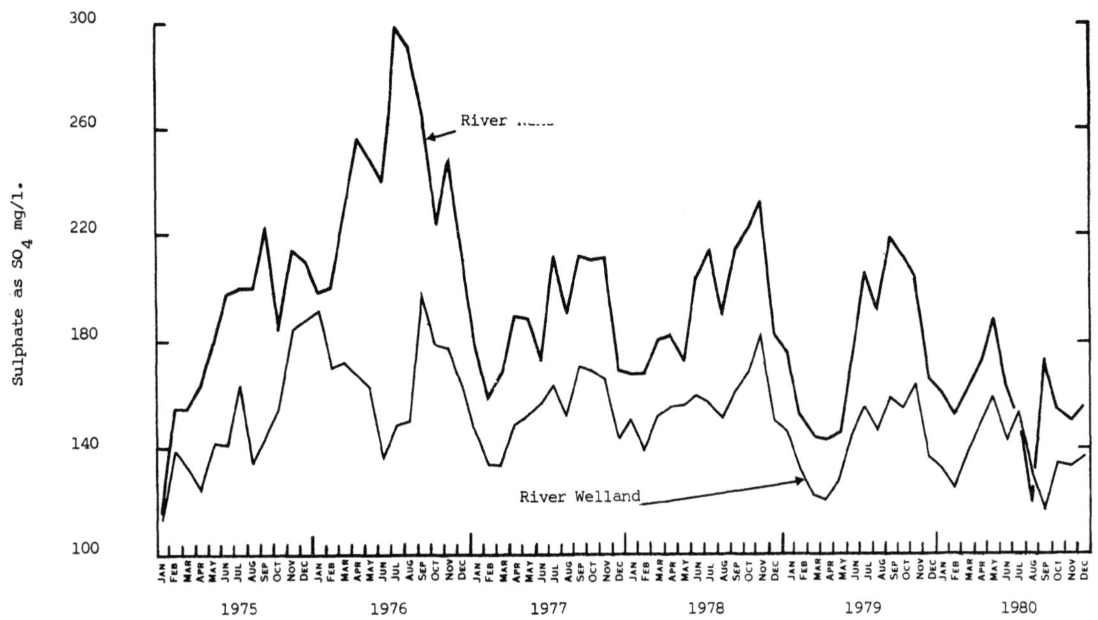

Fig. 3. Mean monthly sulphate concentration in River Nene and River Welland 1975–1980.

recent times) it was decided to pump as much water as possible even though its quality was unsatisfactory with regard to sulphate and latterly nitrate nitrogen (see Figs. 2 and 3). It became necessary at the end of December to stop pumping to ensure that the sulphate and nitrate levels in the reservoir did not exceed the World Health Organisation's recommended value of 200 mg l^{-1} and 11.3 mg l^{-1} respectively (W.H.O. 1971). Because of the high nitrate and sulphate levels in the River Nene it was decided to stop pumping from the River Nene until the autumn of 1977. Pumping from the River Welland continued in a controlled manner to ensure that nitrate nitrogen levels in the reservoir did not exceed 11.3 mg l^{-}.

Since 1977 the only restrictions applied to pumping for water quality control have been as a result of pollution incidents. Though river water quality would have at times been unacceptable for direct supply due to sulphate and/or nitrate nitrogen concentrations it has been suitable for storage purposes.

Reservoir options

Originally provision was made just for the control basin draw-off tower to supply water to the Wing treatment works. The secondary draw-off tower was incorporated as a result of a study (W.R.A. 1971) which concluded that because of its shape, Rutland Water could possibly be managed as if it were two reservoirs, (i) the southern arm and central basin and (ii) the northern arm. This required that the nutrient rich input waters were fed into the southern arm. By managing Rutland Water in this manner it was expected that there would be sufficient difference in the behaviour of the algae in the two 'reservoirs' such that when the central basin water had a difficult algal bloom to treat the northern arm would not be so severely affected. Experience to date has not required the use of the option of the secondary draw-off facility. However it is expected to prove useful in the future when the retention time (volume of reservoir/outflow) is reduced from the present 3.46 y to 1.33 y as a result of the increase in demand for water. One effect of reducing the retention time will be to increase the nutrient loading and as a consequence it is likely that the frequency and extent of algal bloom will also increase (Golterman 1975).

In selecting one of the four draw-off levels at the central basin tower the following points have to be considered:

1) reservoir water level,
2) chemical composition at each draw-off level,

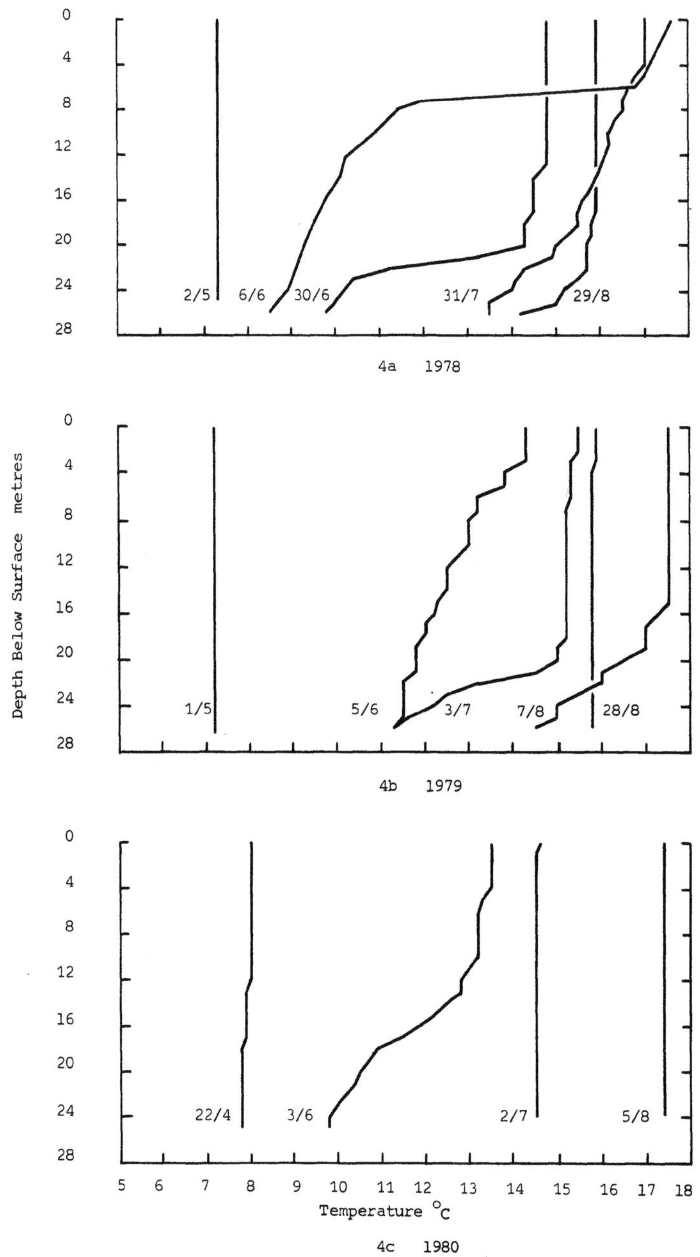

Fig. 4. Depth – temperature profiles, Limnological Tower, Rutland Water 1978–1980.

3) algal density at each draw-off level,
4) water temperature,
5) ease of change over from one level to another.

Initially the bottom draw-off at 55 m O.D. was used to supply water to Wing. It was intended to use this level as much as possible to assist in the summer destratification of the reservoir by removing some of the colder bottom water. However during the summer months of 1977 the chemical composition of this water became unacceptable for the treatment works due to the low dissolved oxygen and higher ammonia iron and manganese concentrations that developed due to stratification effects. To avoid this difficulty the draw-off was changed to 69 m O.D. in the spring of 1978 and has since remained at this position.

Although vertical variation in algal density has been observed (Harper 1978; Ferguson & Harper 1982) it has not been necessary to change the draw-off level for this reason.

The central basin draw-off arrangement is unusual in design with tunnels running into the hillside from the reservoir to the draw-off shaft. As a result any change over from one level to another may require the discharge to waste of stagnant water stored within the tunnel. This operational restriction prevents frequent change of draw-off level unless conditions are severe enough to warrant it.

The destratification system consists of a series of 12 helixor air lift pumps installed in the central basin at a depth of 57 m O.D. and fed by underground pipes with compressed air, from the Empingham Pumping Station. The prime function of this system is to overturn the reservoir water during the summer months and so prevent a thermocline developing with the subsequent problems of oxygen depletion and increase in iron, manganese, ammonia, silica and phosphate in the bottom water. It is also expected (Ferguson & Harper 1982) that the system can be utilised to modify the sequence and extent of algal activity.

The destratification system was commissioned and available for use in 1978. A thermocline was allowed to develop to a greater degree than initially planned so that the Water Research Centre could make an estimate of the efficiency of the system. The air compressors were ran continuously from 6 June until 15 August when the reservoir was relatively uniform in temperature. The behaviour of the thermocline over this period can be seen in Fig. 4. The time taken to achieve mixing was longer than anticipated and as a result the bottom water became slightly depleted in dissolved oxygen. This did not seriously affect the quality of water being drawn off at the 69 m O.D. level.

As a result of the experience of summer 1978 artificial overturn was started at the beginning of May in 1979 and 1980 when a temperature difference in top and bottom water was observed. Apart from maintenance, the air compressor was again run until the middle of August for both years. In each of these years strong north-west winds made a marked contribution to the mixing of the water. The change in the thermocline can be seen in Figs. 4b and 4c.

The destratification systems appear to be underpowered for overturning a full reservoir and as a result it is not possible to ensure that the bottom water at the 55 m O.D. draw-off would be satisfactory for the water treatment works during the summer months. There have been no water quality problems to date in using the draw-off at 69 m O.D. but with a hotter summer than those of 1978–1980 and the absence of a strong north-west wind it is possible that water quality could be adversely affected at this level.

Provision was made for an additional compressor to be installed in the Empingham Pumping Station in the event of the system being underpowered but this option will not be taken up until a study of the performance of the system has been completed and its design details assessed in relation to other destratification systems (Lorenzen 1977).

Monitoring of waters

From January 1975 to August 1976 the river Welland at Tinwell and the River Nene at Wansford were monitored at fortnightly intervals by taking surface grab samples. These samples were analysed for a minimum of 16 parameters which included sulphate and total oxidised nitrogen. From September 1976 the frequency of sampling was increased to weekly intervals.

Sampling of the reservoir underwent a number of changes since filling started in 1975. Initially from April 1975 to March 1976 three sites were monitored at weekly intervals, using a surface grab technique. From April 1976 to March 1980 twelve sites were monitored at the rate of 4.5 sites per week, again using grab samples. During this latter period occasional depth profile samples were taken near the limnological tower, when the reservoir was stratified. All the samples taken up to March 1980 were analysed for a minimum of 22 parameters.

From April 1980 the following sampling programme was implemented: The reservoir draw-off water was sampled weekly and analysed for 24 parameters. The number of sites on the reservoir was increased to twelve with sampling frequency also increased to weekly with only 6 parameters being monitored including dissolved reactive phosphate, silica (molybdate-reactive) and total oxidised nitrogen. The reservoir samples were from the

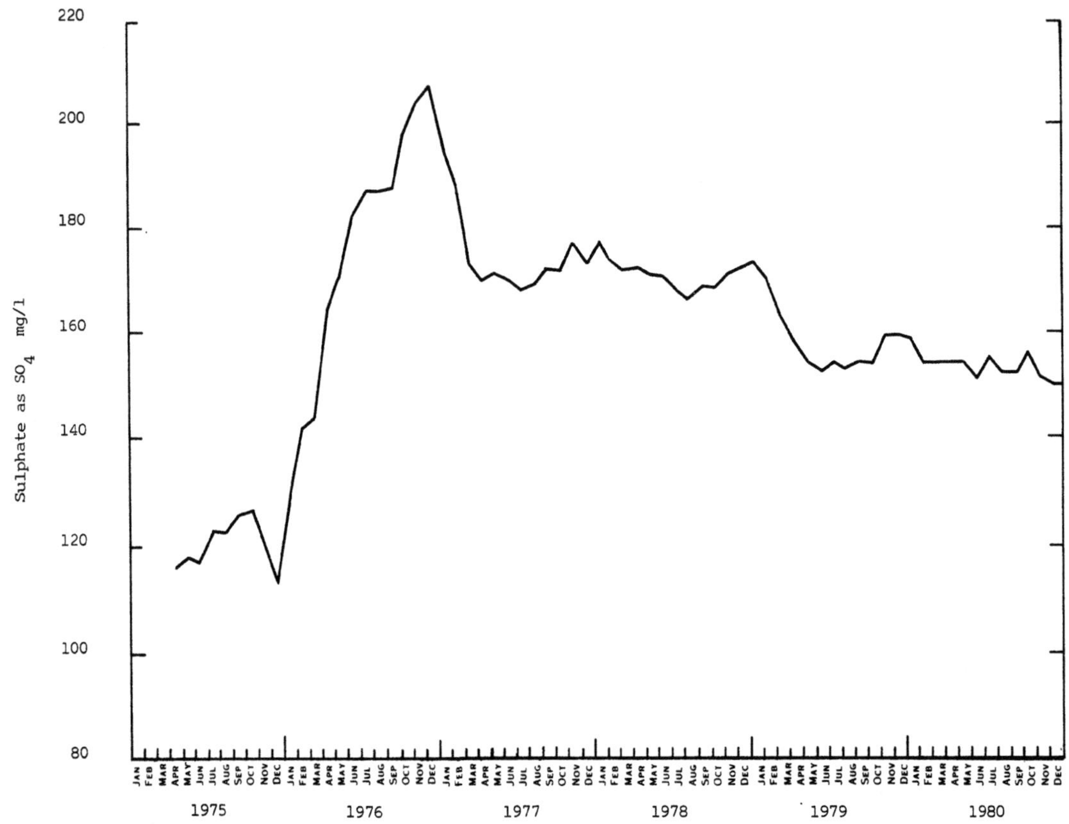

Fig. 5. Rutland Water mean monthly sulphate concentration 1975–1980.

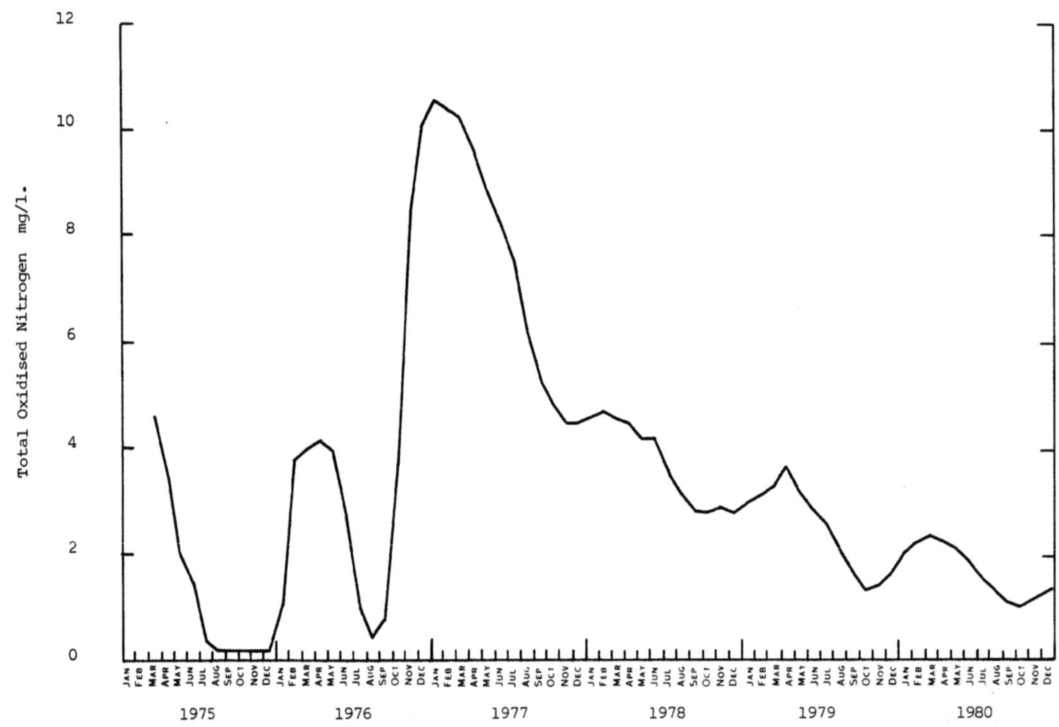

Fig. 6. Rutland Water mean monthly total oxidised nitrogen concentration 1975–1980.

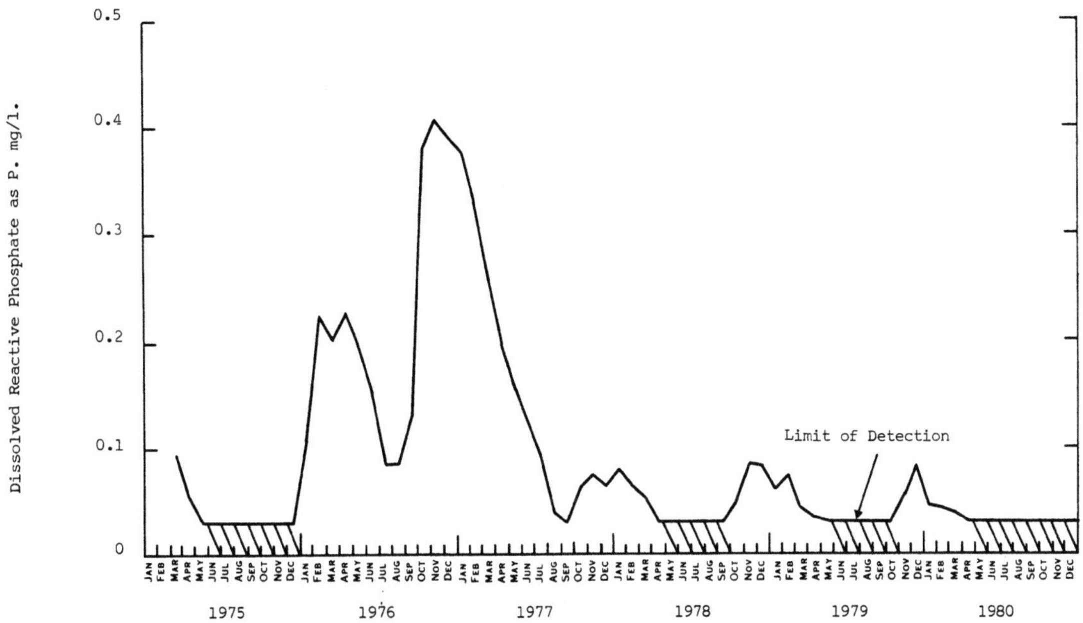

Fig. 7. Rutland Water mean monthly dissolved reactive phosphate concentration 1975–1980.

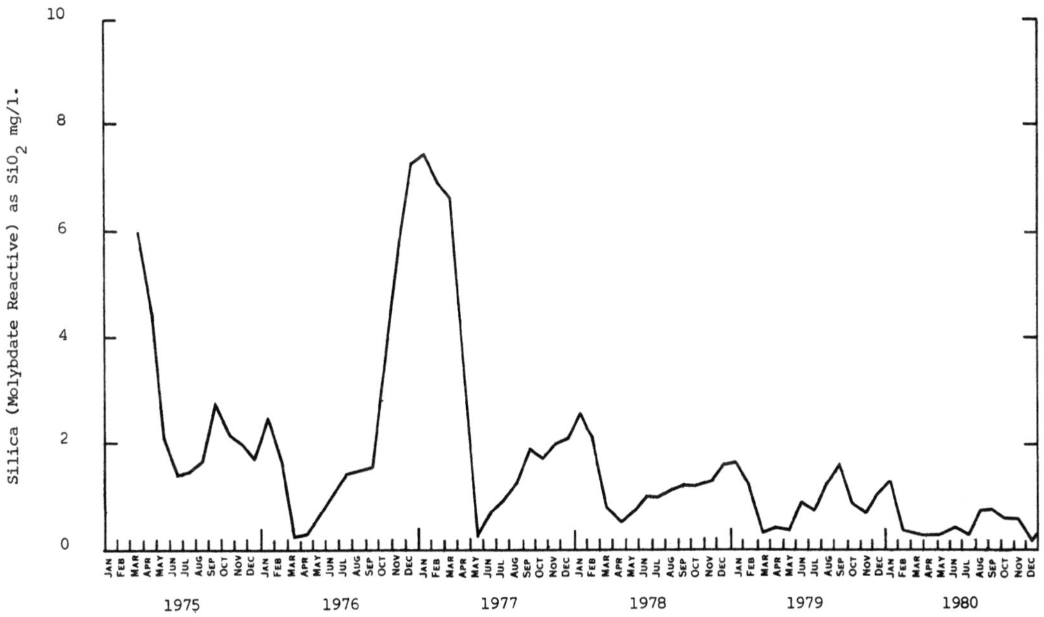

Fig. 8. Rutland Water mean monthly silica (molybdate-reactive) concentration 1975–1980.

surface to 5 m, depth-integrated samples. Depth grab samples were taken at 5 m intervals near the limnological tower at fortnightly intervals during the period mid May to the end of September.

Changes in quality of stored water

As described earlier the quality of water stored in the reservoir at its first filling was as a result of the 1976 drought much more eutrophic in character than expected. Since 1977 there have been marked changes in the chemistry of the stored water and some of these changes can be seen in Figs. 5–8 which are a summary of all the relevant data collected from the reservoir monitoring programmes and presented in the form of unweighted monthly means.

Discussion

One of the mechanisms for the change in water quality is dilution, and this can be seen in Fig. 5 of sulphate concentration, a conservative constituent. This beneficial change appears to have reached a minimum value of 150 mg l^{-1} for sulphate. Sulphate levels will eventually increase as demand for reservoir water increases and more water is abstracted from the sulphate-rich River Nene source.

Denitrification in lakes and reservoirs is a well-recognised process and clearly the process is present in Rutland Water as the change in total oxidised nitrogen (Fig. 6) cannot be explained by dilution and phytoplankton utilisation. The rate of denitrification is affected by the nitrate concentration (Anderson 1977), temperature (Kaplan *et al.* 1977) and organic matter content (Larson 1977). The rate of denitrification achieved in 1977 is unlikely to be repeated in the future even if the nitrate concentration reaches a similar value due to the large amount of decaying vegetation present from the first major flooding of the reservoir. With the long retention time and denitrification it is unlikely that the high nitrate levels present in the input waters will cause any difficulty in the reservoir supply water complying with the World Health Organisation's recommended value of 11.3 mg l^{-1} nitrate-nitrogen for potable waters during non-drought conditions.

Three main mechanisms can be considered to explain the relative rapid decline in phosphate concentration during 1977 (Fig. 7): phosphate uptake by algae, precipitation with lime due to increase in pH caused by algal activity, and absorption on suspended sediment particles. Of these three mechanisms, the first two are likely to have played the major part in removing phosphate as records show that algal activity was at its maximum during 1977. A study (Ferguson 1980) of the phosphate budget for the reservoir showed a gain to it above that from the river inputs. This additional quantity may be due to precipitated phosphate redissolving or desorption from the sediments. Enhanced phosphate levels have been detected in the bottom stratified water during the summer months.

The regular declines in silica (molybdate-reactive) concentration (Fig. 8) are associated with population growths of diatoms, a subject covered in more detail elsewhere (Ferguson 1981). Like phosphate, the silica budget (Ferguson 1980) indicates that silica in the sediment is taken back into solution and again enhanced silica levels have been observed in the bottom stratified water during the summer months.

Conclusion

It has not yet been necessary to use all the options available for water quality control. Increased demand for water will result in a reduction in the retention time of the stored water and an increase in nutrient loading. With these changes the options for water quality control are expected to be fully utilised.

A number of mechanisms for change in the quality of stored water have been identified and some of these have resulted in beneficial changes, e.g. sulphate and nitrate, but others give rise to adverse changes, e.g. return to solution of phosphate and silica associated with the sediments.

Acknowledgements

The views expressed in this paper are those of the author. Grateful acknowledgement is made to colleagues within the Welland and Nene River Division for their assistance in preparing the paper, particularly A. Ferguson. The author thanks the Director of Scientific Services for permission to publish this paper.

References

Anderson, J. M., 1977. Rates of denitrification of undisturbed sediment from six lakes as a function of nitrate concentration, oxygen and temperature. Arch. Hydrobiol. 80: 147–159.

Ferguson, A., 1980. Anglian Water Authority, Welland & Nene River Division. Unpublished communication.

Ferguson, A. & Harper, D. M., 1982. Rutland Water phytoplankton – the development of an asset or a nuisance. In: Harper, D. M. & Bullock, J. A. (eds) Rutland Water – Decade of Change, pp. 117–133. Junk, The Hague.

Golterman, H. L., 1975. Nitrogen and phosphate supplies as causes of eutrophication. In: Ven Te Chow (advisory ed.) Developments in Water Science, 2. Physiological Limnology, pp. 368–378. Elsevier Scientific, Amsterdam.
Harper, D., 1978. Limnology of Rutland Water. Verh. int. Verein. Limnol. 20: 1604–1611.
Kaplan, W. A., Teal, J. M. & Valiela, I., 1977. Denitrification in salt mash sediments: evidence for seasonal temperature selection among populations of denitrifiers. Microb. Ecol. 3: 193–204.
Larson, V., 1977. Nitrogen transformation in lakes. Prog. Wat. Techn. 8: 418–431.
Lorenzen, M. W., 1977. Aeration/circulation keeps algal blooms in check. Wat. Wastes Engng. 14 (10): 69–70 and 72–74; 14 (11): 88–90 and 97.
Water Research Association, 1971. Empingham Reservoir – Biological Considerations. W.R.A., Medmenham, U.K.
World Health Organisation, 1971. International Standards for Drinking Water (3rd edn). Geneva.

Discussion

Mr. Low, introducing his paper, said that it was realised early on that Rutland Water would present problems over water quality because water was being extracted from lowland eutrophic rivers and the reservoir was deep enough for stratification to occur. He outlined the design features incorporated in the construction which include the erection of a limnological tower, establishment of helixor air guns and redirection of sewage effluent into a bunded area.

Initially, natural catchment provided all input from 1975 to 1976. Pumps were installed in early 1976, and poor quality water (e.g. > 23 mg l^{-1} NO_3-N ex river Nene) was pumped in 1976 because of concern over the duration of the 1976 drought. Sulphates from iron pyrites, which are exploited by British Steel, had also affected water quality in the river Nene. The effect of this was dramatic on the reservoir water chemistry, and off-take from the Nene had to be stopped. Phosphates had also risen, but these appeared to precipitate, possibly as calcium oxy-phosphate.

He then discussed the effect of the helixor air guns in destroying a thermocline and affecting water chemistry. No problems had resulted over oxygenation except when wind disturbed the bottom waters reducing oxygen by 60–75% in water at the off-take. He added that certain design features had not been utilised because the reservoir was only supplying one-third of its potential supply. Problems would doubtless multiply as extraction increased.

Mr. Toms opened the discussion by stressing the great value of chemical data on a new reservoir; although much data had been gathered on the chemistry of established reservoirs, little existed on newly established ones. He drew attention to the rapid depletion of phosphates in the water, and suggested that neither the formation of calcium oxy-phosphate nor algal uptake was the most likely mechanism but rather adsorption. Work on other reservoirs had shown that over the first few years phosphate was removed, but thereafter increased, arguing against a simple chemical precipitation, while algal numbers appeared far too low to remove the stated quantity. Adsorption onto newly exposed substrate, which would eventually become saturated, seemed more likely. He further suggested that a decreased retention time would not materially affect phosphate levels.

Mr. Toms further asked whether the use of helixors was considered satisfactory when compared with the other available methods of destratification and went on to point out that, with a three-year retention time, all the water one could use would be available in the epilimnion; it followed that destratification was nugatory as far as water treatment was concerned. Yachtsmen, anglers and others clearly had other requirements when using the reservoir as an amenity.

Mr. Low commented that work had been carried out on the phosphate levels in the sediment by a student at Essex University, and that this point would be borne in mind. The helixors had been rather sluggish in moving the lower waters.

Mr. C. Waller commented on the rather high levels of sulphate being taken into supply and asked if this had caused any problems with hydrogen sulphide formation in sewage treatment. Mr. Low replied that he was unaware of any problems of this nature.

Mr. Morris asked whether recreational activities, which were often conducted close to the dam and off-take point, were affecting water quality. Mr. Low said that none had been noted and he felt any effect would be minimal.

The production of public supply drinking water from Rutland Water at the Wing Treatment Works

D. B. Ford, B. E. Drage & T. J. Roberts
Anglian Water Authority, Northampton Water Division, Pitsford Laboratory, Grange Lane, Pitsford, Northampton, England

Keywords: reservoir, treatment, pulsator, quality, haloforms

Abstract

The paper describes the treatment facilities and processes at the Wing Works where Rutland water is treated to potable standards. It summarises the experience gained in the operation of the works since it opened in 1977 and includes a brief account of an investigation into the production of haloforms.

Introduction

The Wing Treatment Works forms part of the joint Empingham project initiated by the Welland and Nene River Authority and the Mid-Northants Water Board. It was designed and constructed by Degremont Laing Ltd. and was commissioned in June 1977. The finished water is supplied to the Northampton and Peterborough Water Divisions of the Anglian Water Authority and to the Soar Division of the Severn Trent Water Authority. The works is designed for an average output of 205×10^3 m^3 day^{-1} with a peak output of 285×10^3 $m^3 day^{-1}$.

Source of water

Water is taken from Rutland Water via a pumping station at Empingham for treatment to potable standards at the Wing works which is situated on high ground some 7.6 km to the south west of Empingham. The water in the reservoir consists of water from the natural catchment (principally the flow in the river Gwash) plus water pumped from the River Nene at Wansford and the River Welland at Tinwell.

Since the first filling in 1976/77, the proportions of Gwash, Nene and Welland water have been about equal. As the works output increases more Welland and Nene water will be pumped, the latter predominating. Chemical and bacteriological samples of the three component waters from the rivers Welland, Nene and Gwash were examined from 1964 onwards. Prolonged reservoir storage was expected to produce improvements in the chemical, physical and bacteriological characteristics of the river waters but the extent of these changes could not be predicted with certainty. There was also the possibility that raw water might, on rare occasions, have to be drawn direct from the river intakes, e.g. in the event of repairs to or inspection of the valve tower or associated pipework.

Description of the works

A plan showing the block layout of the works with water and chemical flows is given in Fig. 1.

Primary clarification

Three Degremont Pulsator Clarifiers are provided for this process (Figs. 1 and 2). Each pulsator has a three-compartment inlet chamber. Water emerges from the raw water pumping main into the

Hydrobiologia 88, 103–116 (1982). 0018-8158/82/0881-0103/$02.80.

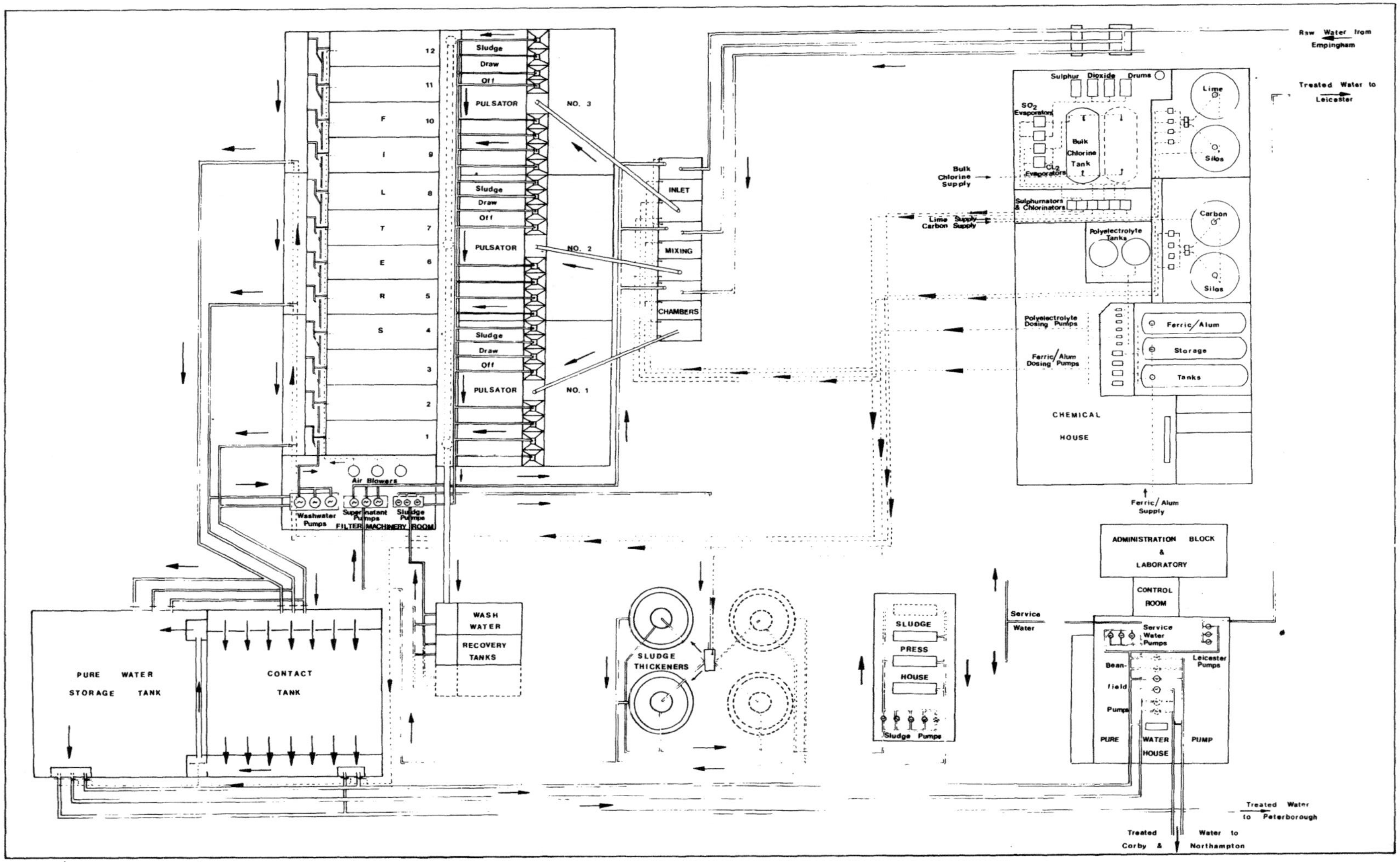

Fig. 1. Plan showing the block layout of the Wing Treatment Works.

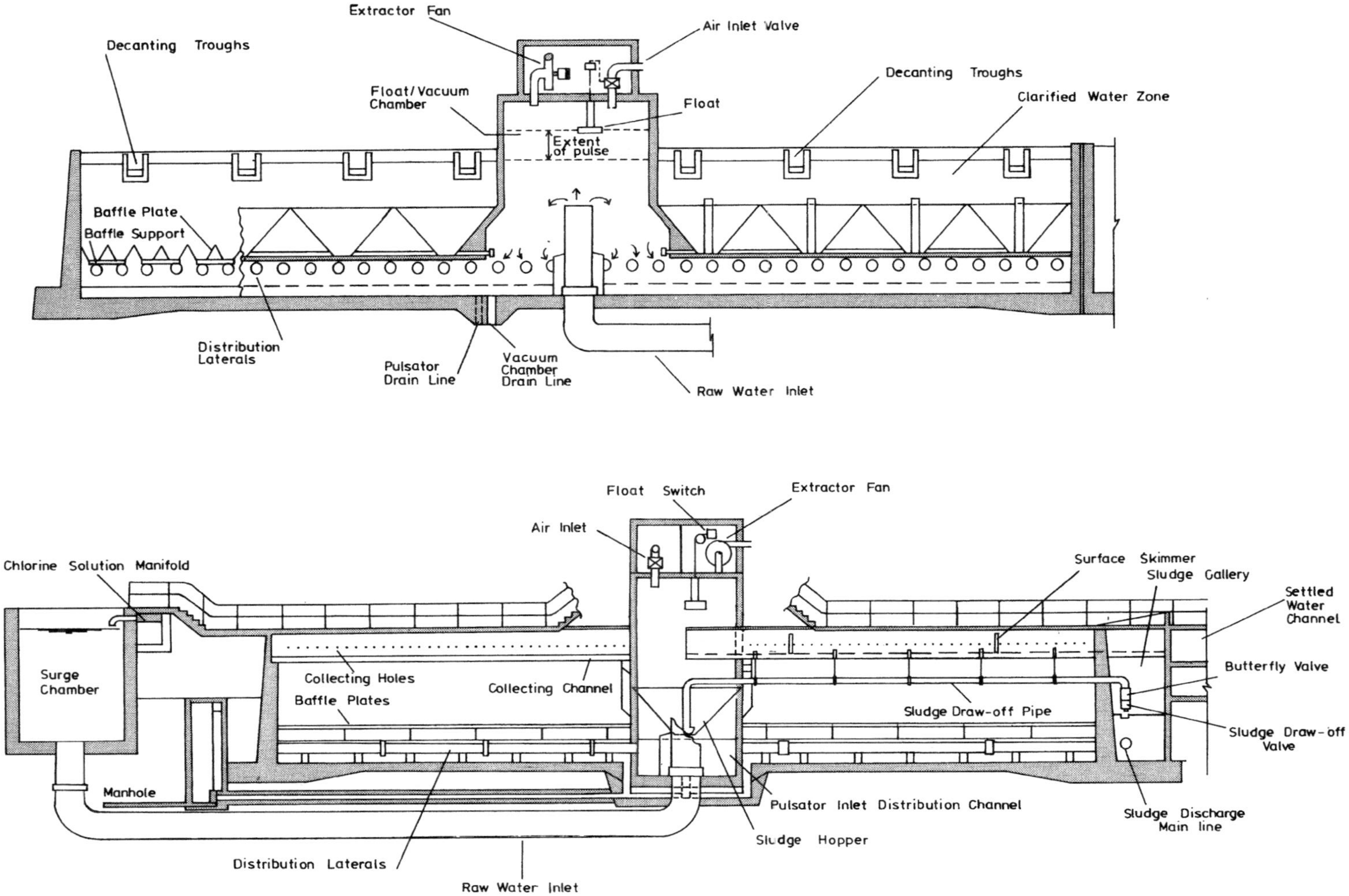

Fig. 2. Cross-sections of the Degremont pulsator units.

first compartment and passes underneath a curtain wall into the second compartment where it is vigorously stirred. After passing over a weir into the third compartment the water enters a 1050 mm diameter pipe and flows to the centre structure of the pulsator. An important function of the inlet chamber unit is the thorough mixing-in of the chemical doses which are applied as follows:

a) First chamber, a marginal dose of chlorine. (This is normally set to give a total chlorine residual of 0.5 mg l^{-1} at the pulsator outlets.)

b) Second (stirrer) chamber. Doses of ferric sulphate and lime.

c) Third chamber. The activated carbon dose and the polyelectrolyte dose.

The pulsator is basically a rectangular tank in which water is introduced near to the bottom and decanted off into troughs at the top. To obtain an equal vertical flow through unit area at all places within the tank the incoming water has to be introduced evenly over the bottom. This is achieved by a system of 400 mm diameter lateral pipes set into and on each side of a broad rectangular sectioned culvert which runs along the middle of the floor of the tank. The laterals, spaced at 1.025 m intervals, have numerous 30 mm diameter holes along the underside and the ends furthest from the culvert are closed.

The incoming water having been dosed with a coagulant, a sludge blanket eventually forms in the lower part of the pulsator, building up to the level of the sludge concentrator hoppers which are situated on top of the inlet culvert. Excess sludge discharges over the edges of the eight sludge hoppers and falls to the bottom where, being in a still-water zone, it concentrates. The sludge is periodically drawn off via the sludge discharge pipework. Although water enters at constant flow, the action of the vacuum chamber and fan at the centre of the pulsator imposes a pulsating effect on the flow of water into the bottom of the tank. The water level in the vacuum chamber rises as air is sucked out by the fan. When it reaches a predetermined height, a valve opens allowing air to enter the chamber and causes the water level to fall rapidly. The valve closes when the water level has fallen to that of the water outside and the next cycle commences as the fan once again begins to extract air.

The downward part of the cycle usually takes about 8 seconds but the time taken for the water level to rise can be varied, as also can the amplitude of the pulse. The upward pulse typically takes about 40 seconds and the maximum rise in water level is about 0.65 m. The pulse helps to maintain the blanket in a homogeneous condition and prevents the channelling of water through it. On each downward pulse excess sludge is eased over the edges of the sludge hoppers. Average rise rates for normal and peak flows are given below. They indicate the effect of having one pulsator out of use at normal flow rate.

Total plant flow 10^3 m^3 day^{-1}	Rise rates m h^{-1}			
	3 pulsators		2 pulsators	
	Blanket	Clarified	Blanket	Clarified
Normal 205	2.82	2.50	4.23	3.74
Peak 285	3.92	3.47	–	–

The highest rise rates likely to be encountered are therefore not at peak load but at normal load if one pulsator is out of use. The settled water decants into launders which run at right angles to and discharge into a common filter inlet channel which runs the whole length of the twelve filters.

Pulsator operation

The maximum design treatment capacity for one pulsator is 95 × 10^3 m^3 day^{-1} and the present chemical dosing maximum allows for 68 × 10^3 m^3 day^{-1} per stream. During plant trials at Wing however, the pulsators were operated for short periods at their maximum hydraulic design of 114 × 10^3 m^3 day^{-1} and with necessary adjustment to chemical dosing plant, continued to produce an acceptable clarified water, using either aluminium sulphate or ferric sulphate as the primary coagulant. Chemical control of the pulsator depends on information from a number of sources; raw and clarified water analyses, laboratory jar tests and condition of the sludge blanket. Degremont recommend three tests for controlling the production and maintenance of a stable sludge blanket. Firstly, laboratory jar tests to determine optimum coagulant doses. Secondly, a settlement test using sludge samples drawn from the top and bottom portions of the pulsator blanket, the results being used for regulation of the pulse. Thirdly, a sludge cohesion test (Test No. 705, Degremont Laing Water Treatment Handbook) the result of which indicates blanket condition and serves as a check for optimum coagulant and polyelectrolyte doses and pulse operation.

It has been found that laboratory jar tests using a normal bench stirrer/flocculator give a good indication of the optimum coagulant dose but not of the optimum polyelectroyte dose.

A sludge settlement test is recommended for pulse regulation, when the settled volumes of samples from the top and bottom of the blanket should be within 5% of each other after a 10 minute settlement period.

The sludge cohesion test depends upon the fact that a layer of sludge subjected to a rising current of water expands to an apparent volume related to the velocity of the water, and the cohesion of the sludge. The numerical expression of this relationship is known as the sludge cohesion coefficient, and may vary from 0.3 in a very fragile flocculate to over 1.0 for a well constituted sludge blanket. For optimum pulsator operation, a figure of 0.8 to 1.0 is desirable.

An opportunity to assess the practical effects of the pulsing action was afforded when the extractor fan on one pulsator was out of action for repairs; the other pulsator was working normally and the chemical doses were the same to both clarifiers. It was found that the settling characteristics of the two sludges were very similar even after five days and the 10-minute settled volumes were all within 5% of each other. The blanket condition in the non-pulsed clarifier was however markedly worse than in the other with fissuring visible at the blanket surface. The clarified water was also worse, turbidity and iron levels rising from 1.2 to 2.4 F.T.U. and 0.74 mg l^{-1} to 1.1 mg l^{-1} respectively. The sludge cohesion coefficient for the pulsator in normal operation was 0.75 whereas the value was 0.39 in the unpulsed blanket.

Filtration

There are twelve Degremont 'V' type rapid gravity filters (Figs. 1 and 3). The main items of filter data are as follows.

Bed size (2 No. per filter): 14.0 × 3.5 m
Filtration area per filter: 98 m^2
Number of nozzles per filter: 5292; spacing 130 × 143 mm
Filtration media: Gravel, sizing 6–10 mm; depth 50 mm
Sand, effective size 0.95 mm; uniformity coefficient 1.5; depth 1.2 m.
Normal filtration rate (205 × 10^3 m^3 day^{-1}, 12 filters working) 7.3 m h^{-1}
Maximum filtration rate (285 × 10^3 m^3 day^{-1}. 11 filters working) 11.0 m h^{-1}
Duration of filter run: 36 h
Maximum head loss: 2 m
Depth of water over sand during run: 1.22 m
Backwash cycle
Maximum draindown discharge volume: 69 m^3
Backwash water flow 1470 m^3 h^{-1} 12 min duration: 294 m^3
Crosswash water flow 698 m^3 h^{-1} 5 min duration: 58 m^3
Typical water per wash: 421 m^3

For the first 6 minutes of the backwash there is a simultaneous application of the air scour, the air flow being at the rate of 5390 m^3 h^{-1}. The air scour then ceases and the crosswash begins and continues for the remainder of the filter wash. The backwash times given above are typical for normal conditions but can be extended to suit more extreme conditions.

Principal features

(i) The sand grading is relatively coarse compared to conventional 16/30 filter sand. (The effective size of the Degremont sand is 0.95 mm compared to about 0.5 mm for the 16/30 sand.) Messrs. Degremont maintain that for filtration rates between 5 and 10 m h^{-1} there is no appreciable difference in filtrate turbidity for sands between 0.5 and 1 mm effective grain size. The coarser media however allows greater flow per unit area for a given head loss. There is also more opportunity for the removed solids to penetrate into the bed, so utilising the available pore space more effectively and preventing the filter from merely acting as a sieve with consequent surface blocking and rapid rise of head loss.

(ii) The depth of sand, 1.2 m, is greater than in previous rapid gravity filter design. This is rendered possible, without incurring unacceptable head loss,

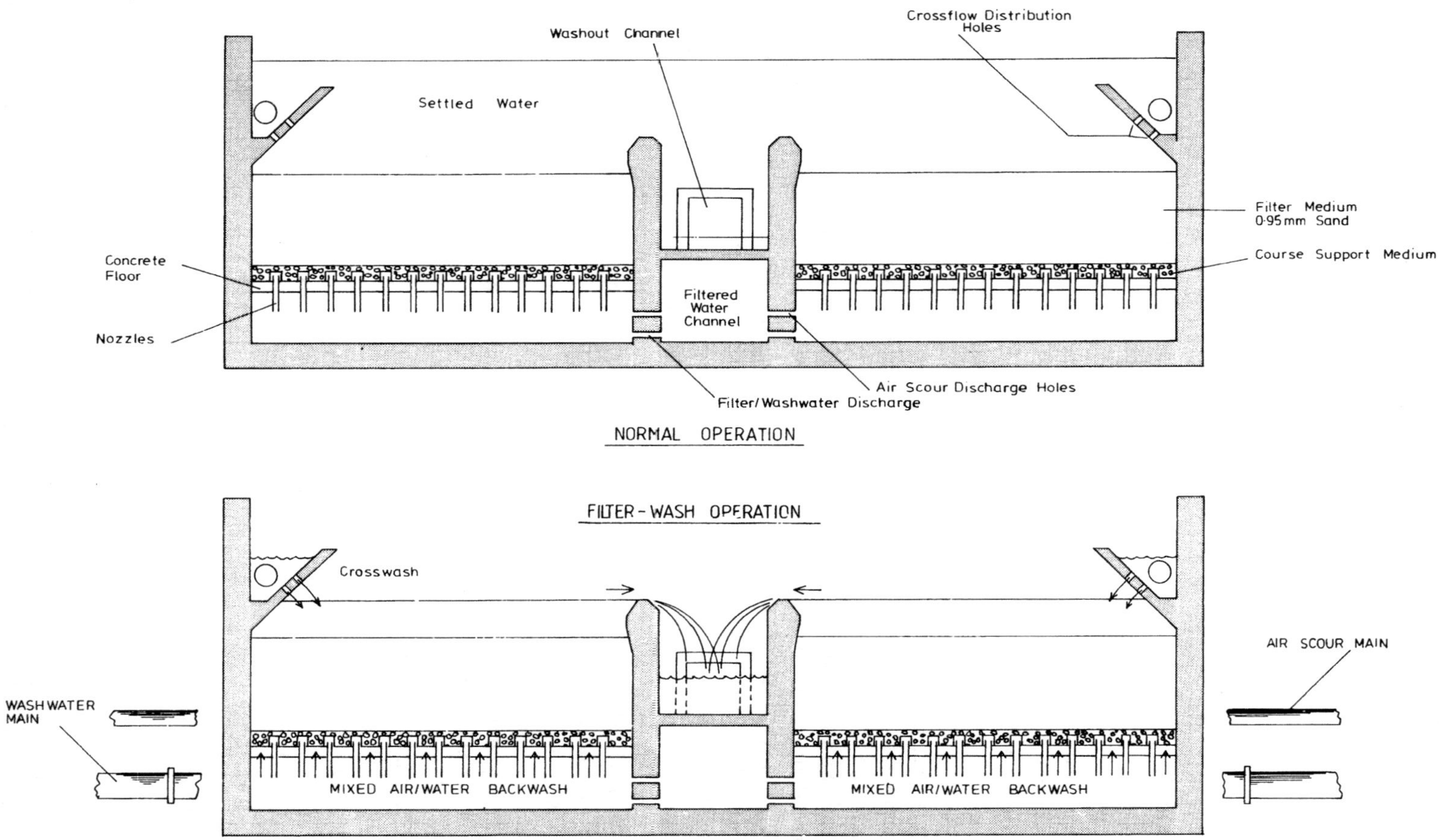

Fig. 3. Cross-section of the Degremont 'Aquazur' filters.

because of the coarser sand. However, it is also rendered desirable because the smaller filtration area would otherwise reduce the total volume of the bed and therefore its dirt holding capacity.

(iii) Washwater requirements are kept to a minimum by the use of a simultaneous air scour and backwash during the first part of the filter washing sequence. This technique also prevents the migration of dirt to the base of the filter and the formation of mud-balls in the sand.

The economy in clean washwater does however mean that an efficient crosswash is needed during the second part of the wash sequence, when the air has stopped, to convey the released dirt across the surface of the filter to the washwater discharge channel. This is achieved by allowing settled water to flow into the perforated V shaped channels running along each long side of the filter.

(iv) The sand-bed is not expanded during backwashing. The simultaneous air scour and waterwash is very efficient and renders bed expansion unnecessary.

(v) Filter outlet control. The dual requirement of maintaining a constant depth of water on top of the filter sand and an equal flow of water through each filter irrespective of head loss is achieved as follows. The water level in the filter inlet channel is measured and compared electrically with a value selected so as to maintain the correct depth of water over the sand in the filters. Any error between these two values is used to adjust electrically the set point of the individual filter outlet valve controllers. The latter receive a signal proportional to the individual filter flow and any error between the set point and the filter flow value is used to open or close the outlet valve.

(vi) The filtered water collecting channel is divided into three sections, the outlets of four filters discharging into each section. Three 1067 mm diameter mains convey the filtrates from each section to the contact tank. (Fig. 1). The post filtration chlorine is dosed separately into each section of the filtered water channel.

Contact and storage

The contact and storage tanks (Fig. 1) lie side by side and are of equal size, 11 000 m^3. The contact tank gives a contact time of approximately one hour at 284×10^3 m^3 day^{-1} flow. Short circuiting effects are minimised by the provision of a screen wall at each end of the tank. The walls are perforated by 22 No. 470 mm diameter holes over each of which is fitted a mushroom-shaped baffle. At the outlet end of the contact tank, the water receives a dose of sulphur dioxide for chlorine residual correction and then passes over a weir into a 2 m diameter pipe which conveys the water into the north end of the storage tank.

Chemical dosing

Electrical signals in the range 4 to 20 mA are generated proportional to the following flow measurements (4 mA corresponds to zero flow).
(i) The raw water flow to each pulsator.
(ii) The flow from each filter outlet.

Summated signals are also produced proportional to the total water flow to the pulsators and to the total filtered water flow.

The above signals are used as the basis of control for much of the chemical dosing and for filter outlet control.

a) *Chlorine dosing.* Storage arrangements for chlorine at the works consist of a single tank bulk storage installation of 30 tonne capacity. A pre-sedimation chlorine dose sufficient to maintain a total chlorine residual of approximately 0.5 mg l^{-1} in the clarified water is added to the raw water in the inlet mixing chambers. This dose is normally between 2.5 and 3.5 mg l^{-1}, but may be greater at times when activated carbon is being dosed for odour control. A post filtration chlorine dose normally about 1 mg l^{-1}, sufficient to maintain a free chlorine residual of 0.8 mg l^{-1} at the end of the contact period, is added to each filtered water stream. The raw water chlorine dose is metered by a single chlorinator which responds to the summated milliamp signal proportional to the total raw water flow to maintain a constant dose at all flows. The split of chlorine between the three pulsators is made manually by the adjustment of the three valves which control the flow of chlorine water from the injector.

The post filtration chlorine dose is again metered from a single chlorinator. The latter is controlled overall both by the summated flow signal from all the filter outlets and by a closed-loop control system which compares the free chlorine residual measured after a limited contact period (about 6 minutes but depending on flow) with a preselected residual level and adjusts the chlorine dose accordingly.

b) Ferric sulphate/aluminium sulphate dosing. These chemicals are stored as the 40% w/w $Fe_2(SO_4)_3$ solution and the 8% w/w Al_2O_3 solution respectively. Three rubber lined steel storage tanks each of 31.5 m^3 are provided, two being used for the liquid ferric and one for the liquid alum.

The undiluted coagulant solution is metered separately for each pulsator by positive displacement, variable speed and stroke dosing pumps. The speed of each pump is governed by the milliamp signal proportional to the flow in the appropriate pulsator. The stroke of each pump is set manually to obtain the appropriate dose of coagulant (normally 5 to 6 mg l^{-1} as Fe or about 3 mg l^{-1} as Al.). The coagulant solution is considerably diluted by 'carrier' water soon after it has been metered and is conveyed to the dosing points in the inlet mixing chambers in the diluted form.

c) Lime dosing. Storage facilities consist of two 20 tonne silos each of which is fitted with a variable speed screw feeder which conveys the lime directly from the bottom of the silo to a 0.30 m^3 slurry mixing tank fitted with a propeller type stirrer. Each of the two slurry mixing tanks is gravity fed with water from a side tank fitted with a ball valve. Lime slurry can be fed from either tank to a bank of four variable-speed Mono pumps (three duty plus one standby) which transfer the slurry to the pulsator inlet chambers.

The reaction between the lime and the bicarbonate alkalinity in the make-up water taking place within the pumps and dosing lines resulted in a crystalline-form of calcium carbonate being deposited tenaciously on all surfaces, particularly at points of high turbulence, and build-ups and blockages could not be prevented even by frequent flushing.

Treatment of the make-up water to remove the bicarbonate ions was seen as the answer to the problem although softening by base exchange was ruled out as this would only have replaced the calcium with sodium leaving the bicarbonate unchanged: on contact with an excess of lime precipitation of calcium carbonate would still have occurred. The treatment adopted was to reduce the pH of the lime make-up water to 4.0 with hydrochloric acid and remove the free CO_2 by aeration. The problem of the blocking of lime dosing lines was effectively solved by the installation of equipment to treat the water passing to the lime mixers in this manner.

The original intention to dose lime slurry to the three filtrate streams for pH correction to pHs + 0.2 was abandoned as it gave rise to unacceptable turbidity in the final water. As ferric sulphate is used as coagulant for most of the time, the lime dose can be applied at the coagulation stage (not so if alum is being used); however it has been found that there is still a need for further pH adjustment upwards at the final water stage. It was therefore decided to install caustic soda dosing for final water pH correction and at the time of writing, this equipment is about to be commissioned. It will be capable of effecting the full pH correction even when alum is being used without the presedimentation lime dose.

The possibility that alum coagulation might be required at times of high raw water pH due to algal activity, necessitating the use of sulphuric acid for downward pH correction, has been given some attention in the works laboratory. Heavy algal blooms were induced by the illumination of 10 l samples of raw water and increases in pH were obtained. The acid doses required to reduce the pH to 7.6 were determined experimentally. It was found that doses increased up to initial pH values of 9.2 but declined at higher pH values. The sulphuric acid dose at pH 9.2 was 49 mg l^{-1} and at pH 10.2 it was 27 mg l^{-1}. It was found that the alum dose itself was sufficient to further reduce the pH to 7.3, the assumed optimum value for alum coagulation.

The doses of acid indicated by this investigation are high and the conditions which would call for their use have so far not been encountered. Consideration of the acid bulk storage facilities which would be required is therefore in abeyance until a clear need is indicated.

d) Powdered activated carbon. Storage, dosing and dose control facilities for this material are similar to those for lime. The make-up water is not however treated, dosing lines being kept clear by regular flushing and occasional acid dosing.

Powdered activated carbon (P.A.C.) – normally Ness Anthrasorb CC100 – has been dosed at Wing each year since 1977 for taste and odour control. Dosing levels have varied between 5 and 25 mg l^{-1} and dosing has normally been necessary between May and October but exceptionally up until

December 1980. Raw water and final water Threshold Odour Numbers (T.O.N.) are determined at 40 °C and P.A.C. is dosed to reduce the level of the raw water T.O.N. (which has been as high as 132) to a figure of 8 or less in the final water. It has been our experience that reducing T.O.N. to this level considerably reduces the number of consumer complaints attributable to odour.

e) Polyelectrolyte. Polyelectrolyte can be dosed to the main water treatment stream and to the sludge thickeners, normal doses being 0.01 mg l^{-1} and 0.2 g kg^{-1} dry sludge solids respectively. The chemical is made up as a 0.1% solution in 4.8 m^3 mixing tanks.

f) Sulphur dioxide. This chemical is used to reduce the residual of chlorine in the water leaving the contact tank. It is controlled by a closed loop residual controller which takes its sample at the point where the water enters the clear water storage tank.

g) Ammonium sulphate. An ammonium sulphate storage, make-up and dosing system is at present being installed in order to convert the portion of the final chlorine residual present as free chlorine into a combined residual to ensure the maintenance of a chlorine residual further into the distribution system.

Wastewater and sludge treatment

Overflows to waste exist at various places in the plant but these are only intended for use in the case of difficulty or emergency and the works is designed so that there is no effluent discharge resulting from any of the treatment processes. The only treatment waste which leaves the site is the pressed ferric hydroxide sludge which is transported to the Pilton tip some 2 km from the works.

The following wastewater and sludge treatment processes are provided.

(i) *Washwater recovery.* There are two settling tanks, each of 600 m^3 capacity, into which the filter wash water is discharged. The volume of washwater from a normal filter wash is about 421 m^3, comprising 294 m^3 from the washwater pumps, 58 m^3 of crosswash water and 69 m^3 of draindown water. After 2 hours settlement, the supernatant water is decanted off through a floating arm draw off and gravitates to a sump underneath the filter machinery house. This reclaimed water is then pumped back to the head of the works. The settled sludge is discharged to a sludge sump from which it is pumped to the inlets of the sludge thickeners.

(ii) *Pulsator sludge.* The contents of the pulsator sludge hoppers are bled off at intervals and drain to the same sludge sump mentioned in (i) above.

(iii) *Sludge conditioning and thickening.* This operation is carried out in two cylindrical sludge thickening tanks, each 148 m^2 in area and 4 m deep with a peripheral clear water decanting trough. The sludge thickeners were designed to produce 6% sludge solids with a loading of 18.5 kg m^{-2} day^{-1}. At present, the solids loading is approximately 7 kg m^{-2} day^{-1}. The bottom of each tanks slopes towards a central sump, towards which the settled sludge is gradually moved by angled scrapers fitted to a slowly rotating diagonal arm.

Polyelectrolyte is added at the inlet to the thickeners at a normal dose of 0.2 g polyelectrolyte (Magnafloc LT25) per kilo of sludge solids. The thickeners operate well at the low solids loadings to give a sludge of 5–6%, and their successful operation depends on the careful regulation of the polymer dose and thickened sludge draw-off rate (and consequently the level of thickened sludge in the thickener) in order to prevent either over-pressing or over-thickening.

(iv) *Filter presses.* The three filter presses, by Manor Engineering Ltd., have a design filter cake capacity of 5 m^3 each. There are 79 plates per press, each measuring 2 × 1.5 m, and the normal thickness of the filter cake is 25 mm. The pumps which transfer the thickened sludge from the thickeners to the presses are specially controlled so as to ensure a steady pressure of 7 bar within the press. The press run is terminated after a predetermined period, usually about 6 hours, and the press is then opened. As the plates are mechanically separated, the filter cake drops into a trailer positioned underneath the press. The filter cake has a solids content of between 25 and 30% (dried at 180 °C) and is disposed of at the Pilton tip about 2 km from the plant. Current daily sludge production is about 8 to 12 tonne day^{-1} (depending on whether powdered activated carbon is in use), the corresponding dry weight being 2.2 to 3.2 tonne day^{-1}. Sludge production will increase pro rata as the volume of water treated increases.

Table 1. Average analysis of Wing raw and final water 1977–1980.
Units are mg l^{-1} unless otherwise stated

	Raw water			*Final water*		
	Max.	Min.	Mean	Max.	Min.	Mean
Colour (Hazen)	30	<5	10	12.5	<5	<5
Turbidity (Formazin)	6.5	0.25	1.7	2.3	0.10	0.22
pH	8.8	7.5	8.2	8.3	6.9	7.7
Conductivity (μS cm^{-1} 25 °C)	1040	510	780	1140	520	808
Total dissolved solids	655	399	533	661	402	548
P.V. (4 h, 27 °C mg l^{-1} O)	2.9	1.6	2.1	1.5	0.71	1.1
Ammonia F. & S, (as N)	0.18	0.01.	0.03	0.13	<0.01	0.03
Ammonia Albd. (as N)	0.46	0.19	0.30	0.42	0.08	0.20
Nitrite (as N)	0.125	0.001	0.015	0.005	<0.001	<0.001
Nitrate (as N)	9.7	1.1	3.9	9.2	1.4	3.9
Total hardness (as $CaCO_3$)	372	285	325	392	290	335
Alkalinity (as $CaCO_3$)	176	116	155	166	100	142
Calcium as Ca	133	101	114	142	102	119
Magnesium as Mg	18.5	4.9	10.9	18.0	4.8	10.6
Sodium as Na	55	32	41	54	32	41
Potassium as K	9.7	5.7	7.9	9.1	5.7	7.9
Phosphate soluble as P	0.125	<0.001	0.039	0.080	<0.001	0.003
Fluoride as F	0.38	0.30	0.36	0.47	0.30	0.39
Chloride as Cl	68	51	59	76	52	63
Sulphate (as SO_4)	177	124	152	188	141	166
Silicate (as SiO_2)	6.0	0.3	1.5	4.0	0.3	1.2
Iron (total)	0.30	<0.01	0.03	0.38	<0.01	0.02
Manganese (total)	0.22	<0.02	0.02	0.08	<0.02	0.02
Aluminium as Al	0.01	<0.01	<0.01	0.10	<0.01	0.04
Zinc as Zn	0.04	<0.02	<0.02	0.05	<0.02	<0.02
Copper as Cu	0.04	<0.02	<0.02	0.11	<0.02	<0.02
Lead as Pb	<0.03	<0.005	*	<0.03	<0.002	*
T.O.C. as C	7.3	5.7	6.3	5.8	4.2	5.2
Chlorophyll 'a' (μg l^{-1})	112	0.7	12.4	–	–	–
Metals:						
Cadmium as Cd	<0.01	<0.01	<0.01	<0.01	<0.01	<0.01
Chromium as Cr	<0.03	<0.02	*	<0.03	<0.02	*
Cobalt as Co	<0.02	<0.02	<0.02	<0.03	<0.02	*
Mercury as Hg	<0.001	<0.0001	*	<0.001	<0.0001	*
Arsenic as As	<0.002	<0.002	<0.002	<0.002	<0.002	<0.002
Selenium as Se	<0.001	<0.001	<0.001	<0.001	<0.001	<0.001
Pesticides (μg l^{-1}):						
Y.B.H.C.	0.027	0.018	0.021	0.015	0.007	0.011
Dieldrin	<0.005	<0.002	*	<0.005	<0.002	*
Aldrin	<0.002	<0.001	*	<0.002	<0.001	*
Endrin	<0.005	<0.002	*	<0.005	<0.002	*
p.p′ D.D.E.	<0.005	<0.002	*	<0.005	<0.002	*
p.p′ D.D.T.	<0.010	<0.0.005	*	<0.010	<0.005	*
Haloforms (μg l^{-1}):						
Chloroforms	<0.8	<0.8	<0.8	19	15	17
Dichlorobromomethane	<0.4	<0.4	<0.4	27	20	24
Dibromochloromethane	<0.8	<0.8	<0.8	20	16	18
Tribromomethane	<2.0	<2.0	<2.0	<2.0	<2.0	<2.0
Total haloforms	<4.0	<4.0	<4.0	65	52	58
Polyaromatic hydrocarbons						None detected <40 ng l^{-1}.

* Different limits of detection in the determinations. None detected in any test.

Plant performance and aspects of treatment

Analytical data

The extensive analytical data which has been obtained on the Wing raw and final waters since treatment began in 1977 is summarised in Table 1.

Those parameters reflecting the efficiency of coagulation and filtration namely colour, turbidity and residual iron have all been low with a very high degree of consistency. Nitrate and sulphate were both fairly high when the reservoir was first filled owing to the severe drought in 1976 and pumping from the River Nene was in fact curtailed as Nene water had both the highest nitrate and sulphate, (Low 1982).

The nitrate level has steadily declined from its initial level of between 9 and 10 mg l^{-1} N and has now levelled off at between 1 and 2 mg l^{-1}. Sulphate has also eased to lower values as the sulphate content of the pumped river water has fallen. The use of ferric sulphate at current doses of about 5 to 6 mg l^{-1} as Fe unavoidably increases the sulphate level through treatment by about 14 mg l^{-1}.

At the time of writing, the total hardness of the raw water is about 305 mg l^{-1}. The use of lime in the coagulation process raises the hardness through treatment to about 325 mg l^{-1}.

Examination of raw and final water samples for the six most persistent organochlorine pesticides has revealed minute traces of only one – gamma B.H.C. This compound is routinely found in most lowland surface waters and the concentration found is well below the acceptable level. The figures so far obtained suggest about a 50% removal by the treatment process.

Trihalogenated methanes (haloforms) in Wing water

Attention has recently been focussed on the possibility of trace quantities of haloforms being present in drinking waters. Haloforms can be produced when the chlorine used for disinfection reacts with humic substances in the water. The haloforms most frequently found are chloroform ($CHCl_3$), dichlorobromomethane ($CHCl_2Br$), dibromochloromethane ($CHBr_2Cl$) and bromoform ($CHBr_3$).

At the time of writing no formal limit for haloforms in drinking water exists in the U.K.; nevertheless it is obviously desirable to minimise haloform production during and after treatment if this can be achieved without prejudicing the disinfection process. An investigation into haloform formation in the water at various stages of treatment at Wing and in the distribution system is therefore being undertaken, the results obtained to date being illustrated graphically in Figs. 4 to 6. They indicate the following:

(i) Trace quantities of haloforms are produced in the treatment process at Wing. The predominant one is dichlorobromomethane; chloroform and dibromochloromethane are also produced at lower and roughly similar concentration but bromoform has not been detected.

(ii) With the present regime of prechlorination, haloform production within the works is about 90% complete by the end of the clarification process.

(iii) There is evidence of further haloform production in the water after it leaves the works but in much smaller quantity than within the works.

(iv) The use of a monochloramine instead of a free chlorine residual would much reduce haloform production. Commissioning of the ammonium sulphate dosing plant in the near future will ensure that the final water will not contain a free chlorine residual and this should curtail haloform production in the distribution system. However, more work is required to assess the effectiveness of chloramine disinfection within the works before this could be safely adopted as a haloform-limiting measure.

(v) Although no formal limit exists, a tentative guide level of 100 $\mu g\ l^{-1}$ for total haloforms has been suggested. None of the samples examined so far in this investigation has exceeded this level.

Acknowledgements

The authors gratefully acknowledge the help given by colleagues in the preparation of this paper and thank the Directors of Operations and Scientific Services and the Anglian Water Authority for permitting publication. The views expressed by the authors in this paper are not necessarily those of the Anglian Water Authority.

References

Low, N. F., 1982. Raw water quality: problems and management. In: Harper D. M. & Bullock, J. A. (Eds) pp. 93–101. Junk, The Hague.

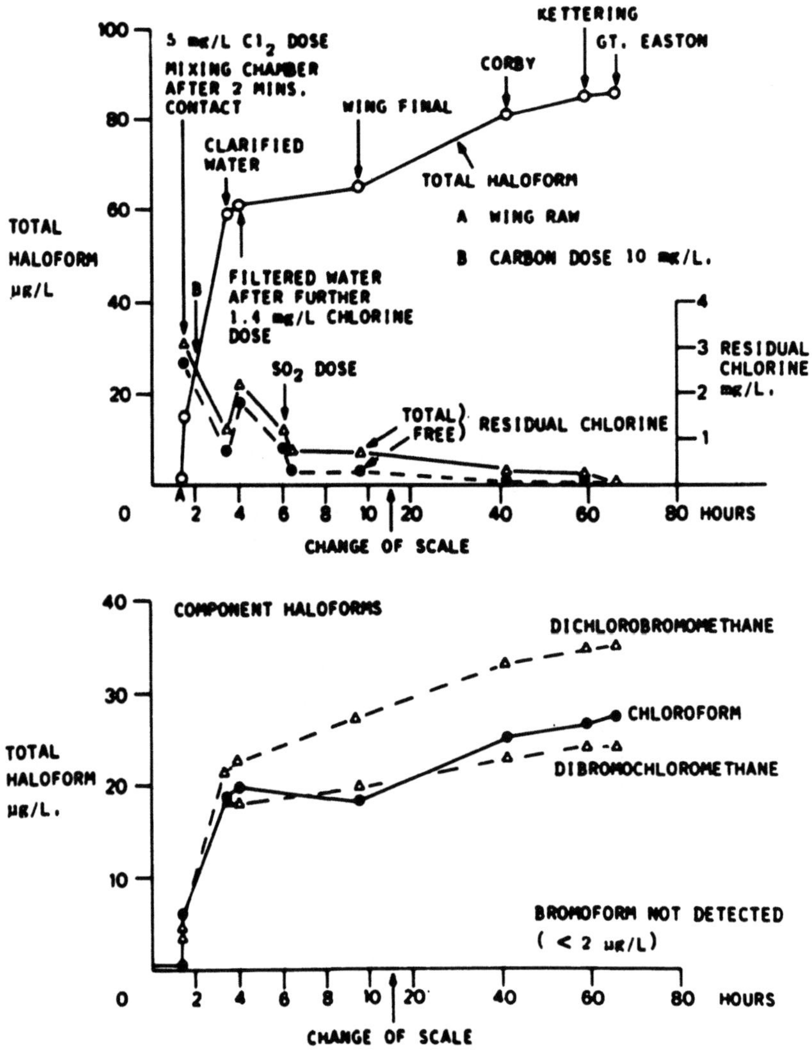

Fig. 4. Total haloform concentration (upper graph) and component haloforms (lower graph) through Wing Treatment Works and distribution.

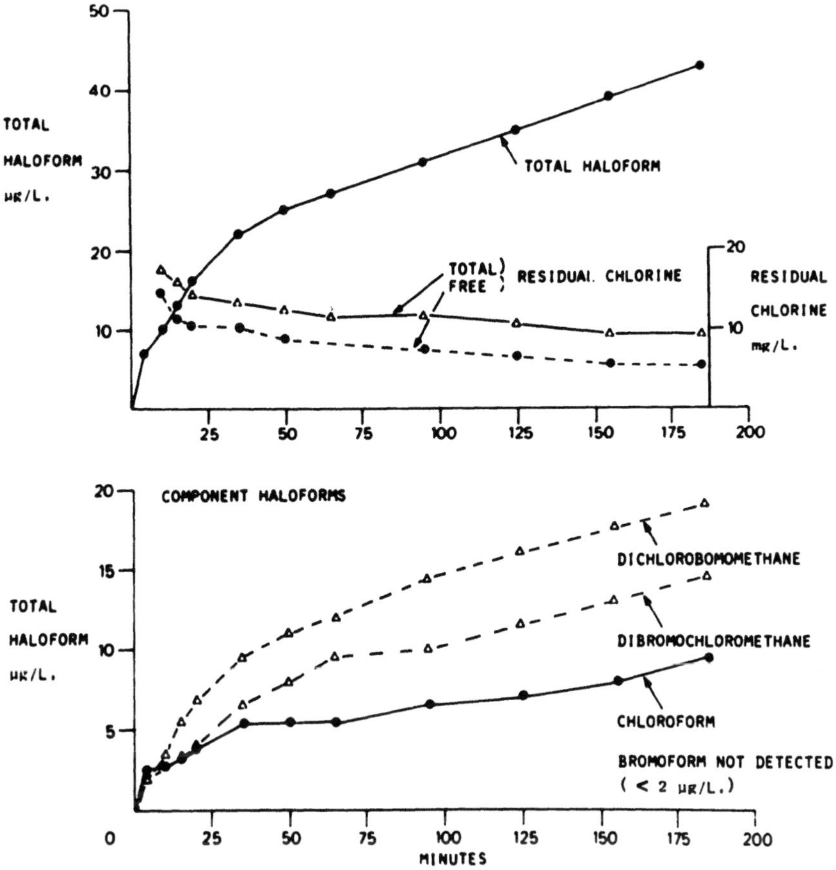

Fig. 5. Total haloform concentrations (upper graph) and component haloforms (lower graph) during pre-chlorination (chlorine dose 3 mg l^{-1}).

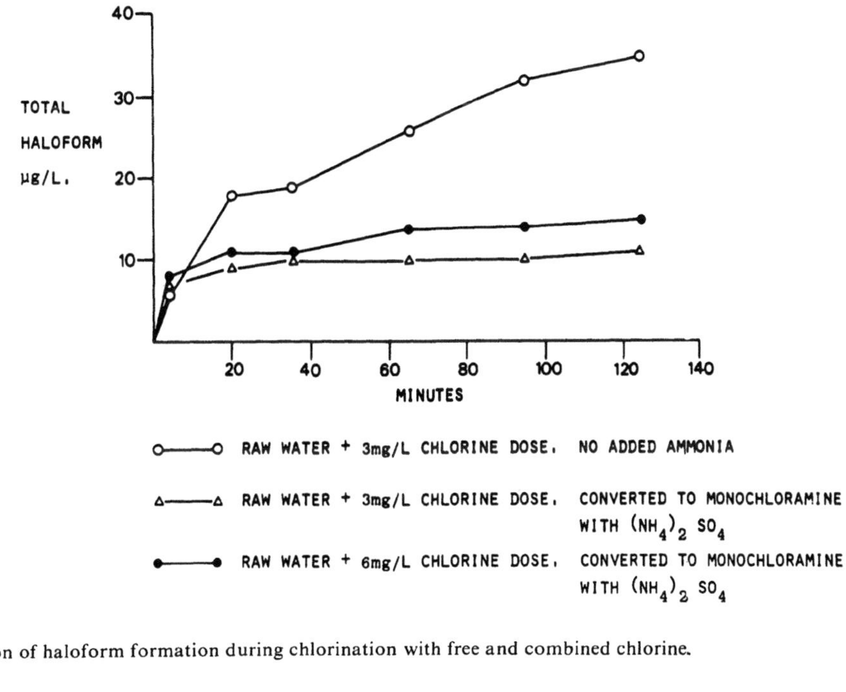

Fig. 6. Comparison of haloform formation during chlorination with free and combined chlorine.

Discussion

Mr. Ford stated that water treatment works were designed with little or no prior knowledge of the chemical or biological problems with which they must cope. He considered that one of the most striking features of the Wing plant was that it had, to date, been able to treat all water drawn from the reservoir without any problems. He drew particular attention to pulsator clarification, and gave details of the major processes involved.

Mr. Robertson then commented on the performance of the plant, drawing particular attention to the consistently good quality of the water (conforming in virtually all respects to W.H.O. European Standards and E.E.C. directives 1980), particularly in those parameters reflecting filtration and coagulation. The only notable departure from standards had been when a filter malfunction caused iron residues to reach 0.36 mg l^{-1}.

Mr. Drage outlined the study of the trihalogenated methanes (haloforms) which appeared following chlorination. Four major factors affected their formation: quantity of chlorine, contact time, occurrence of carbon dosing, and form of chlorine (free or combined). He offered evidence that combined chlorine residuals reduced haloform incidence, but were less effective disinfectants than free chlorine. If control of the formation of haloforms was needed, free chlorine disinfection followed by combination of the residuals seemed a promising solution.

Mr. Gardiner opened the discussion by enquiring why caustic soda treatment had not been preferred to lime in the initial operation of the plant. He went on to ask when pre-treatment at the Empingham off-take would be implemented. Regarding the use of the pulsator, he asked if supernatant water was recycled and what effect this had on operation of the pulsator. He questioned whether the use of polyelectrolytes at normal flows was necessary and suggested that other alternatives might have been considered.

In water treatment, he asked why the quantities of iron and manganese were presented as joint values. He also drew attention to data presented in Table 3 where values for alkalinity/chloride balance suggested that the specification that water should not be supplied that would remove zinc from brass taps had been ignored; he asked if there was any evidence for this occurring. On the matter of the specifications again, he was disturbed by the failure to reduce organics by 80% as had been specified, but noted that the additional criterion of a maximum load of 1.5 mg l^{-1} savoured of being a convenient 'let-out' clause.

Regarding the removal of manganese, he pointed out that free chlorine at 0.7 mg l^{-1} or more will oxidise manganese and enable it to be precipitated at the clarification stage. This could be extended to applying chlorine between clarification and filtration when the precipitated manganese would be filtered off and would then catalyse further manganese. This could reduce the quantity of chlorine and hence the problems of haloforms. In closing, he warned against delaying a decision to instal an acid doser to reduce pH, and hence algal growth.

Mr. Ford said that when he was appointed, the works were nearing completion and one of the features he criticised was the lime post-treatment. Attempts to make it work failed, and attempts to add it with ferric sulphate at coagulation presented problems. Caustic soda would be used for final pH trimming. Turning to pretreatment at the off-take point, he said that this had proved difficult because of the head of reservoir water.

The use of polyelectrolyte was recommended by the contractors, but it was difficult to discern any real benefit to water quality. Other aspects which had been built into the system included the combined iron/manganese value, the specification for 'dezincification' and the potentially contradictory requirements for organic reduction. He noted the comments on the control of manganese and elaborated the problems of the acid doser.

Mr. Waller welcomed the study on haloforms and asked about the increase in haloform concentration when there was no free chlorine present. Mr. Drage replied that he could not comment. Mr. Warren enquired what arrangements had been made for cleaning the raw water main and what would happen to water used in cleaning. Mr. Ford replied that a twin mains system with several cross connections would be installed soon so that the mains could be cleaned in sections. Cleaning water would be diverted into an empty pulsator.

Mr. Youngman asked what was the nature of the taste and odour which necessitated the frequent use of powdered activated carbon, and to what source this was attributed. Mr. Ford replied that it was the usual earthy, musty odour. This did not *always* coincide with algal blooms.

Rutland Water phytoplankton: the development of an asset or a nuisance?

A. J. D. Ferguson & D. M. Harper[1]
Anglian Water Authority, Welland and Nene Division, Oundle, Peterborough, England
[1] *Present address: Department of Adult Education, University of Leicester, Leicester, England*

Keywords: phytoplankton, reservoir, succession, abundance, control

Abstract

The development of a seasonal succession of phytoplankton species in Rutland Water is described. The factors controlling their succession and abundance are discussed in relation to changes in physical and chemical variables. At present, input of water from the catchment and by pumping from the Rivers Welland and Nene, has the greatest influence on phytoplankton abundance.

Introduction

The benefits which can be derived from storing water in impoundments before treatment for public supply are manifestly greatly in lowland areas where the high degree of instability of river systems is accentuated by eutrophication. Both the quality and quantity of river water in lowland areas are unpredictable and hence difficult to manage and often costly to treat. Impounding the water results in a greater degree of stability and consequently predictability (cf. Figs. 1, 2a and 2b). However, in large reservoirs such as Rutland Water, high or increasing nutrient loading can result in a greater frequency of algal blooms, some of which may cause treatment problems. This process is evident in relatively oligotrophic lakes such as Blelham Tarn in the English Lake District (Lund 1979) and other areas (Kalff & Knochel 1978). The main difficulties encountered when treating algal blooms in British reservoirs are recorded (Collingwood 1976) as being through filter blockages, reduced filter runs, taste and odour problems and through variable pH. The possibility of encountering similar problems at Rutland Water was therefore foreseen and after consultation with the Water Research Centre (W.R.A. 1971; Lack & Collingwood 1975) the design features previously described (Harper 1978), such as the provision of two draw-off towers with multiple draw-off levels and the limnological tower, were included to help deal with these problems. The Helexor type air guns were also included to aid destratification and thus reduce water quality problems associated with an anoxic hypolimnion (Cooley & Harris 1954; Lorenzen & Mitchell 1975; Ridley 1970).

A further possible control over the reservoir environment was provided by the ability to change the timing and extent of pumping water into the reservoir. However, other major considerations such as water quality and quantity in the rivers and economic constraints imposed by electricity tariff charges may be overriding.

Consequently there are a number of management decisions which have to be made in order to use the available control and abstraction facilities effectively. Many of these decisions relate directly or indirectly to the algal community and therefore much of the biological and nutrient chemistry work has been directed towards describing the fundamental characteristics of the plankton populations. It is evident (Harris 1980; Tapps 1978) that few mathematical models which have been developed to describe plankton changes in relation to en-

Hydrobiologia 88, 117–133 (1982). 0018-8158/82/0882-0117/$03.40.

vironmental variations are universally applicable although some (e.g. Steel 1972, 1975; Tapps *op. cit.)* provide very useful indicators of general trends and the likely importance of particular variables. The main difficulty found in applying models which have been developed for other areas (Nicholls & Dillon 1978) arises from the complexity of the system and problems in defining some variables such as the effects of morphometry or the physiological conditions of the algae. This complexity is amply demonstrated by the diversity of plankton and rapid changes in species dominance. The philosophy behind the approach to managing Rutland Water in terms of biological processes has been to begin developing predictive relationships based on observed responses of algae to changes in environmental conditions rather than trying to develop more complex and often more demanding deterministic models.

The timescales of management decisions can vary from years to hours but, in most cases, the predictability of the timing, extent and spatial distribution of algal blooms is of foremost importance and it is the degree of this predictability which forms the basis of this paper. Predictability of spatial distribution is discussed separately.

Methods

Temperature, fluorescence and oxygen percentage saturation can be continuously monitored at 2 m depth intervals on the limnological tower. The results are transmitted by telemetry to the Control Room below the dam and read on visual display units. This provides instantaneous information on thermocline formation and on vertical distribution of algae. Calibration of this instrumentation has to be carried out at regular intervals, particularly fluorescence against chlorophyll *a* and algal counts, because the degree of fluorescence can vary between species and with physiological condition of the algae (Harris 1980).

A full range of meteorological factors are continuously monitored by a Didcot Automatic weather station. Total photosynthetic radiation is measured by a Li-Cor Instruments Quantum Sensor confined to 400–700 μm waveband of the visible spectrum and read at daily intervals.

Chemical and algal sampling is carried out at weekly intervals at each of thirteen sites round the reservoir, as previously described (Harper 1978). A 5 m integrated tube sampler is used for describing horizontal distributions and a 3 l bottle sampler is used for observations on vertical distributions.

The combination of routine weekly sampling on the reservoir and data from the limnological tower and weather station provides a high degree of flexibility in terms of timescales on which observations can be made.

Results and discussion

The development of the phytoplankton community and species succession

Details of the early establishment of phytoplankton in Rutland Water between 1975 and 1977 have been described (Harper *op. cit.*). In 1975 about 20 species of diatoms dominated, most of which were more characteristic of lowland rivers but, by the end of that year, common lentic water phytoplankton species such as *Stephanodiscus astraea* and *Aphanizomenon flos-aquae* had become established. However, only about 30 more algal species were recorded up to mid 1977 compared with over 100 by 1980.

An increasing stabilisation of phytoplankton successional dominance has occurred and is reflected in Fig. 1, which shows changes in mean chlorophyll *a* concentration at weekly intervals since the reservoir's impoundment in 1975. This illustrates how the amplitude in oscillation of phytoplankton abundance has decreased since the reservoir was filled (see Fig. 3) in 1977 and that the fluctuations are much smaller than those found in the rivers (Figs. 2a and 2b). The higher concentration of phytoplankton and greater rates of change in the first three years were associated with high concentrations of nutrients in the catchment and in the Nene and Welland rivers. Details of changes in nutrient levels have been described (Low 1982) which show a steady decline in major nutrient ion concentrations (e.g. annual maxima of NO_3 from 7.5 to 1 mg.l^{-1}). Thus, although many of the physical variables which affect phytoplankton concentrations are likely to show similar degrees of variation to those already experienced, changes in nutrient ions will be less predictable, particularly

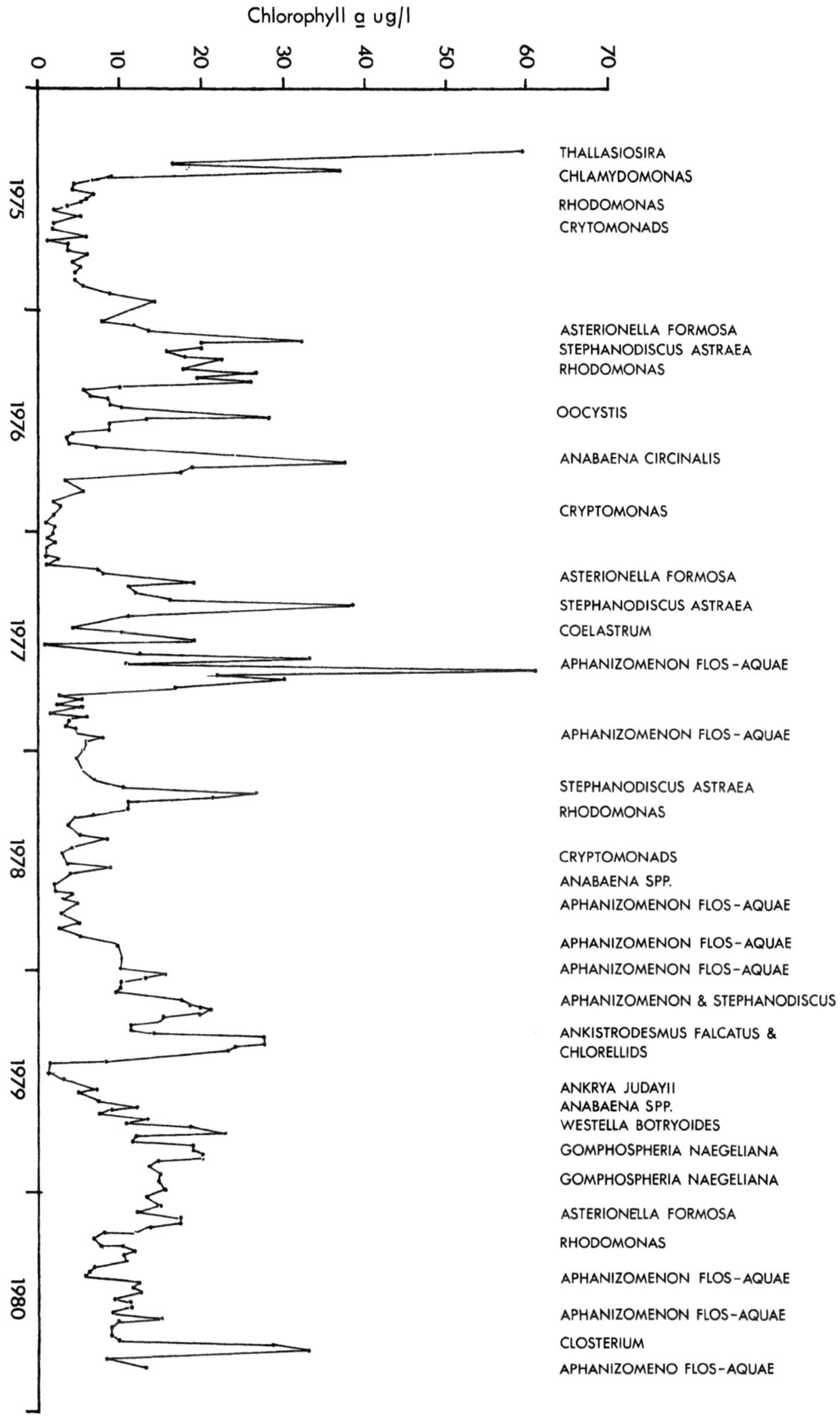

Fig. 1. Rutland Water - mean chlorophyll *a* 1975-1980.

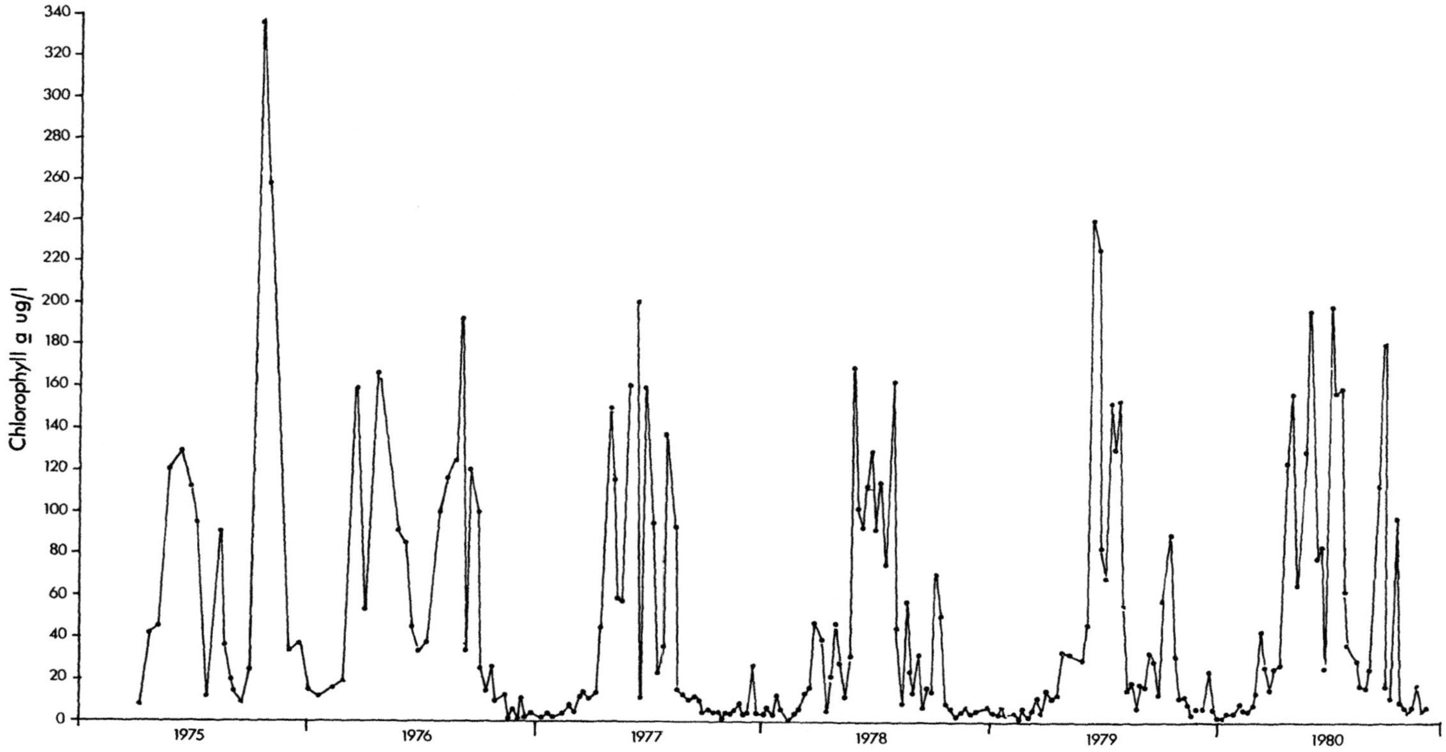

Fig. 2a. River Nene at Wansford - chlorophyll *a* 1975–1980.

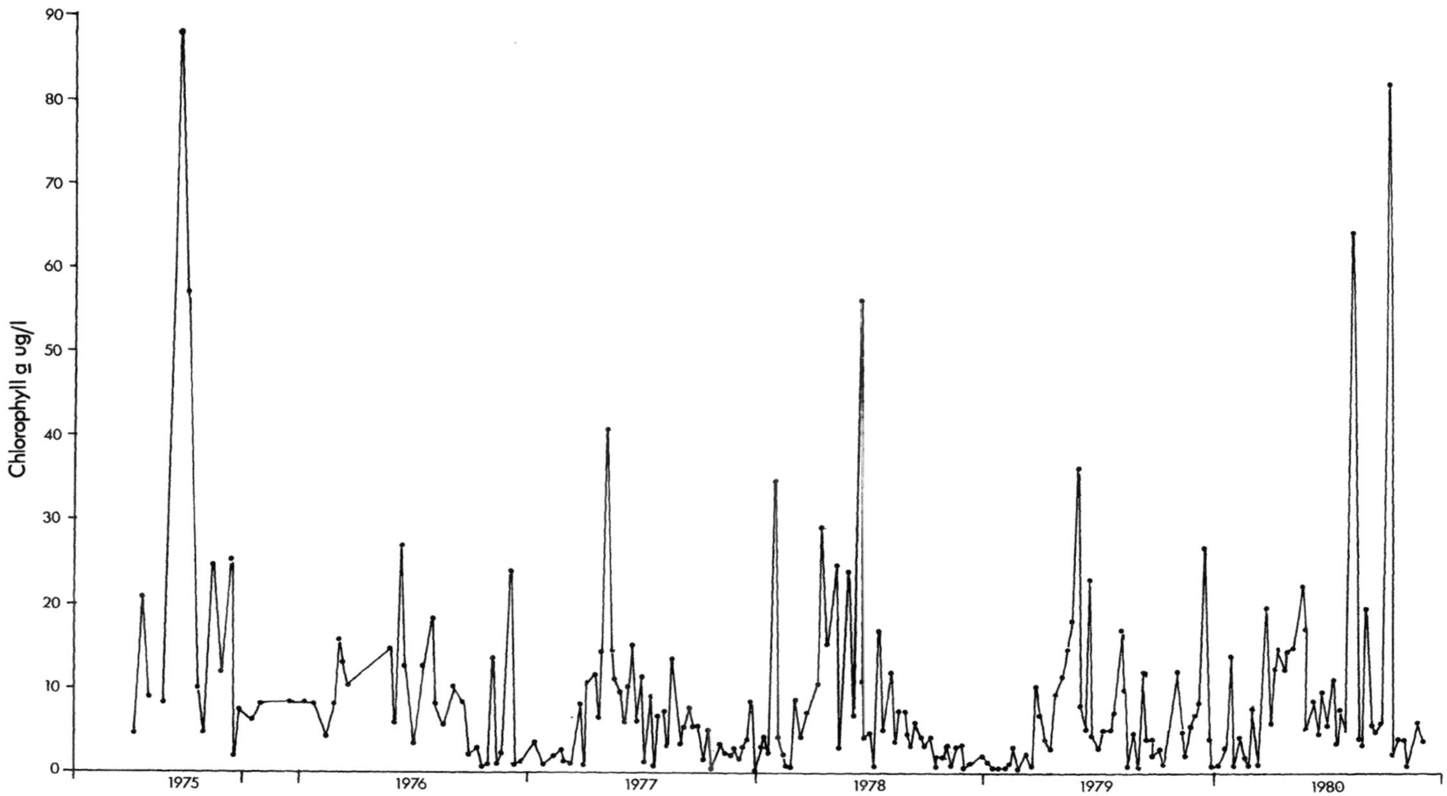

Fig. 2b. River Welland at Tinwell – chlorophyll *a* 1975–1980.

Fig. 3. Rutland Water – reservoir water level – 1975–1980.

when the reservoir goes into full supply and the retention time is reduced.

The rate at which phytoplankton populations showed a species succession characteristic of other lowland reservoirs such as Grafham (Toms *et al.* 1975) and Farmoor (Youngman 1975) reservoirs and of temperate lowland lakes (Hutchinson 1967) has been quite remarkable. The species succession over the past few years (Fig. 1) has begun with a spring diatom bloom usually dominated by *Stephanodiscus astraea* and *Asterionella formosa* which have become silica limited, followed by increases in unicellular algae such as *Ankistrodesmus falcatus, Rhodomonas minuta* and *Cryptomonas* species. Changes in dominance during the summer have been more rapid and less predictable with self buoyant blue green algae such as *Anabaena circinalis* or *A. flos-aquae* apparently having a competitive advantage over non-buoyant species during intermittent periods of stratification. An autumn peak of diatoms has usually occurred but winter dominance by blue green algae such as *Gomphosphaeria naegeliana* and *Aphanizomenon flos-aqua* has been a rather unusual feature.

Some species such as the cryptomonads are present in relatively low concentrations all year and become dominant for short periods while others such as *Ceratium hirudinella, Cosmarium* species and *Scenedesmus quadricauda* are well established but have so far never been dominant. The most difficult algae are those which appear in the reservoir and then rapidly rise to a dominant position. For example, *Westella botryoides* was absent until mid July 1979 but dominated the plankton by 7 August 1979 and in August 1980 a *Closterium* species which had hitherto gone unnoticed or was absent became dominant by 16 September. One hopes that events such as these will become less frequent as the reservoir stabilises (Baxter 1977).

The predictability of algal population dynamics

The above data provide information on the species likely to be dominant in general terms, but predictions of the actual timing and extent of algal blooms will be a much more exacting and improbable task (Harris 1980). The possible extent of algal blooms which can be expected can be calculated from phosphate/chlorophyll relationships based on data collected from many different water bodies (Schindler 1978; Nicholls & Dillon 1978) or on more complex models which take account of the relationships between algal growth, photosynthesis and light climate (Talling 1971; Steel 1975). As we are dealing with a single water body and data collected over long periods we should be able to modify relationships observed and thus increase the predictive capability.

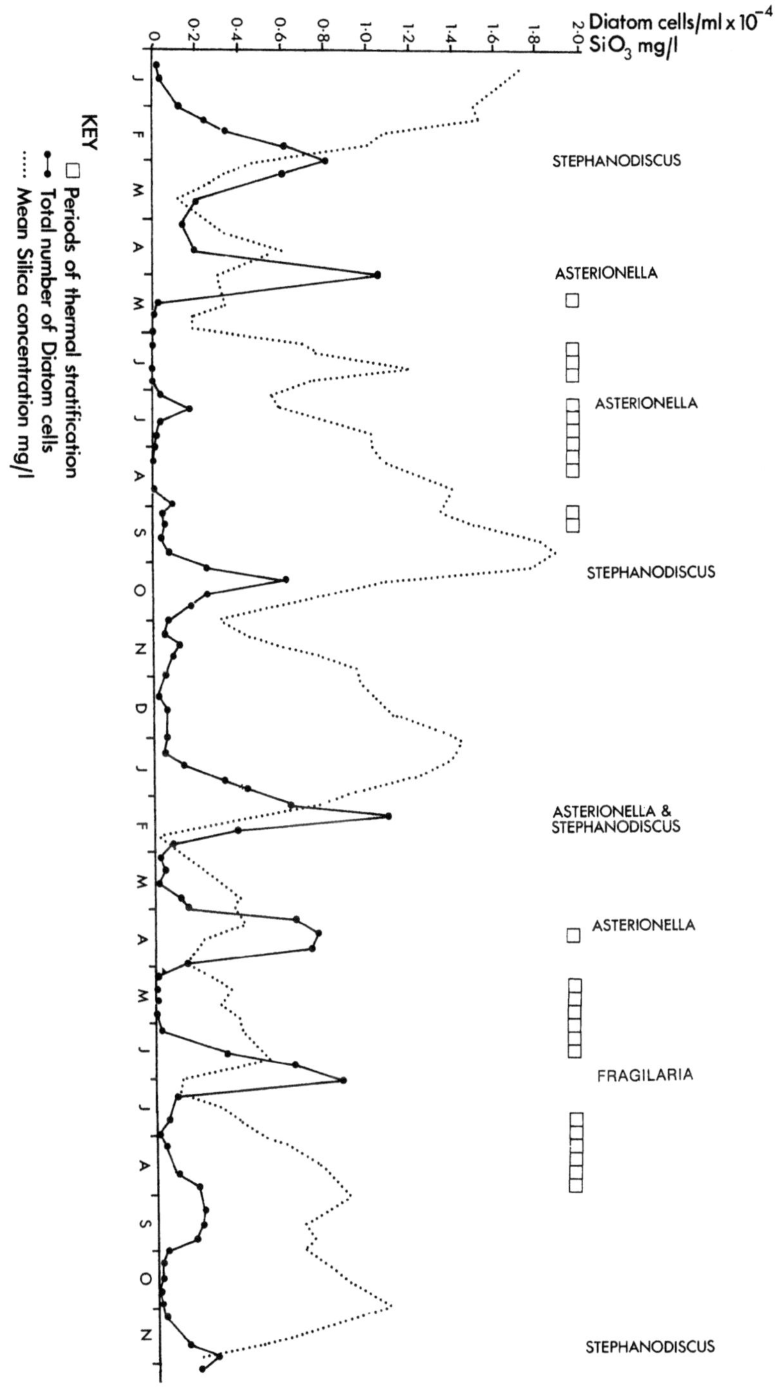

Fig. 4. Rutland Water silica/diatom interactions.

Table 1. Volumes and volume ratios relative to *Asterionella formosa.*

Species	Volume μm^3	Volume ratio
Asterionella formosa	704.1	1
Stephanodiscus astraea	13060.1	18.55
Fragilaria crotonensis	1317.0	1.87
Stephanodiscus hantzschii	67.3	0.096

The most predictable system in Rutland Water are the silica/diatom interactions, which serve as an example. Some complex deterministic models of silica/diatom relationships have been developed (e.g. Baumert 1979) but require more precise and complex information than is available with limited resources and may not be applicable to Rutland Water. Data from Rutland Water collected in 1979 and 1980 show (Fig. 4) a very strong negative correlation between cell numbers and silica concentration and a relatively predictable succession of dominant species. This figure also shows periods of relatively stable thermal stratification as this factor seems to play a major role in limiting diatom populations (Reynolds *et al.* 1982). The other main limitations are light and temperature during the winter.

The variation in numbers of cells produced in relation to the amount of silica available and hence difficulty in making predictions results mainly from the differing cell sizes and the sedimentation characteristics of different species. For example, the volume ratio of *Stephanodiscus astraea* and *S. hanzschii* from Rutland Water is 1:175. Information on precise silica requirements of Rutland Water species is lacking so volume ratios shown in Table 1 have been used to convert different species cell numbers to standard units equivalent to a single *Asterionella formosa* cell of 704 μm^3. Average dimensions were used to calculate the volumes. It is interesting to note that so far there has been little intraspecific temporal variation in cell size. Data from eight diatom peaks shown in Fig. 4 were converted to equivalent cell number units taken to have been produced by the observed depletion of silica. These observations are illustrated in Fig. 5, and are best related by the power curve:

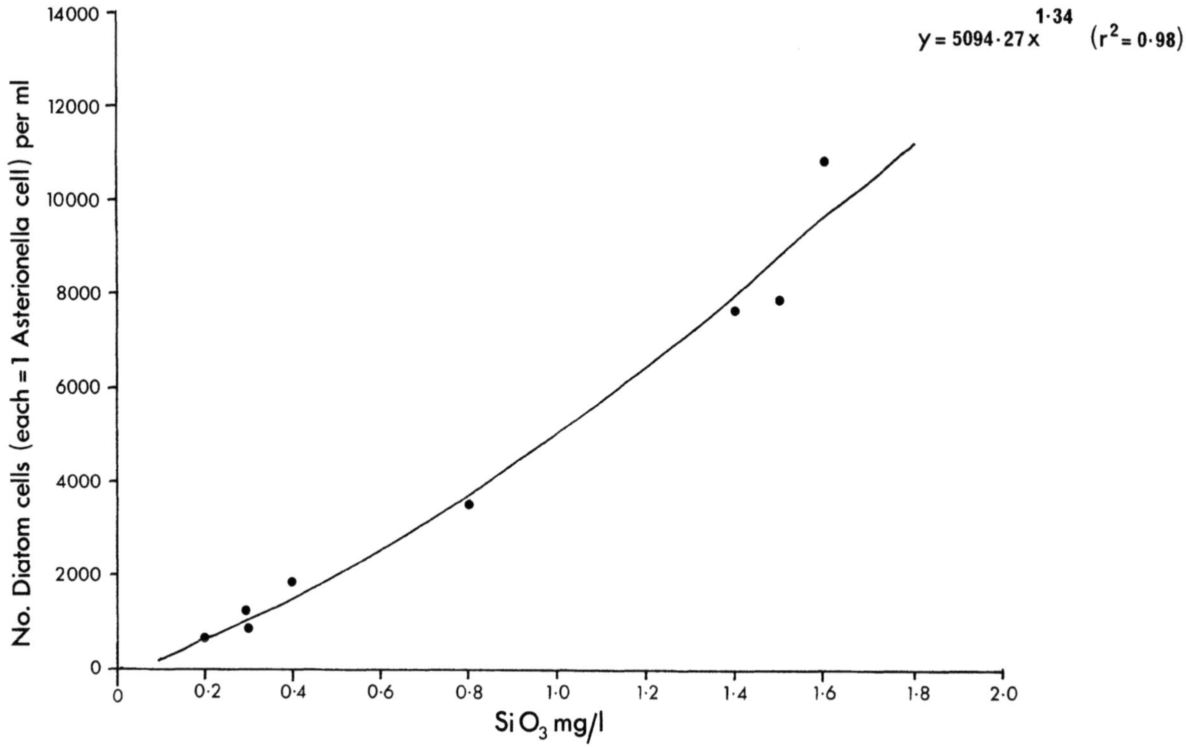

Fig. 5. The relationship between the number of diatom cells found (each equivalent to an *Asterionella formosa* cell of 704 μm^3) and observed silica depletion.

$$y = 4578x^{1.39} \ (r^2 = 0.87) \quad (1)$$

where y is the standing crop of diatom cells each equivalent to an *Asterionella formosa* cell of 704 um^3 and x is silica depletion in mg/l. The standing crop is the net production of cells and is equivalent to gross production less cells lost due to sedimentation, grazing and senescence (Reynolds *et al.* 1982). In terms of water treatment the maximum net production and timing are most important.

A method for predicting the timing and extent of diatom peaks is illustrated in Fig. 6. The initial concentration of silica is used to calculate maximum net production by using equation (1) and cell counts from weekly observations are used to predict the net growth rate using the exponential equation

$$N_t = N_0 e^{k_n t} \quad (2)$$

where N_0 and N_t are the initial and final observed populations after time t. k is the net growth constant k_n where:

$$k = \frac{\ln N_t - \ln N_0}{t} \quad (3)$$

Extrapolation of the growth towards the maximum expected population provides an estimate of the timing of the peak population.

If any other factors such as stratification or parasitisation become limiting the growth curve is apparently irretrievably interrupted and both predictions become over-estimates. An example of a population limited by stratification is the *Asterionella formosa* population from July 1979 and for

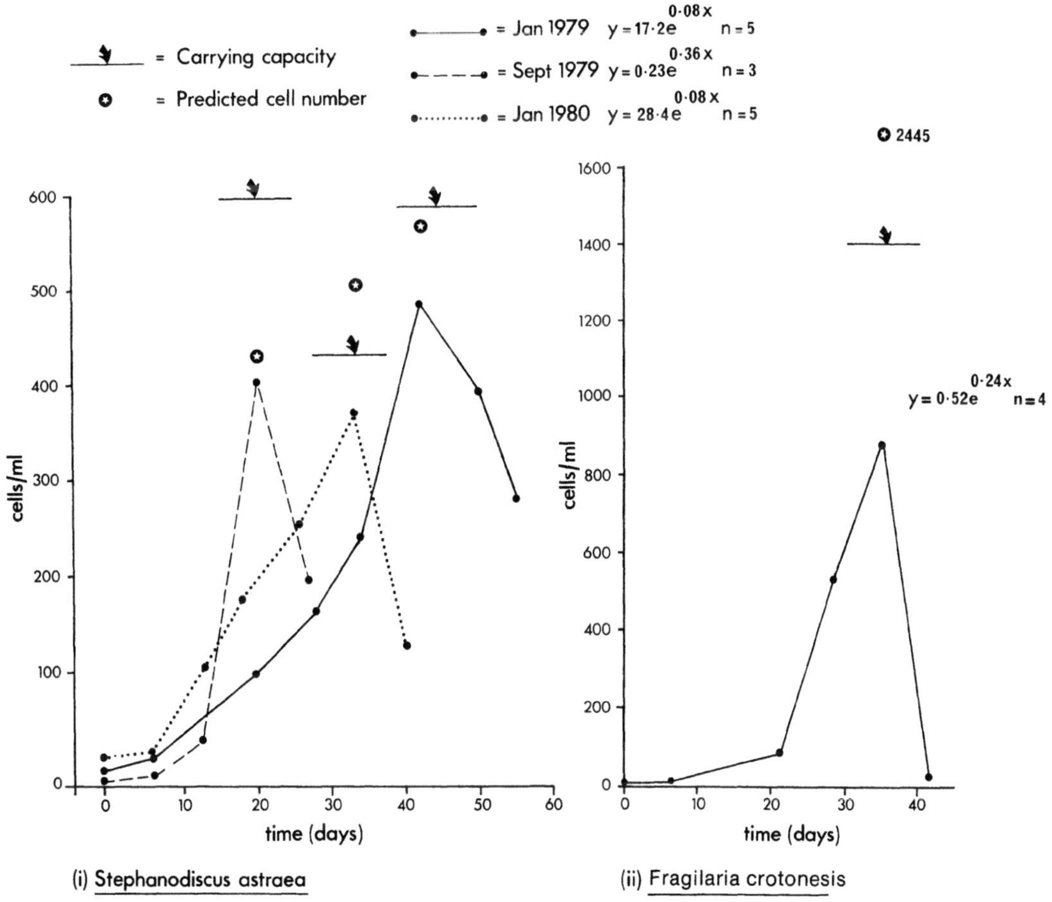

Fig. 6. Observed, predicted and maximum expected cell numbers for (i) *Stephanodiscus astraea* (ii) *Fragilaria crotonensis.*

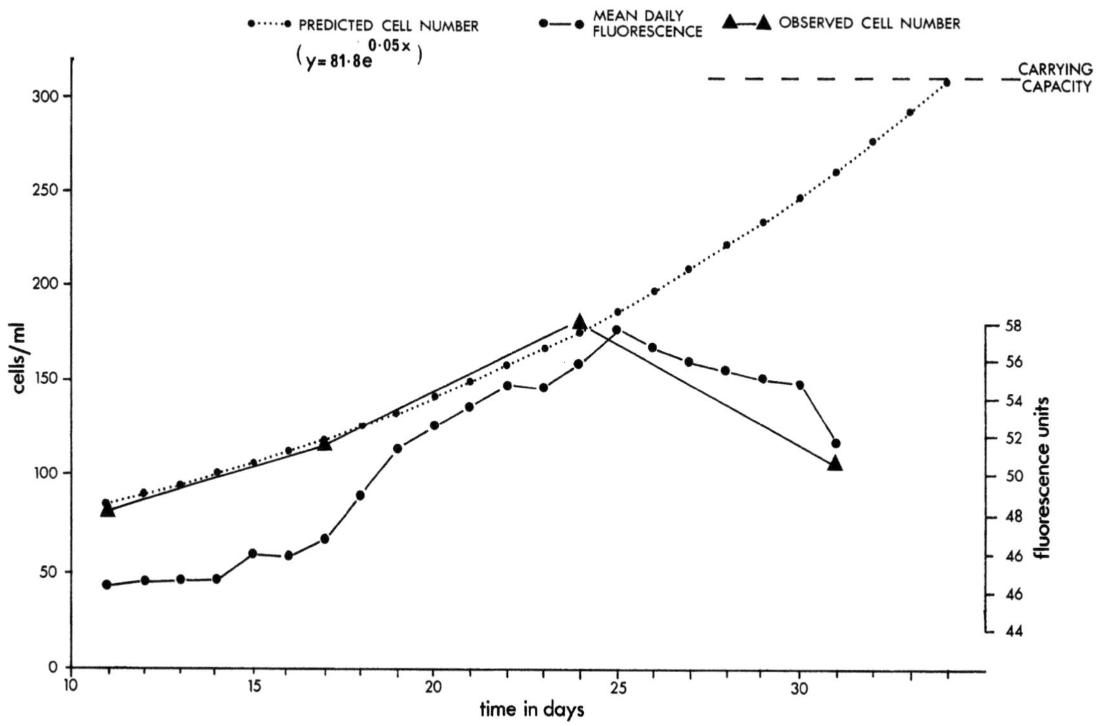

Fig. 7. Observed and predicted cell numbers for *Stephanodiscus astraea* November to December 1980. Mean daily fluorescence is also shown (chlorophyll *a* $\mu g/l = 0.33 \times$ fluorescence $+ 0.037$ at this time).

less clear reasons by the last peak in 1980 which was dominated by *Stephanodiscus astraea* (Fig. 7). The advantage of increased resolution obtained by daily readings of fluorescence is illustrated in this second example (Fig. 7) which gives an indication of the day when growth was interrupted. The reasons why loss processes exceeded the growth rate at this time are not obvious but the onset of calmer conditions and a drop in air temperature of 13 °C between 24 and 25 November may have been a sufficient 'shock' (cf. Round 1971; Reynolds 1980a).

Our confidence in the predictability of the timing, extent and dominance of diatoms peaks is thus relatively high when weekly and daily observations are used. The accuracy of these predictions should increase with continued observations on

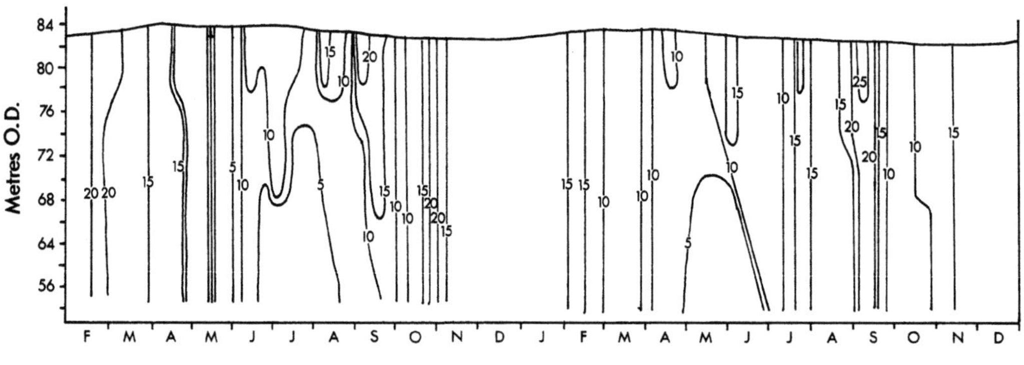

Fig. 8. Depth – time diagram of chlorophyll *a* concentrations at the Limnological Tower in 1979 and 1980 (isopleths at 5 $\mu g/l$ Chl *a* intervals).

factors likely to interrupt growth and the introduction of constants to modify any extrapolations (e.g. equation (4)). Perhaps the most encouraging aspect of this so far relatively simplistic approach is the possibility of applying similar but more complex methods to less predictable groups. Currently our main predictive capability for many species relies on observed changes in growth pattern, e.g. from exponential to linear growth, and experience gained from past data on species succession. The period from November to March has been found to be the most stable and predictable period with few algal species dominated by blue green algae. Figure 8 illustrates the temporal and spatial heterogeneity during this period in the winter of 1979/80 when chlorophyll *a* levels only varied between 12 and 17 μg/l between 1 November and 1 March. *Gomphosphaeria naegeliana* was dominant during this period. However, data for 1980/81, which include fluorescence readings, indicate that some vertical stratification has taken place during the winter while *Aphanizomenon flos-aquae* dominated. Daily data from the fluorimeter (Fig. 9) confirm that chlorophyll *a* fluctuations are minimal during the winter.

The predictability of spatial distributions

This subject is discussed in detail separately (Myers *et al.* 1982) with particular reference to abstraction during heavy algal blooms using different draw-off points. However, two points are of particular relevance.

Firstly, consideration of spatial distributions of the algae has had to be made when calculating growth rates because variations in horizontal distributions have been noted which could cause some considerable error.

Secondly, variations in spatial distributions have been accentuated when algal densities were high. For example in 1977 chlorophyll *a* concentrations varied from 19.8 μg/l in the north arm to 162.6 μg/l near the dam. Studies of this spatial distribution will therefore become increasingly important as demand for abstraction rises, the retention time is reduced and algal densities increase.

Control and management of algal populations

We have seen how the reduction in fluctuation of environmental variables brought about by impounding the water has introduced a greater stability to phytoplankton populations and thus increased both their predictability and management. However, we also know (Collingwood 1971) that treatment problems are not always proportional to the quantity of phytoplankton because the quality may be equally or more important. Some species are known to have particularly abhorrent physical or chemical characteristics in terms of treatment problems. Thus, it would be very useful to be able to alter the species dominance in a controlled manner. This might be achieved in several ways. A change in nutrient status would almost certainly result in an alteration in the pattern of species succession (Reynolds 1980a), possibly in a predictable manner.

This could be achieved by phosphate removal or relatively short interruptions in pumping during the winter. The former would provide an overall change in algal species succession, while the latter could cause a change in succession by interrupting the growth of the dominant species. Winter populations are generally steady, so unless the dominant blue green was particularly troublesome, rapid change might not be advantageous to the treatment works which usually prefers a consistent raw water quality.

Interference with grazing effects on succession either by husbanding or by reducing zooplankton numbers through fish predation has been suggested as a management weapon (Reynolds 1980a) but the former would be impractical with such a large water body and the latter would require very high stocking rates. It has been shown (Gliwicz & Preis 1977) that a biomass of between 400 and 900 kg fresh weight of fish per hectare would be required to reduce zooplankton by any significant amounts. Current trout densities in Rutland Water are about 20–25 kg fresh weight per hectare (Moore 1982).

All these methods are worthy of consideration but the most easily altered physical factor is the degree of stratification; consequently, particular attention has been paid to the relationships between algae and the mixed depth (Z_m) with a view to controlling algal concentration and/or succession. The mixed depth is evidently related to both thermal input and losses and to the circulation caused by wind speed and direction (cf. Stephen *et al.* 1976). The results in Fig. 10 illustrate daily

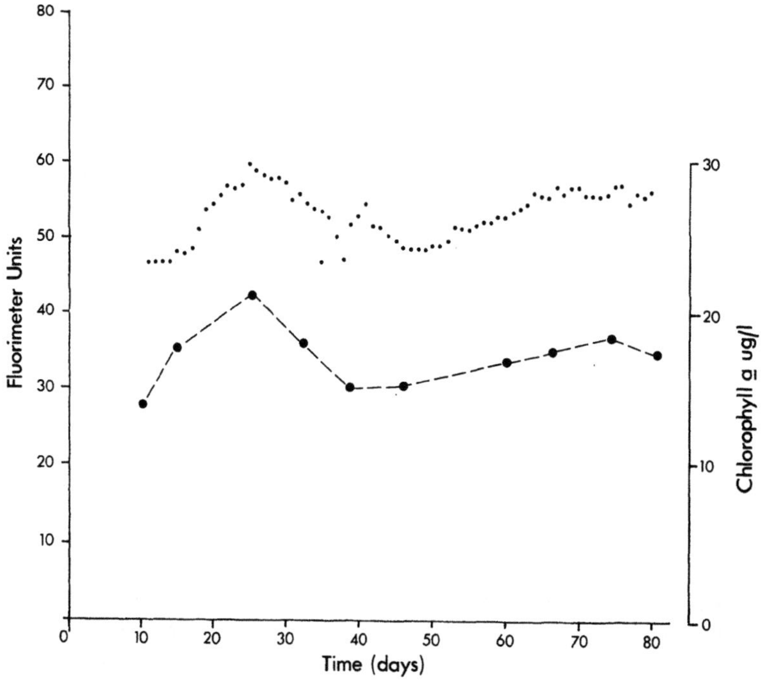

Fig. 9. Mean daily fluorescence readings from the Limnological Tower (11 November 1980 to 20 January 1981). Chlorophyll *a* concentrations are shown for comparison (.---------.).

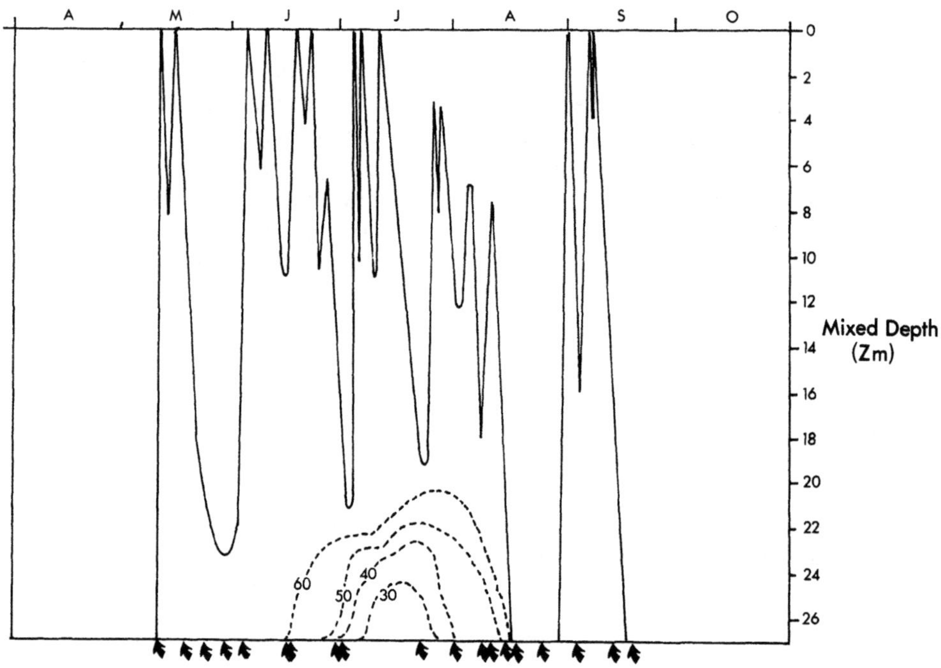

Fig. 10. Daily changes in mixed depth (Z_m - solid line); oxygen percentage saturation when less than 60% (broken line); days when wind speed exceeded 3.5 m/s (↑). Data from the Limnological Tower and Met Station 1979.

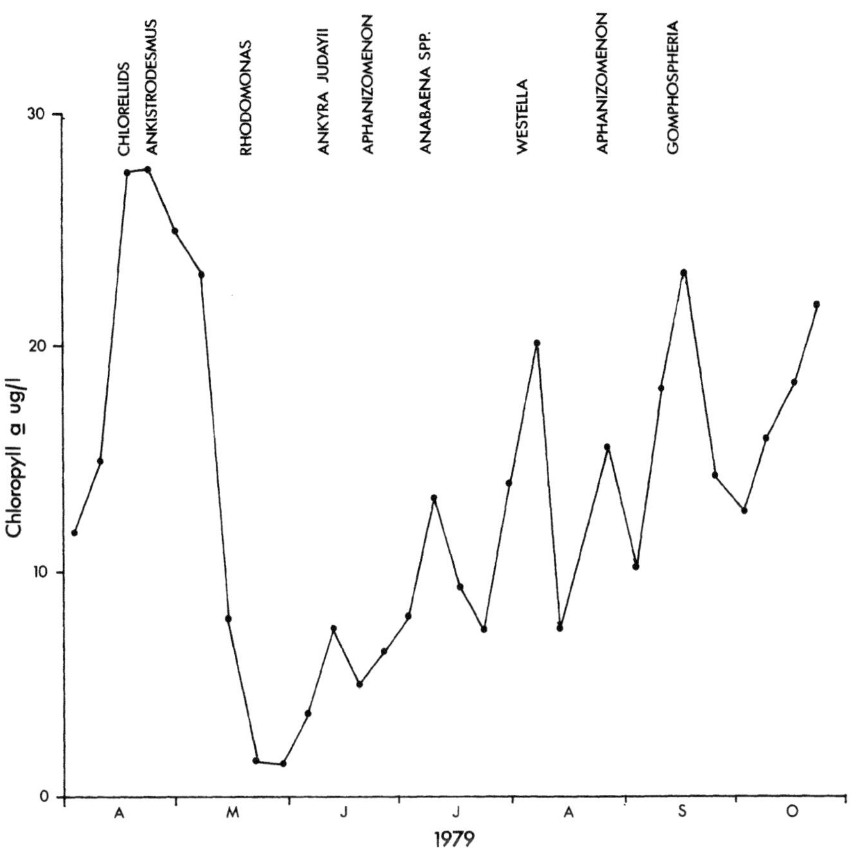

Fig. 11. Chlorophyll *a* concentrations and dominant algae during April to October 1979 (cf. Fig. 10).

changes in mixed depth at the Limnological Tower. This figure also shows oxygen depletion where the percentage saturation was less than 60% and periods when the wind speed exceeded 3.5 m/s. Changes in chlorophyll *a* are shown in Fig. 11 and major periods of change and associated algal species are shown in Table 2. The destratification equipment was in operation throughout this period.

It is evident from these data that the combination of high winds and destratification prevents the formation of a stable thermocline. However, Fig. 10 also indicates that there is considerable variation

Table 2. Periods associated with changes in algal species-dominance in 1979 (summer only).

Period	Date	Stratification	Algal species
(1)	12 May - 4 June	12 - 18 May then destratified	Chlorellids, *Ankistrodesmus*
(2)	4 June - 23 June	Mainly stratified	*Rhodomonas, Cryptomonas, Ankyra*
(3)	23 June - 30 July	Mainly stratified	*Aphanizomenon, Anabaena, Rhodomonas*
(4)	30 July - 31 August	Mainly destratified	*Westella, Anabaena*
(5)	31 August - 16 September	Mainly destratified	*Aphanizomenon, Gomphosphaeria*
(6)	16 September on	Destratified	*Gomphosphaeria, Anabaena*

in mixed depth (Z_m) and relatively long periods in June and July when thermal stratification exists. Examination of the relationships between the mixed depth algal concentrations and species succession mathematically has not produced any predictable relationships, probably because data on algal concentrations was only collected at weekly intervals. The data from 1979 can therefore only be used to indicate general trends and the possible outcome of future investigations using daily data on fluorescence as the basis for the algal component. Table 2 shows six main phases which can be identified from changes in the algal population and mixed depth (Figs. 10 and 11). This suggests that, as others have found (Reynolds 1980b), non buoyant species such as the chlorellids, *Ankistrodesmus falcatus* and later *Westella botryoides* all tend to dominate in unstratified water or where the mixed depth is large and that these types of algae are often adversely affected by periods of thermal stratification even when they are short (e.g. 12–18 May). This is also true for the diatom population such as *Fragilaria crotonensis* as seen earlier (Fig. 4). The actual mechanisms by which stratification interrupts growth of these species is unclear and may not be the same for all species. A shock change in the environment may be sufficient (Round 1971) or, equally likely, the growth rate is suppressed by a reduction in mixed depth and losses due to sedimentation increase, so exceeding the growth rate (Reynolds *et al.* 1982), i.e. in the equation:

$$k_n = k' - (k_d + k_s + k_g) \qquad (4)$$

(equation 6 in Reynolds *et al.* in press)

k_s (gross sedimentation rate constant) increases and k' (growth rate constant) decreases so that:

$$k' < (k_d + k_s + k_g)$$

and k_n declines. The effect of grazing (k_g) must also be taken into account, particularly for smaller green algae such as *Ankistrodesmus falcatus* (Ferguson *et al.* in press; Reynolds 1980a). However, during this period the grazing pressure would have been low (k_d = death rate constant).

The data in Table 2 also show that mobile species such as the cryptomonads and *Rhodomonas* and buoyant species such as the blue green algae appear to have a competitive advantage over non buoyant and non mobile species in Rutland Water during 1979. Initially the cryptomonads and *Rhodomonas* dominate (Period (2)) probably because they have a potentially higher growth rate than the blue greens. The blue green algae then become dominant, until progressive destratification in August results in an increase in the non buoyant *Westella botryoides*.

It is evident from the results of this study and those of others (e.g. Steel 1975; Stephen *et al.* 1976; Reynolds 1980a; Fast *et al.* 1973) that careful examination of loss rates and growth rates combined with constant examination of change in the environment will lead to a better understanding of the effects of changes in stratification so that controlled alterations in succession and possibly of abundance may be possible in the summer period, as has been suggested (Reynolds 1980a).

Conclusions

The demand for abstraction of water from Rutland Water is almost certain to increase in the future. This would result in a reduction in the retention time, increase in nutrient loading and a corresponding increase in maximum and mean chlorophyll *a* concentrations (e.g. Schindler *et al.* 1978). A change in the species succession will also occur (Reynolds 1980b) and fluctuations in algal densities will probably be more frequent. It seems logical therefore to observe the system now in order to determine what factors control the algal populations, how fast they can grow and where they are most likely to be found, in preparation for the problems which we will probably experience. At present we are evidently deriving full benefit from the development of algal populations through nutrient reduction in a system which is essentially cyclic. In the future they may become more of a nuisance as their abundance increases. By the time this arrives we hope to have a sufficient understanding of the system to be able to use the control of inputs, destratification and the abstraction points effectively and efficiently.

Acknowledgements

The views expressed in this paper are those of the authors.

The authors would like to thank Mr. A. W. Davies, Director of Scientific Services, Anglian Water Authority, for his permission to publish this paper. Other helpful comments on the manuscript came from Mr. P. R. Langford (Manager of the Welland and Nene River Division) and from Mr. R. E. Field and Mr. N. F. Low. We would also like to thank Mrs. P. C. Race, Mr. M. Healey and Mr. P. Springthorpe for their assistance, and Mrs. D Thurlby for typing the manuscript.

References

Baxter, R. M., 1977. Environmental effects of dams and impoundments. Ann. Rev. Ecol. Syst. 8: 255–283.

Baumert, H., 1971. System analysis of silica-diatom interaction. Int. Revue ges. Hydrobiol. 64: 457–473.

Cooley, P. & Harris, S. L., 1954. The prevention of stratification in reservoirs. J. Instn Wat. Engrs 8: 517–537.

Collingwood, R. W., 1976. A survey of eutrophication in Britain with particular reference to its effects on water supplies. In: Eutrophication of lakes and reservoirs – Institution of Public Health Engineers Symposium. B 1–33.

Collingwood, R. W., 1971. The effect of algal growth on the quality of treated water. In: James, A. & Evison, L. (Eds) Biological Indicators of Water Quality, chap. 10, pp. 1–19. John Wiley.

Fast, A. W., Moss, B. & Wetzel, R. A., 1973. Effects of artificial aeration on the chemistry and algae of two Michigan lakes. Wat. Resour. Res. 9: 624–627.

Ferguson, A. J. D., Thompson, J. M. & Reynolds, C.S., 1982. Structure and dynamics of zooplankton communities in closed systems, with special reference to the algal food supply. Paper submitted to J. Plankton Res.

Gliwicz, Z. M. & Preis, A., 1977. Can planktoniverous fish keep in check planktonic crustacean populations? A test of size-efficiency hypothesis. Ecol. Pol. Ser. A. 25: 567–591.

Harper, D. M., 1978. Limnology of Rutland Water. Verh. int. Ver. Limnol. 20: 1604–1611.

Harris, G. P., 1980. Temporal and spatial scales in phytoplankton ecology. Mechanisms, methods, models and management. Can. J. Fish. aquat. Sci. 37: 877–900.

Hutchinson, G. E., 1967. A Treatise on Limnology, Vol. 2. John Wiley.

Kalff, J. & Knoechel, R., 1978. Phytoplankton and their dynamics in oligotrophic and eutrophic lakes. Ann. Rev. Ecol. Syst. 9: 475–495.

Lack, T. J. & Collingwood, R. W., 1975. The control of reservoir water quality by engineering methods. In: The Effects of Storage on Water Quality. W. R. C. Symposium Paper 18: 1–26.

Low, N. F., 1982. Raw water quality – problems and management. In: Harper, D. M. & Bullock, J. A. (Eds) Rutland Water – Decade of change, pp. 93–101. Junk, The Hague.

Lorensen, M. W. & Mitchell, R., 1975. An evaluation of artificial destratification for control of algal blooms. J. Aer. Wat. Auth. 67: 373–376.

Lund, J. W. G., 1979. Changes in the phytoplankton of an English lake 1945–1977. Hydrobiol. J. 14: 6–21.

Moore, D. E., 1982. Establishing and maintaining the trout fishery. In: Harper, D. M. & Bullock, J. A. (Eds) Rutland Water – Decade of change, pp. 179–189. Junk, The Hague.

Nicholls, K. H. & Dillon, P. J., 1978. An evaluation of phosphorus – chlorophyll – phytoplankton relationships in lakes. Int. Revue. ges. Hydrobiol. 63: 141–154.

Myers, E., Oldham, R. & Ferguson, A. J. D., 1982. Circulation patterns: Their investigation and potential manipulation. In: Harper, D. M. & Bullock, J. A. (Eds) Rutland Water – Decade of change, pp. 67–77. Junk, The Hague.

Reynolds, C. S., 1980a. Processes controlling the quantities of biogenic materials in lakes and reservoirs subject to cultural eutrophication. Pollut. Rep. Dep. Environ. U.K. No. 8: 45–62.

Reynolds, C. S., 1980b. Phytoplankton assemblages and their periodicity in stratifying lake systems. b. Holarctic ecology 3: 141–159.

Reynolds, C. S., Thompson, J. M., Ferguson, A. J. D. & Wiseman, S. W., 1982. Loss processes in the plankton dynamics of phytoplankton maintained in closed systems. Paper submitted to J. Plankton Res.

Round, F. E., 1971. The growth and succession of algal populations in freshwaters. Mitt. int. Verein. theor. angeuw. Limnol. 19: 70–99.

Ridley, J. E., 1970. The biology and management of eutrophic reservoirs. J. Soc. Wat. Treat. Exam. 19: 374–399.

Schindler, D. W., Fee, E. J. & Ruszczynski, T., 1978. Phosphorus input and its consequences for phytoplankton standing crop and production in the experimental lakes area and in similar lakes. J. Fish. Res. Bd Can. 35: 190–194.

Steel, J. A., 1972. The application of fundamental limnological research in water supply. Symp. Zool. Soc. Lond. 29: 41–67.

Steel, J. A., 1975. The management of Thames valley reservoirs. In: The effects of Storage on Water Quality – W. R. C. Symposium Paper 14: 1–35.

Stephen, H., Skoglund, T. & Megard, R. O., 1976. Wind control of algal growth in eutrophic lakes. Proc. Am. Soc. Engrs – J. environ. Eng Divn 102: 1201–1213.

Steichen, J. M., Gorton, J. E. & Rice, C. E., 1971. The effect of lake destratification on water quality. J. Am. Wat. Auth. 71: 219–225.

Talling, J. F., 1971. The underwater light climate as a factor in the production of freshwater phytoplankton. Mitt. int. Verein. theor. angeuw. Limnol. 19: 215–243.

Tapps, J. S., 1978. Eutrophication analysis with simple and complex models. J. W.P.C.F. 50: 484–492.

Toms, I. P., Wood, G. & Owens, M., 1975. Grafham Water: some effects of the impoundment of nutrient rich river water. In: The Effects of Storage on Water Quality – W.R.C. Symposium Paper 6: 1–33.

Water Research Association, 1971. Empingham Reservoir – biological considerations. Technical Report, W.R.A., Medmenham, U.K.

Youngman, R. E., 1975. Observations on Farmoor, a eutrophic reservoir in the upper Thames valley, during 1965–1973. In: The Effects of Storage on Water Quality – a W.R.C. Symposium Paper 7: 1–15.

Discussion

Dr. Ferguson introduced the paper, explaining that the development of algae was of major interest, particularly because water pumped from both the rivers Welland and Nene was very nutrient rich. The levels of chlorophyll in the reservoir had initially oscillated greatly, but more recently these had become damped with increasing stability. He speculated that this was the result of a long retention time (currently 3 years) and the consequent depletion of nutrients. This resulted in the algae becoming nutrient limited, thus increasing stability. He suggested that in the future the substrate would become nutrient saturated and that there would be a consequent rise in the general phosphate level. In addition, retention time would decrease with greater water extraction. As a result of both these factors, the algal densities would increase.

He felt that the present monitoring programme would improve prediction of algal levels and the efficiency of treatment processes.

He drew attention to the recent installation of a Turner Fluorimeter which could measure chlorophyll levels at regular intervals. Although such methods had obvious faults, he considered that coupled with normal weekly assays, much more detailed information on fluctuations in algal density would accrue. In closing, he drew attention to the relation between silicate concentration and diatom density.

Mr. Youngman opened the discussion by commenting on the wisdom of the Anglian Water Authority in appointing a biologist to the programme at such an early stage. He believed that the knowledge gained would be of great value in the future. The concept of continued monitoring, coupled with active management, was relatively new, and offered better possibilities for effective water supply.

He noted that the authors had sampled 13 sites weekly over the six years, and asked if this was the ideal number of sites. On the matter of chlorophyll, Mr. Youngman asked whether the data were presented as chlorophyll *a*, or were they uncorrected for the phaeopigments? If the former was the case, he found the winter values for the rivers very surprising. He enquired how frequently the fluorimeter was recalibrated.

He considered the use of *Asterionella* units to quantify diatom presence (i.e. expressing each individual of each species as equivalent to so many *Asterionella*) as of doubtful merit. He pointed to work on *Stephanodiscus astrea* which showed that the species went through a two year cycle during which the diameter changed from 60 to 20 μm. He further considered that the use of surface area could improve the prediction of silica uptake. Dr. Harper replied that he believed that 13 sites were adequate for phytoplankton, but not for zooplankton. The horizontal variation was now much less than in 1977, but the authors intended to maintain the intensity of sampling in case heterogeneity reoccurred. He confirmed that the data were uncorrected for phaeopigments.

Dr. Ferguson stated that the fluorimeter had been installed in November 1980 and was currently recalibrated fortnightly but this frequency would be increased in the summer months.

He added that the methods were being constantly reviewed and the suggestions on the prediction of silica uptake were most helpful.

Mr. F. Perrin asked whether colorimetric methods were being used to measure silicates and, if so, whether consideration had been given to the possibility that depletion was due to their conversion to a form not susceptible to such assay.

Dr. Ferguson replied that he was aware of this possibility, but that resources did not permit the use of alternative assays.

Mr. Toms said that the concern of management biologists lay in the interaction of various algae with the treatment process, e.g. *Tribonema* was up to 100 times as important as *Asterionella* (w/w), and asked what was the likelihood of these other filter-blocking genera appearing in the reservoir.

He considered the single silica model had limitations because it considered only one horizon of a total situation and enquired what would be the effect of temperature, light penetration, grazing, etc. He also expressed doubts that a reduction in retention time would materially affect the situation. There was a possibility that the grazing intensity would increase because there would be more overwintering food from the organic material pumped from the rivers. He added that, properly used, a fluorimeter was a valuable tool.

Dr. Ferguson commented that with increasing knowledge it was becoming possible to predict which algae species would appear in which water bodies. He felt the more work that was done, the sooner would prediction be more accurate. Naturally, since algal counts were made weekly, 'problem' species were readily discerned. He considered that the predictive equation would not necessarily help management, but that working with the collected data certainly offered the possibility of improving management. He believed that this was a realistic approach.

Zooplankton and their relationships with water quality and fisheries

D. M. Harper[1] & A. J. D. Ferguson
Anglian Water Authority, Welland and Nene River Division, Oundle, Peterborough, England
[1] *Present address: Department of Adult Education, University of Leicester, Leicester, England*

Keywords: zooplankton dynamics, algal succession, *Daphnia, Diaptomus,* fish predation, Rutland Water

Abstract

Zooplankton population structure in Rutland Water changed over the first five years of the reservoir's existence. *Daphnia pulex,* initially dominant was replaced by *D. hyalina* in late 1975 and since 1979 this latter cladoceran has coexisted for part of each year with *Bosmina longirostris.* Population fluctuations of all the main crustacean zooplankton – *D. hyalina, Cyclops* spp. and *Diaptomus gracilis* – were erratic in the first three years (1975–1977) but have since become more regular. The relationships of the zooplankton with their potential algal food supply and with their potential predators, the fish population, are discussed.

Introduction

Because of its large size and unusual shape, consideration was given in the early planning stages of Rutland Water to the limnological implications of impoundment (W.R.A. 1971; W.N.R.A. 1972) and as a result the reservoir incorporated a number of design features which it was hoped would aid management for water supply (Harper 1978). It was also recommended (W.R.A. 1971) that a full time biologist be appointed to the reservoir scientific staff and the first was in post by early 1975. This latter point is of some practical importance, because only if at least one full time biologist is in post can attention be paid to the zooplankton community. Although Duncan (1975) has shown the importance of zooplankton in reservoirs and Nilssen (1978) their role in the succession of algae, few studies in this country have had reservoir management as a major consideration (e.g. Jones *et al.* 1979) and even fewer have been carried out by water authority scientists themselves. Consequently our understanding of the role of zooplankton in reservoir management is still at an early stage.

It was the intention of this study to monitor the changes in the zooplankton community from the beginning of impoundment, in order to relate them to the other environmental factors operating in the reservoir and to use the data thus obtained in a predictive fashion as experience is gained during the early years when the reservoir is taken up to its full yield.

Methods

Zooplankton were sampled with a 2 litre Friedinger sample bottle from April 1975 to January 1979 and from then on with a Patalas 10 litre zooplankton sampler. The usual precautions over avoidance of capture (Smyly 1968) were taken. Initially samples were taken from a central point but as the reservoir expanded, sampling was increased to a maximum of 13 sites, sampled at 0.5 m depth, together with a depth profile at the limnological tower. Sampling was weekly, with the depth profile fortnightly until early 1979, when the depth profile was usually taken weekly. Samples

Hydrobiologia 88, 135–145 (1982). 0018-8158/82/0882-0135/$02.20.

were returned to the laboratory, filtered through phytoplankton netting and preserved in 70% alcohol with added glycerol. Identification and sizing were carried out in a circular perspex trough under a stereoscopic microscope with magnification from 10 to 80×. All crustaceans were routinely counted but rotifers could not be due to inadequate time available. Where sub-sampling of the Patalas samples was necessary, these were taken after thorough mixing in a magnetically stirred polygonal perspex container. The counts distinguished the copepodite stages of *Diaptomus gracilis*, and artificially designated size classes of *Daphnia hyalina*. These followed Ferguson *et al.* (1982), viz. I - 1.0 mm; II - 1.0–1.3 mm; III - 1.3–1.6 mm; IV - 1.6–1.9 mm; V - 1.9 mm.

Results and discussion

Species composition and origin

The reservoir began to fill in February 1975 and increased rapidly in size through March and April to reach an area of over one hundred hectares, about 5 m deep in that summer. By early April, when sampling commenced, populations of planktonic crustaceans were well established. They may have had their origin in the sluggish stretches of the river Gwash, or a series of medieval fishponds on the course of the north Gwash.

Initially the zooplankton was overwhelmingly dominated by cyclopoid copepods of several species (the major one being *Cyclops abyssorum* Sars),

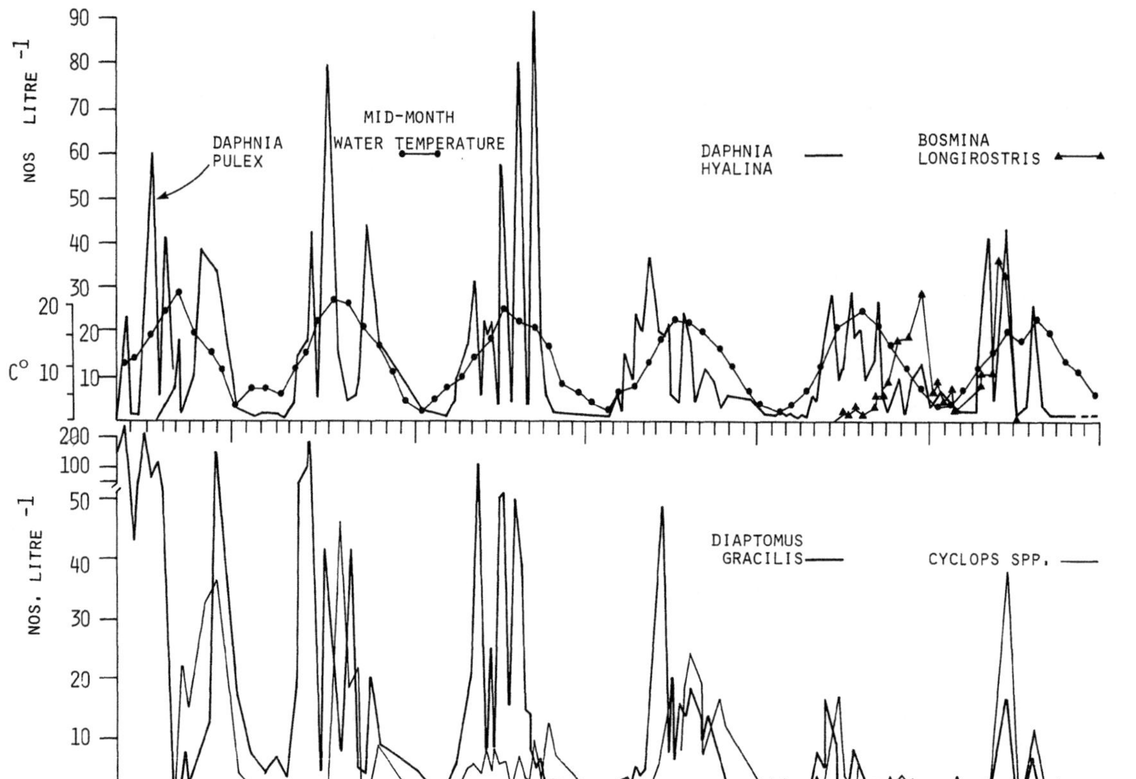

Fig. 1. Changes in the density of the main species of crustacean zooplankton in the central basin of Rutland Water 1975–1980 (mean of four samples).

reaching numbers in excess of 250 l^{-1} at the end of April and again in early June (Fig. 1). The cladoceran *Daphnia pulex* (De Geer) was also present from the first samples, reaching a maximum of 60 l^{-1} in mid-June, with a smaller peak a month later. Zooplankton numbers were generally low by early August, but the next three months saw the appearance and upsurge of the calanoid copepod *Diaptomus gracilis* Sars together with a second species of *Daphnia, D. hyalina* var. *lacustris* Sars. This latter species, smaller than *D. pulex*, completely replaced it in the plankton from August 1975 onwards. Both the *D. gracilis* and *D. hyalina* populations peaked in late September–October and were followed by a peak of *C. abyssorum* a few weeks later. *Cyclops abyssorum* has generally been considered to be a predator (Fryer 1957) although some studies have suggested an omnivorous diet based upon gut analyses and its high density in the plankton (Johnson & Walker 1974). Its late summer increase, following that of the herbivorous species, suggests a predatory relationship, but its spring population increases must have been sustained by other food – algae, detritus or rotifers.

For the next 5 years, 1976–80, these same three species – *D. hyalina, D. gracilis* and *C. abyssorum* – dominated the plankton. At least one other species of *Cyclops, C. leukartii* was regularly recorded together with occasional individuals of *Chydorus sphaericus, Ceriodaphnia quadrangula,* and *Eurytemora affinis*. The cladoceran *Bosmina longirostris* was also only recorded rarely up until mid-1979, but became abundant from October 1979 to June 1980, the standing crop equalling or exceeding that of *D. hyalina*.

This changing pattern of cladoceran dominance is of considerable interest, because it confirms similar observations made on other new reservoirs (e.g. Munro & Bailey 1980); in addition changes in the size of the dominant filter feeding cladocerans from larger to smaller forms in both natural lakes and reservoirs have been attributed to the effects of size-selective predation by fish (Brooks & Dodson 1965; Duncan 1975; Hall *et al.* 1975). In Rutland Water, fingerling trout were stocked to a high density in early 1975 (Moore 1982), and a high proportion of their diet in this first year consisted of zooplankton (Harper 1982). Their feeding would have been maximal during July and August so it is likely that size selective predation during this time resulted in the competitive exclusion of *D. pulex* by the smaller bodied *D. hyalina*. A comparison made in October 1975 between the *Daphnia* in the reservoir and in trout stomachs showed that whilst the mean size in the reservoir was 1.34 mm, that in trout stomachs was 1.58 mm, a significant difference (comparison of means, $p < 0.05$).

After the spring of 1976, *Daphnia* ceased to be found in significant quantities in trout stomachs, and it is assumed that the increase in size of the trout coupled with the enlargement of the reservoir (hence decreased trout density) relieved the predation pressure on the *Daphnia* population. The present density of trout (around 20–25 kg ha^{-1}) is not high enough to have a significant effect upon zooplankton grazers (Gliwitz & Prejs 1977).

All coarse fish had been removed from the reservoir area in early 1975 and so populations of such species as roach (*Rutilus rutilus*), which have probably originated from individuals pumped in from the rivers Welland or Nene, were slow to become established. Large shoals of cyprinid fry have been observed in the reservoir shallows each summer for the last two years (1979–80) now and they have become an important component in the diet of trout (Warlow & Oldham 1982). Since coarse fish, particularly in their early stages, are planktivorous, it is feasible that predation pressure on *Daphnia* has increased in the last two years, resulting in the observed increase in *Bosmina* and its coexistence with *Daphnia*. Munro & Bailey (1980) reported the replacement of *D. pulex* by *D. longispina* (similar in size to *D. hyalina*) within the second year of Bough Beech reservoir in Kent, and the dominance of *B. longirostris* in the same autumn. They attributed this to the effects of predation by stocked trout and indigenous coarse fish. Since coarse fish had not been removed from this impoundment as they were from Rutland Water, the greater speed with which the same changes were recorded is of interest.

In the five years 1976–1980 the seasonal fluctuations of the three major zooplankton species have been broadly similar, but the most striking feature is the reduction in oscillation which has taken place, particularly in the population changes of *D. hyalina* and *C. abyssorum* and in the three years since 1977. This is a characteristic which we have also observed for the phytoplankton (Ferguson & Harper 1982). There are three possible reasons for this.

Firstly, it is in line with the common ecological observation that populations only slowly come into balance with their resources in a new ecosystem, stabilisation following several 'overshoots' of the carrying capacity of the environment (Rickleffs 1978). This is particularly true for the *Cyclops* populations (Fig. 1).

Secondly, during the periods of rapid filling of the reservoir in the first three years (1975–77), rapid changes in water chemistry occurred which have not been repeated due to the present long retention time in the reservoir (Low 1982). Reduction in nutrient levels has reduced phytoplankton oscillations and thus resulted in lower peak populations of zooplankton.

Finally, the large temporal fluctuations apparent in Fig. 1 are partly due to the very uneven horizontal distribution of most zooplankters, which was most pronounced in the second quarter of 1977 (see below) and which has not yet been repeated.

Population changes in relation to potential food supply

Since the middle of 1978 the herbivorous species have been dominant in the zooplankton and this is clearly the stable state compared to that existing earlier. For the first three summers and four spring periods in the reservoir, numbers of *Cyclops abyssorum* were considerably in excess of potential food organisms at times, and frequently showed increases earlier than the herbivores. This paradox has already been mentioned, and we are presently attempting to explore it by examining the population structure and productivity of the cyclopoids and rotifers.

The dominant herbivore species is *D. hyalina*. Although its population oscillations were greater in magnitude in the early years, the underlying pattern of population change has been broadly similar in each year, with reproductive peaks fairly consistent in occurrence differing only in precise timing and numbers of offspring produced. This can be seen by

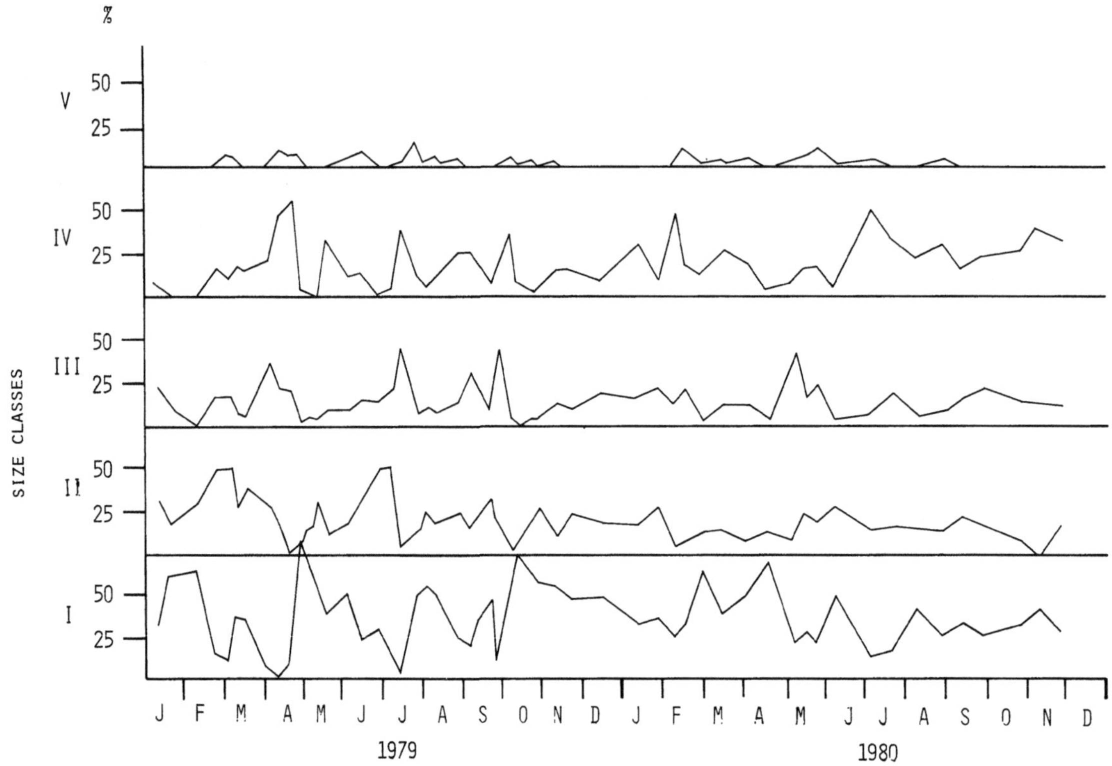

Fig. 2. Changes in the size structure of *Daphnia hyalina* populations expressed as a percentage of total density 1979–1980.

comparison of Fig. 2, which shows the changes in population structure during 1979 and 1980 expressed as a percentage of standing crop density, with the appropriate part of Fig. 1 which shows the actual changes in standing crop.

The typical pattern of seasonal change was for there to be three to four major peaks – usually around May, July and September – between which numbers fell back to low levels similar to those of winter. In this the reservoir was different from other eutrophic ones studied (e.g. George & Edwards 1974; Jones *et al.* 1979), where the populations of *D. hyalina*, once built up in the spring, remained at a plateau all summer, from which further outbursts occurred, before they fell back down to winter levels at the end of the year. Perhaps this is another aspect of the instability of the present Rutland Water zooplankton.

A close examination of the population changes of *D. hyalina* over the past six years reveals the similarities and individual differences from which additional investigations can be planned and progress towards a predictive model explored (Fig. 3).

Usually the spring diatom bloom, the first algal increase of the year, consisted of species palatable to *D. hyalina* (Ferguson *et al.* 1982) which stimulated an increase in fecundity, mean brood size and hence egg production. It was followed by an increase in *Daphnia* which was usually occurring as the bloom was declining through silica depletion (Ferguson & Harper 1982). In the two years when this March increase did not take place (1976 and 1979), the spring water temperatures were lower than usual, for instance only 2 °C at the beginning of March 1979 compared with at least 5 °C in the normal years. This would have extended egg development times considerably (Botrell *et al.* 1975).

The extent of the first increase of *Daphnia*, when it did occur, was only approximately related to the magnitude of the spring diatom bloom; thus it was slightly lower in 1980 than 1977 or 1978. When the

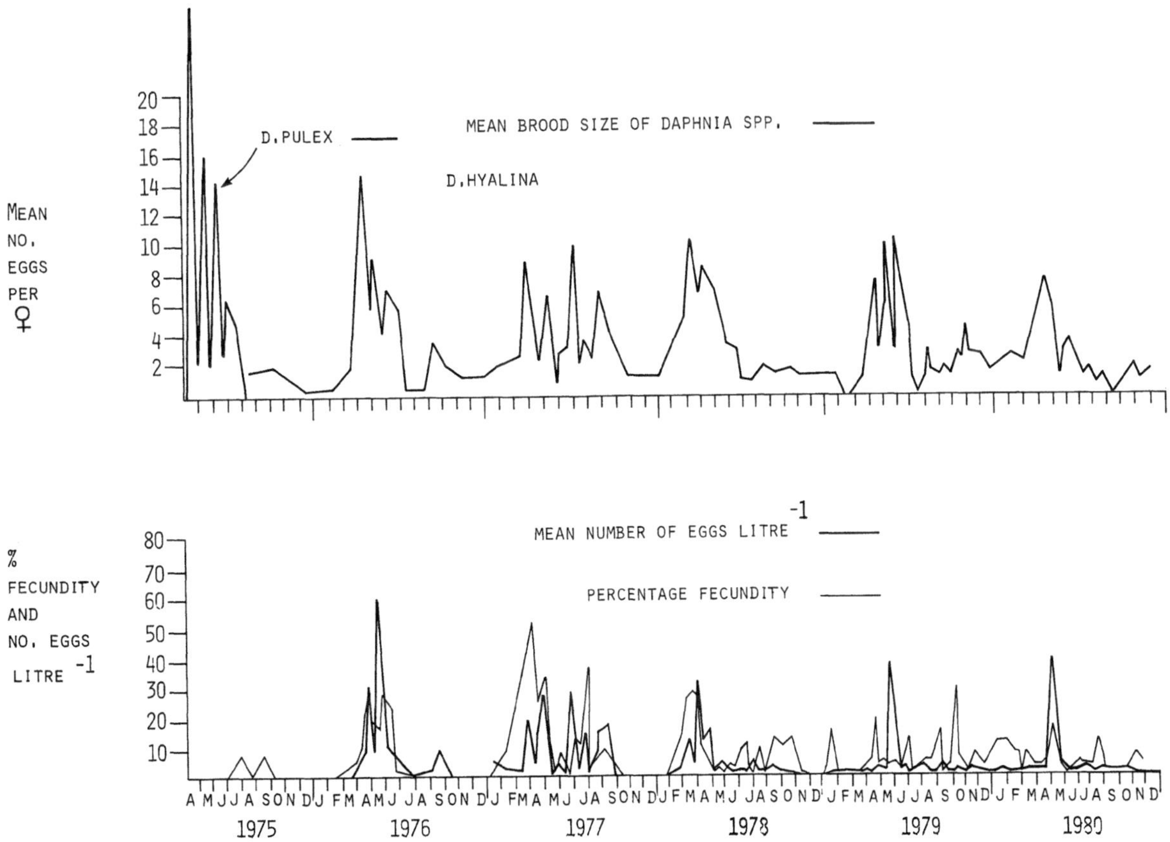

Fig. 3. Reproductive statistics of *Daphnia hyalina* 1979–1980.

increase in *Daphnia* was delayed by low spring temperatures in 1976 and 1979 it was not related to the spring diatom bloom at all, but merged into the May upswing in numbers which followed.

Rapid increase took place in all years in May and June; in each year they were coincident with or closely followed algal increases dominated by palatable species. Thus *Rhodomonas* dominated in 1976 and 1978, *Coelastrum* in 1977, *Ankistrodesmus* in 1979 and *Aphanizomenon* (but with *Rhodomonas* sub-dominant) in 1980. In the two years with a cold March and delayed *Daphnia* growth, the post-diatom algal peak was higher than in the other three years and this could be a result of reduced grazing pressure in the immediate aftermath of the diatom bloom. In all years this early summer peak of numbers could not be sustained and was followed by a dramatic decline. Minima were reached in June in the cold spring years and July in other years.

Where a rapid recovery of algae followed this zooplankton trough, a rapid increase in *Daphnia* followed. Thus in 1976 an *Oocystis* bloom in late June was followed by a rapid zooplankton outburst which crashed in July. In 1977 a bloom of mixed *Aphanizomenon, Rhodomonas* and *Cryptomonas* developed rapidly in July, fluctuated through August and declined in September. The *Daphnia* population showed two pulses separated by a trough; one in mid-August, the other in mid-September, which followed these two chlorophyll peaks.

In the last three years there have been no dramatic increases of algae during July–September and no correspondingly rapid changes in *Daphnia* populations. Each year has had an increase in late July (1979) or August (1978 and 1980). These late summer peaks occurred during low, oscillating or slowly rising levels of chlorophyll, made up from a dominant blue-green alga (*Aphanizomenon* or *Gomphosphaeria*) and associated, more palatable species such as *Rhodomonas* and *Cryptomonas*. There is thus no clear link between standing crop levels of grazer and potential food, probably because of the mixture of algal species and the rapidity of change at summer temperature. Analysis of the percentage size structure of the population over the last two years shows how similar they

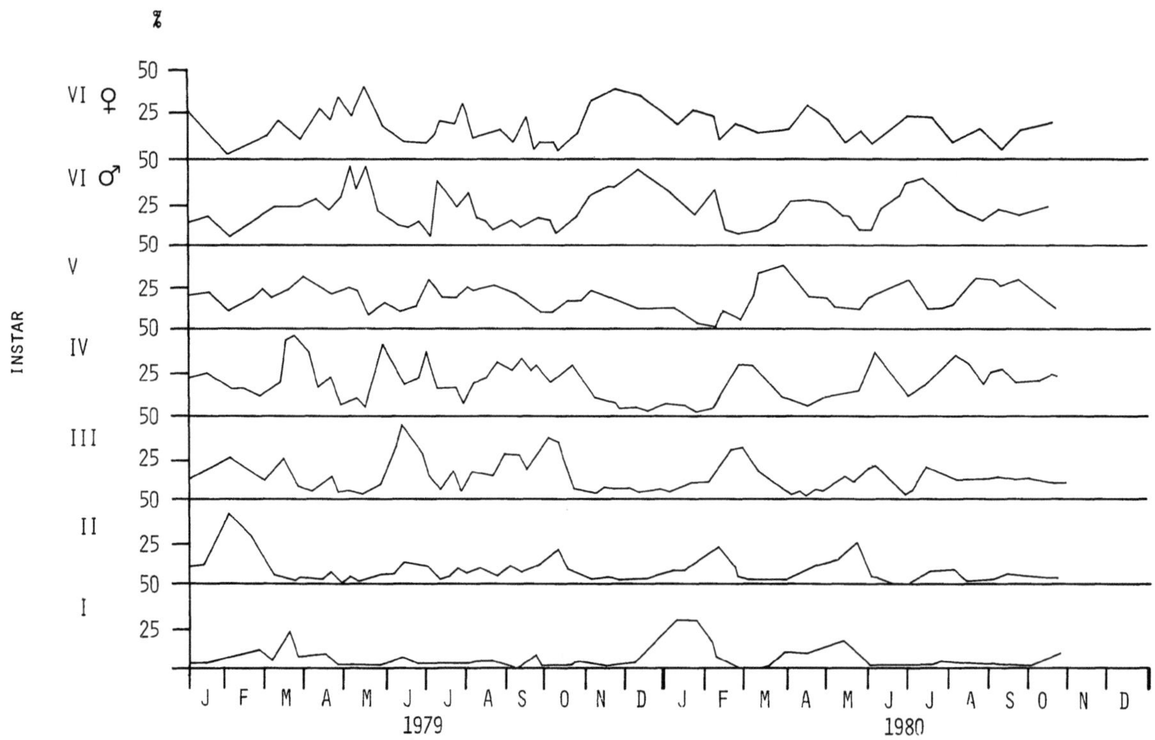

Fig. 4. Changes in the size structure of *Diaptomus gracilis* populations expressed as a percentage of total density 1979–1980.

are. We are presently analysing the relationship between zooplankton change and algal species counts to try to obtain a clearer picture of events during the summer months.

The population dynamics of *Diaptomus gracilis* seem to be less closely linked with algal changes. Firstly, this is a sexually reproducing species whose population oscillations are slower; instar analysis (Fig. 4) shows that only three generations have occurred each year by comparison with at least six in *Daphnia*. The first generation was produced very early in the year from overwintering adults and it was largely the maturation of this brood in April which then produced the second generation which was the major increase in population seen in May–June. This second generation, maturing in August, itself only gave rise to a small third generation but it was the survivors of this brood which formed the bulk of the small overwintering population, either as late stage copepodites or adults.

There was no noticeable numerical response to the spring diatom blooms in any years; usually small rises which occurred at about the same time. The main May–June increase was variable, occurring at the same time as the increase in *D. hyalina* in some years or during a trough of them in others.

Further analysis of these data will be carried out by size measurements of adults, which will facilitate the identification of generations, and comparisons of the algal species composition with the periods of egg production.

Horizontal distribution of zooplankton

It was anticipated at the design stage of this

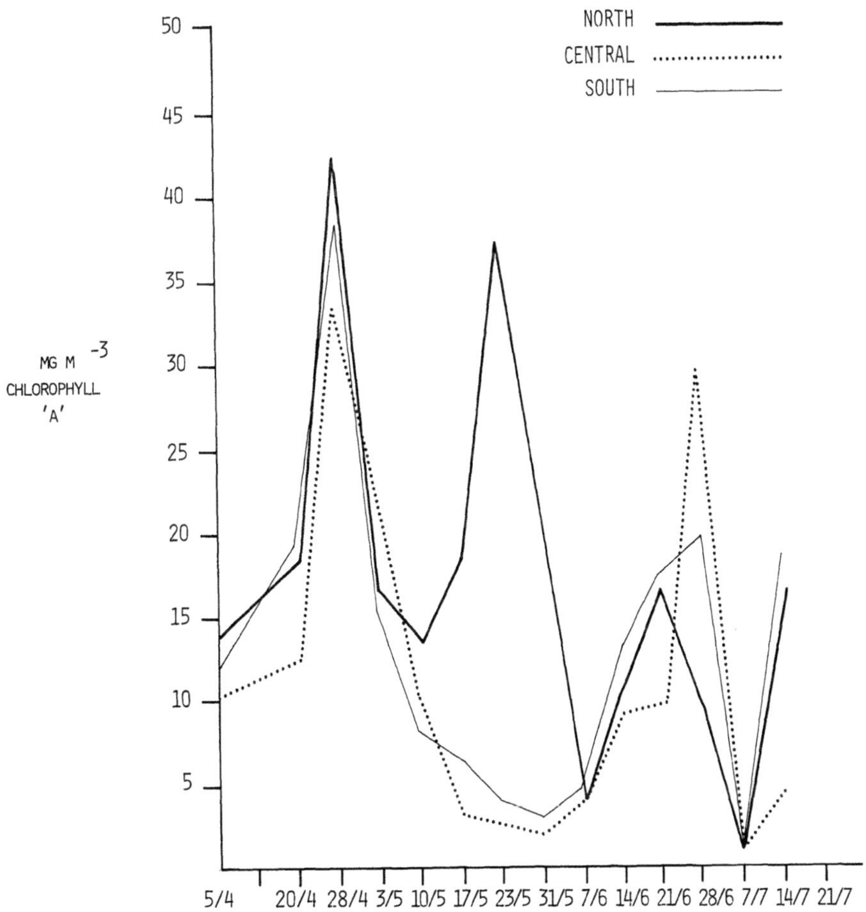

Fig. 5. Changes in the mean density of chlorophyll *a* in the three sections of Rutland Water 1977 (mean of 3–5 samples).

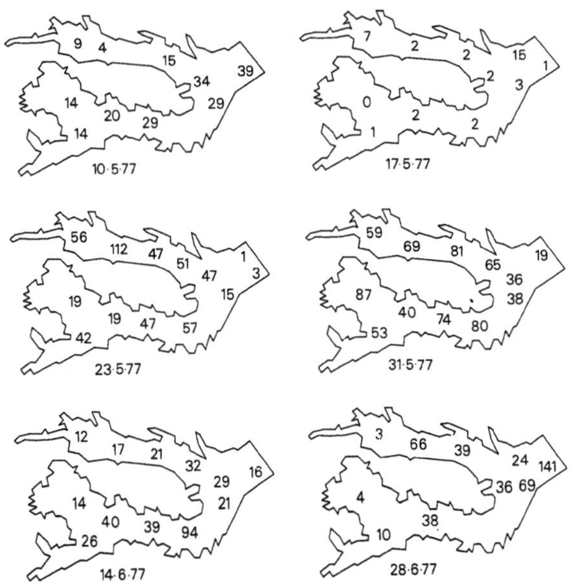

Fig. 6. Changes in the density of *Daphnia hyalina* in Rutland Water 1977 (density expressed as numbers l^{-1} from single 2-l samples).

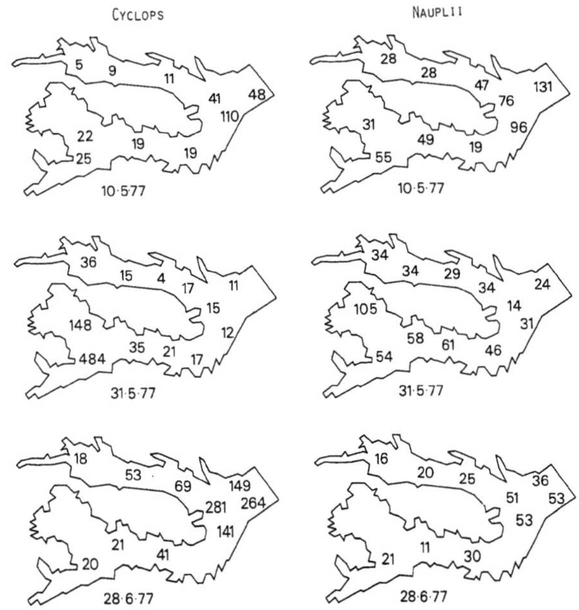

Fig. 7. Changes in the density of *Cyclops* spp. and total nauplii in Rutland Water 1977 (density expressed as numbers l^{-1} from single 2 l samples).

reservoir that considerable horizontal differences might result (W.R.A. 1971), and Harper (1978) showed how these differences were realised in the first year of the reservoir. The horizontal differences in algal populations which occurred in 1977, and which have not been repeated in such a dramatic fashion, were not just due to the short-term effects of wind distribution because there were blooms of completely different species for a time in the south arm (probably as a result of the input of water pumped from the Welland). The horizontal distribution of zooplankton was also examined over this period April–July 1977 to see if there was any evidence of longer-term differences within the reservoir.

The results show that clear differences developed between the three arms, with some relationship apparent between zooplankton and phytoplankton populations.

Figure 5 shows the changes in chlorophyll *a* which illustrate the major difference when populations of *Scenedesmus* and chlorellids developed in the south arm independently in late May, between the April diatom peak and the late June increase of *Aphanizomenon* and cryptomonads which were fairly similar throughout the reservoir.

Populations of *Daphnia* had risen in early April and peaked a fortnight after the spring diatom bloom. They then declined to low levels (mean 3 l^{-1}) on 17 May (Fig. 6).

A week later, chlorophyll reached a maximum mean of 35 mg m^{-3} in the south arm but remained very low elsewhere. The *Daphnia* population began to reproduce again following its crash, although the upsurge which occurred in the north arm was as great as that in the south (both significantly greater than the central basin) despite the large differences in chlorophyll between the two arms. Perhaps surprisingly *Cyclops* and nauplii were the ones which showed a dramatic increase in the south arm only (Fig. 7), particularly at the far end.

Algal density was at its lowest on 7 June throughout the reservoir and then rose steadily to peak towards the end of the month and fall back to almost undetectable levels of chlorophyll on 7 July. Once more, zooplankton numbers followed the algal changes, and this time both *Daphnia* and *Cyclops* numbers were at their highest in the central basin and lowest in the south arm, which was consistent with the differences in peak algal density. The density of *Cyclops* in the central basin was so great, however, that it could not have arisen by reproduction alone.

Throughout this period the densities of *Diap-*

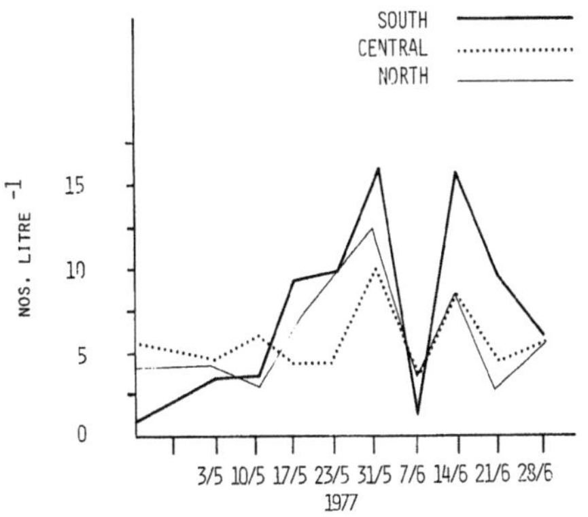

Fig. 8. Changes in the density of *Diaptomus gracilis* in the three sections of Rutland Water 1977 (mean of 3-5 samples).

tomus were not high enough to show any significant differences between different areas of the reservoir (Fig. 8).

Thus in both phytoplankton and major zooplankton groups distinct differences were apparent in horizontal distribution during late spring and early summer 1977. The algal differences probably are due to the reservoir design – location of input jets – but the zooplankton differences cannot at present be fully explained: they appear to be due in part to feeding relationships and in part to redistribution within the reservoir by active or passive means. However, the differences are real, and the practical value is that the very highest densities have not been recorded in the north arm and central basin simultaneously. Initially it was thought that these horizontal differences might have been due to some combination of the stratification of the reservoir in 1977 and discontinuous vertical distribution in the zooplankton. Figure 9 compares vertical distribution of *D. hyalina* during the summer of 1977 with 1980, when there was little stratification, and shows that the patterns were essentially similar. Further work in attempting to explain the horizontal distribution of the zooplankton will involve a more detailed analysis of algal cell density, rotifer populations and wind patterns.

The pattern of zooplankton population change is now similar to that of other lowland reservoirs such as Farmoor (Jones *et al.* 1979) or Eglwys Nynydd (George & Edwards 1974) although with lower standing crops maintained in Rutland during the summer.

The seasonal changes in the populations of the main grazer, *Daphnia hyalina,* are becoming predictable, but the effect of grazing upon the algae and their succession during the summer is not yet clear. Other workers have made similar remarks (Duncan 1975; Reynolds 1979) and trying to understand the processes will be a main feature of our

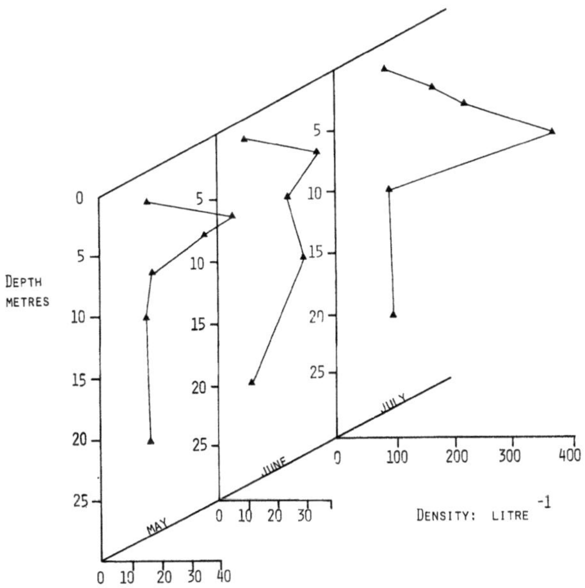

Fig. 9. Changes in the density of *Daphnia hyalina* with depth at the Limnological Tower 1977.

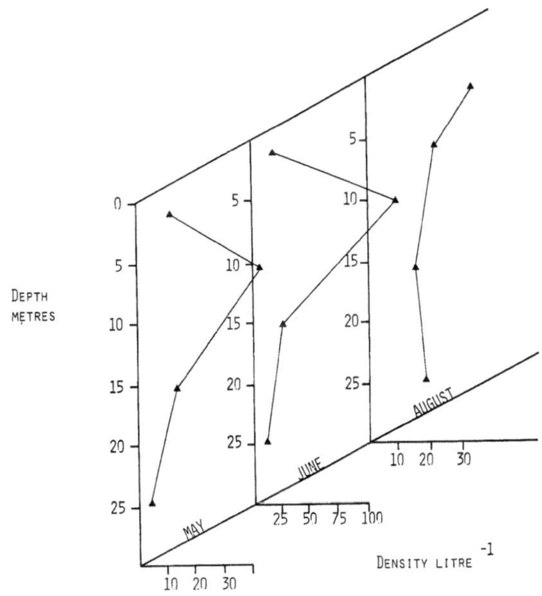

Fig. 10. Changes in the density of *Daphnia hyalina* with depth at the Limnological Tower 1980.

future studies. It is anticipated that a clearer understanding of grazing pressure will be of value to the future management of the reservoir, particularly in relation to the range of draw-off options. The development of our zooplankton studies will involve attempting simple modelling of zooplankton growth, and more detailed analyses of the population structures of different species, including rotifers. As Duncan (1975) has pointed out, such detailed studies of plankton interactions in reservoirs is facilitated by close cooperation between the management and research organisations.

Acknowledgements

The authors wish to thank the Anglian Water Authority for permission to publish this paper, but the opinions expressed are the authors' own. Mrs P. C. Race assisted with the sampling and many colleagues in the Water Authority and University contributed helpful discussion of the results. The senior author is grateful to Professor MacGregor for facilities in the Zoology Department, Leicester University, where part of the work was carried out.

References

Bottrell, H. H., Duncan, A., Gliwicz, Z. M., Grygierek, E., Herzig, A., Hillbricht-Ilkowska, A., Kurasawa, A., Larson, P. & Weglenska, T., 1976. A review of some problems in zooplankton production studies. Norw. J. Zool. 24: 419–456.

Brooks, J. L. & Dodson, S., 1965. Predation, body size and composition of the plankton. Science 150: 28–85.

Duncan, A., 1975. The importance of zooplankton in the ecology of reservoirs. In: The effects of Storage on Water Quality, pp. 247–272; Water Research Centre.

Ferguson, A., Thompson, J. M. & Reynolds, C. S., 1982. Structure and dynamics of zooplankton communities maintained in closed systems, with special reference to the algal food supply. J. Plankton Res. (in press).

Ferguson, A. & Harper, D. M., 1982. Rutland Water phytoplankton – the development of an asset or a nuisance? In: Harper, D. M. & Bullock, J. A. (Eds) Rutland Water – Decade of Change, pp. 117–133. Junk, The Hague.

Fryer, G., 1957. The food of some freshwater cyclopoid copepods and its ecological significance. J. Anim. Ecol. 26: 263–286.

George, D. & Edwards, R., 1974. Population dynamics and production of Daphnia hyalina in a eutrophic reservoir. Freshwat. Biol. 4: 445–465.

Gliwitz, Z. M. & Prejs, A., 1977. Can planktivorous fish keep in check planktonic crustacean populations? A test of the size-efficiency hypothesis in typical Polish lakes. Ekol. Pol. 25: 567–591.

Hall, D. J., Threlkeld, S. T., Burns, C. W. & Crowley, P. H., 1976. The size-efficiency hypothesis and the structure of zooplankton communities. Ann. Rev. Ecol. Syst. 7: 177–208.

Harper, D., 1978. Limnology of Rutland Water. Verh. int. Ver. Limnol. 20: 1604–1611.

Harper, D., 1982. The feeding of trout during the filling phase of Rutland Water. In: Harper, D. M. & Bullock, S. A. (Eds) Rutland Water – Decade of Change, pp. 191–198. Junk, The Hague.

Johnson, D. & Walker, A. F., 1974. The zooplankton of Loch Leven, Kinross. Proc. R. Soc. Edinb. B, 74: 285–294.

Jones, H. R., Lack, T. J. & Jones, C. S., 1979. Population dynamics and production of Daphnia hyalina var. lacustris in Farmoor – I: A shallow eutrophic reservoir. J. Plankton Res. 1: 45–65.

Low, N. F., 1982. Raw water quality – problems and management. In: Harper, D. M. & Bullock, J. A. (Eds) Rutland Water – Decade of Change, pp. 93–101. Junk, The Hague.

Moore, D. E., 1982. Establishing and maintaining the Trout fishery. In: Harper, D. M. & Bullock, J. A. (Eds) Rutland Water – Decade of Change, pp. 179–189. Junk, The Hague.

Munro, I. G. & Bailey, R. G., 1980. Early composition and biomass of the crustacean zooplankton in Bough Beech reservoir, south east England. Freshwat. Biol. 10: 85–96.

Nilsson, J. P., 1978. Eutrophication, minute algae and inefficient grazers. Mem. Ist. Ital. Idrobiol. 36: 121–138.

Reynolds, C. S., 1980. Phytoplankton assemblages and their periodicity in stratifying lake systems. Holarctic Ecol. 3: 141–159.

Rickleffs, R. E., 1973. Ecology. Nelson, London.

Smyly, W. J. P., 1968. Some observations on the effect of sampling technique under different conditions on the numbers of freshwater planktonic Entomostraca and Rotifera caught by a water bottle. J. nat. Hist. 2: 569–572.

Warlow, A. & Oldham, R., 1982. Temporal variations in the diet of brown trout and rainbow trout in Rutland Water. In: Harper, D. M. & Bullock, J. A. (Eds) Rutland Water – Decade of Change, pp. 199–206. Junk, The Hague.

Water Research Association, 1971. Empingham Reservoir – Biological Considerations. W.R.A., Medmenham, U.K.

Welland and Nene River Authority, 1972. Empingham Reservoir – Stratification and Circulation Models. Unpublished report W.N.R.A., Oundle, Peterborough, U.K.

Discussion

Dr. Harper introduced the paper, explaining that the joint authorship stemmed from his having been Biologist, AWA, responsible for Rutland Water and that Dr. Ferguson had succeeded him in this role. He explained that the current paper constituted a descriptive background to more detailed experimental studies.

The salient characteristics of zooplankton development had been a change in species composition and a markedly reduced oscillation in abundances, reflecting less dramatic oscillations in phytoplankton density as the reservoir matured. He expanded further on the detailed studies of the important zooplankton species, *Daphnia hyalina*, indicating the problems in estimating abundance stemming in part from variable horizontal distribution.

Dr. Oldham, opening the discussion, commented on the striking annual modulation of two species of *Daphnia* which conformed to that of the temperature modulation. Superimposed on this annual pattern there were a series of 'troughs' in the graph – or alternatively of minor 'peaks'. He speculated that these anomalies could stem from variation in the sampling procedure. Noting that the authors had commented on the uneven horizontal distribution, he suggested that even at much smaller intervals this would continue, and enquired whether the authors had any evidence of the distribution over a smaller scale.

He noted that, in Fig. 2, there appear to be 11 major peaks of production in *Daphnia hyalina*. These should coincide with the peaks in Fig. 1, but only four peaks did so coincide, again suggesting uneven sampling.

Dr. Oldham commented that there was good evidence for motile and less motile plankton being segregated in response to wind activity. Motile forms which maintain position near the surface accumulate on the down-wind shore since they resist the downward circulatory movement. Forms that maintain themselves at greater depths remain on the up-wind shore. Work on *D. hyalina* further suggested that in Langmuir cells, motility responses result in markedly discontinuous distribution. The fact that high values in the June and July counts occurred at opposite ends of the reservoir for zooplankton and phytoplankton appeared to support this general pattern.

Dr. Oldham commented that the pattern of colonisation seemed to follow the same trends as noted in other groups. There was an initial phase dominated by riverine species, a second phase with opportunistic species dominating, a third period of increasing diversity and a fourth signalling stabilisation.

He then commented on the contrast between *D. hyalina* in trout stomachs with those in the reservoir, suggesting that the difference arose from inadequate sampling and that the trout were giving a better sample than the bottle. In closing, he questioned the value of attempting to make a predictive model of zooplankton and asked what value such studies could have in reservoir management.

Dr. Harper agreed that more frequent samples were desirable. He pointed out that the values in Fig. 2 were expressed as percentages to show the underlying pattern of reproduction which would only appear in Fig. 1 if reproduction were successful.

He pointed out that there was a general problem of unevenness in zooplankton sampling. Nonetheless, if one examined the 1977 figures, the violent oscillations were due to high reproduction followed by high juvenile mortality. This did not occur in subsequent years, and consequently more stable figures were obtained. He considered that, although there might well be problems with uneven sampling, this would only affect absolute values but not overall changes.

Turning to the contrasting mean size of *D. hyalina* in trout stomachs and in samples, Dr. Harper explained that samples were taken with horizontal net hauls and several hundred individuals were measured; he believed this provided an adequate sample. Also, the number of *D. hyalina* bearing ephippia was quite high in the trout but was close to zero in the plankton samples suggesting, as others have shown, that trout are taking large and obvious individuals as prey.

Dr. Ferguson said that he did not believe that sufficient was known of zooplankton to dismiss further studies upon it and that, when there was a more complete understanding of the interaction between phyto- and zooplankton, the true value of the work would emerge. He added that there had already been trouble with zooplankton at the treatment plant. There was a strong likelihood that action and advice would be needed in the future. Finally, he considered that it was better to have some knowledge of zooplanktonic population fluctuations and distribution than none at all.

Benthic invertebrates of Rutland Water

J. A. Bullock, F. Clark & S. Ison
Zoology Department, University of Leicester, Leicester LE1 7RH, England

Keywords: reservoir, benthos, invertebrates, colonisation, abundance

Abstract

The colonisation of Rutland Water by benthic invertebrates has been monitored from 1976 when the reservoir was first filled. A modified Ekman grab was used to sample populations quantitatively while false substrates and dredge samples were used to obtain supplementary data. Hand net and hand-sorting were used to study the littoral zone.

The dominant taxa in the reservoir are *Asellus aquaticus, A. meridianus, Gammarus pulex, Crangonyx pseudogracilis, Helobdella stagnalis, Erpobdella octoculata, Polycelis tenuis,* larval Chironomidae and Oligochaeta. Other species which appeared abundantly, such as *Limnaea pereger* and *Planorbis albus,* have now disappeared and it is suggested that the generally low levels of gastropods might account for the poor performance of species such as *Dugesia lugubris* and *Glossiphonia complanata.*

Other aspects of abundance are discussed, as are the effects of maturation of the system.

Introduction

The study of the benthic invertebrates of Rutland Water, Leicestershire, was undertaken initially to discover which species would transfer from the feeder streams to the reservoir and to monitor the appearance of new species. Thereafter, we set out to study the changes in abundance of the benthos and to build up a picture of the energetics of selected major species and to investigate interspecific interactions. The objective in this account is to give an overview of the study to date while more detailed studies will be reported later by individual investigators.

Studies on the feeder streams

Sampling began on the main stream of the Gwash and on the northern stream, which runs through Oakham, in late 1974. Samples were taken from two major sites and three minor sites, with several sampling stations at each site. The three minor sites were abandoned in 1975 because of either flooding or major disturbance, but the two major sites were maintained until October 1976. At this time, the last sampling stations on the floor of the reservoir were submerged. Sampling upstream of the reservoir on the main stream at Brooke Priory and, less satisfactorily, just above the reservoir on the northern stream began in January 1976 and continued until 1980. Samples were taken with a hand net, either by kick sampling in riffles or by sweeping the net through the water, amongst vegetation and along the bank in deeper areas. The species list comprises the following taxa:

Tricladida		3 species
Mollusca:	Gastropoda	9 species
	Bivalvia	2 genera
Annelida:	Hirudinea	6 species
	Oligochaeta	unstudied

Hydrobiologia 88, 147-156 (1982). 0018-8158/82/0882-0147/$02.00.

Crustacea:	Isopoda	2 species
	Amphipoda	1 specie
Insecta:	Plecoptera	2 species
	Odonata	2 species
	Ephemeroptera	8 species
	Hemiptera	13 species
	Trichoptera	12 species
	Diptera	5 families
	Coleoptera	31 species (at least)

This list is provisional, as some specimens still require identification, and others require checking, but provides some idea of the diversity of forms present. A full account of this study is being prepared, but it should be noted that the fauna of the streams changed during the course of this study as a result of the extensive drought in 1976. At this time the flow in the main stream was negligible and that in the northern stream, which received the Oakham treated sewage effluent, was markedly reduced. As a consequence, many lotic species disappeared including both Plecoptera species, several Ephemeroptera, many Trichoptera, such as the four species of Hydropsychidae, and many Coleoptera (e.g. Elmidae), while many other species dwindled greatly. In so far as colonists for the reservoir were concerned, the decrease in numbers was probably more important than were the outright losses, since the latter are clearly intolerant of lentic conditions.

Studies on the reservoir

Sampling of the reservoir began in 1976 using an Ekman grab, but results were unsatisfactory, mainly because the bottom was too hard to give a satisfactory sample but also because the inflatable dinghy which was used provided an unstable platform. In August 1977, an 11 ft fibre glass dory was obtained and a serious programme of grab sampling began in September 1977. A subsequent step has been to modify the grab to close on impact on reaching the bottom (Rawson 1947). This has meant that grabs can be taken successfully in heavy waves and with the boat drifting.

Three other methods of collecting have been used. A simple bottom dredge was used to collect material for laboratory experiment. This gives some idea of relative abundance but is unsatisfactory for quantitative estimates.

Eight sets of false substrates, each set comprising a basket of rocks, a roll of plastic netting and a series of hardboard plates, were used for a limited period. Although these proved very attractive to some species, e.g. *Dendrocoelum lacteum* (Muller), and a species of Hydridae, this method could not be considered as adding much to our knowledge of the reservoir's fauna. (Also the marker buoys disappeared one by one and, although some substrates and buoys were replaced, it was eventually decided that the returns were not commensurate to the effort.)

Finally, samples were collected from the littoral zone, principally by hand net but also, in rocky areas, by hand searching over a specific area. These did increase our knowledge of the fauna, and complemented the grab samples, although this has become less valuable as the water level has fluctuated in response to draw-down and pumping.

Results from grab samples

The major taxa which have been encountered regularly throughout the study are:

Platyhelminthes	*Dugesia lugubris* (Schmidt)
Hirudinea	*Helobdella stagnalis* (L.)
	Erpobdella octoculata (L.)
	Glossiphonia complanata (L.)
Gastropoda	*Valvata piscinalis* (Muller)
Crustacea	*Gammarus pulex* (L.)
	Asellus aquaticus (L.)
	A. meridianus (Racovitza)

Two other major taxa have been present in large numbers. Larvae of the Chironomidae are abundant and have been studied by Dr. A. Brown (Brown & Oldham 1982). Aquatic Oligochaeta are also abundant, although we do not consider our methods particularly suitable for this taxon. For this reason, and because of taxonomic problems, we prefer to regard them as a single taxon.

Several further species have been recorded intermittently throughout the study, but generally at low densities. These are:

Platyhelminthes	*Dendrocoelum lacteum* (Muller)
Hirudinea	*Piscicola geometra* (L.)
Gastropoda	*Bithynia tentaculata* (L.)
Arachnida	Hydracarina

Several species were abundant early in the study but have since declined. The two most prominent species were the gastropods *Limnaea pereger* (Muller) and *Planorbis albus* (Müller), both of which had effectively disappeared before the end of 1978.

Recruits to the reservoir have been *Polycelis tenuis* (Ijima) and *Crangonyx pseudogracilis* Bousfield, both of which attained high densities in 1980 and are maintaining their numbers in early 1981. In addition, the caddis, *Oecetis ochracea* (Curtis), appeared in fair numbers in the summer of 1980.

Hand-net results

The results from hand-net samples reflect in general the species taken in the grabs but there have been a number of additional species which occurred in large numbers, albeit for a relatively short period of time. *Limnaea pereger* persists in the littoral although at low densities while the caddis *Agraylea multipunctata* (Curtis) was abundant from August to October 1979. The small corixids *Sigara dorsalis* (Leach), *S. falleni* (Fieber), and *Callicorixa praeusta* (Fieber) also appeared intermittently as did the beetle, *Deronectes depressus* (F.) and the may-flies, *Cloeon dipterum (L.) and Caenis horaria* (L.). The leech, *Theromyzon tessulatum* (Muller) was frequent in 1979 but less frequent in 1980 despite the increase in wild-fowl on the reservoir.

Dredge and false substrate results

As indicated earlier, these two methods give little information on abundance because of the attractiveness of the substrates to some species, e.g. *Dendrocoelum lacteum*, and of the uncertainty of area and efficiency of the dredge. Nonetheless they have a function in showing what species are present, and thus serve as a check on the grab programme. They also have a value in contributing to the knowledge of population structure.

In the main, it is true to say that no species appeared in either substrates or dredges which did not also appear in the grabs, except for the hydroid which appeared abundantly on rocky substrates but has not been recorded elsewhere.

The grab programme

From September 1977 to October 1979, a regular grabbing programme was maintained at monthly intervals (conditions permitting) along four transects (Fig. 1). These were:

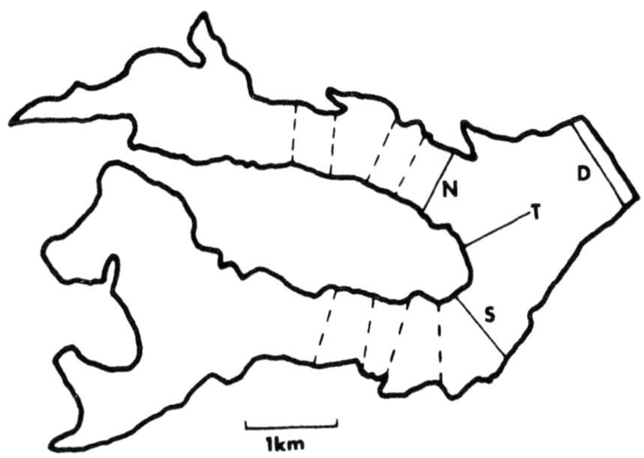

Fig. 1. Outline map of the reservoir showing the original four transects (D = Dam, T = Tower, N = North arm and S = South arm) and the new transects (broken lines).

A. Dam transect which ran NW - SE more or less parallel to the dam.
B. South arm transect running N - S across the south arm a short distance up the arm.
C. North arm transect which was similar to the south arm transect, and roughly opposite to it in the north arm.

All three ran from shallow water to deep and thence to shallow, with usually 12 grabs, equally spaced, on each run.

D. Tower transect which ran from the shallows at the tip of the peninsula out to deep water near the limnological tower, and comprised ten samples.

This sampling pattern has now changed. Monthly samples are now taken on four transects (each of six grab samples) on both the north and south arms with the most easterly of each following the line of the original 'Arm' transect, while the 'Dam' and 'Tower' transects are run three times a year, in March, July and November.

On collection, each sample is placed in a separate small container, labelled and sealed for transport to the laboratory. The samples are sorted by hand within two days of collection and all material is stored in 70% ethanol with the exception of the triclads which are kept in water. Specimens of *Polycelis* are dissected to confirm the identification (cf. Reynoldson 1978); other triclads are recorded immediately because of difficulties with preservation. The remaining material is subsequently identified to species with the exception of Diptera larvae, Oligochaeta and Hydracarina.

Distribution of species

Crustacea

Asellus aquaticus and *Asellus meridianus* (Fig. 2). The study of the feeder streams showed *A. aquaticus* present in the northern stream and *A. meridianus* dominant in the southern one. As the reservoir filled, *A. aquaticus* spread to occupy both the entire north arm of the reservoir and the main pool above the dam. *A. meridianus* remained within the south arm, appearing in grabs from the south arm transect only.

The population density of *A. aquaticus* partly reflects this fact in that, estimated as a mean density over all samples, the density had reached a high

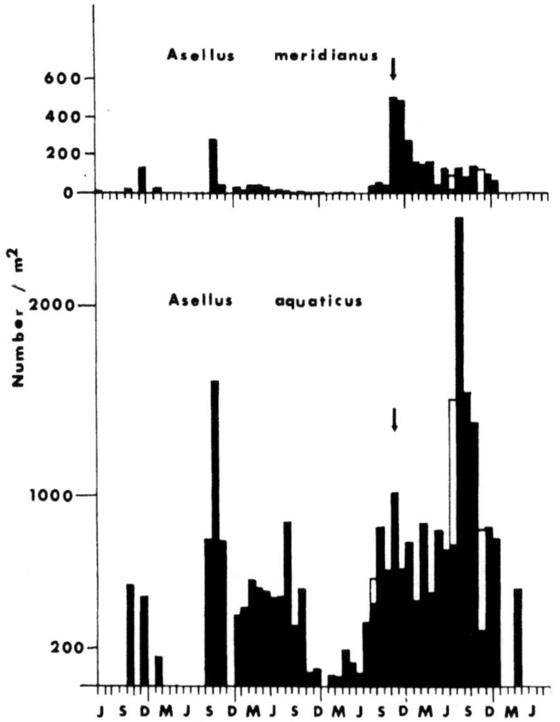

Fig. 2. Population fluctuations in *Asellus aquaticus* and *A. meridianus.* Note that the arrow above November 1979 indicates when the change in sampling pattern occurred. In months when samples were taken from the Dam and Tower transects as well as on the new transects, the mean densities in the former are indicated by the black column, and the latter by the white. (N.B. The columns are nested, cf. Fig. 3.) Also, the same convention is used to show the values obtained on consecutive days of sampling in August 1979.

level in the autumn of 1977 and then regressed with slight peaks in autumn 1978 before crashing in that winter only to recover in the autumn of 1979 and maintain a high level through the winter of 1979–80. Despite the change in sampling patterns, the figures indicate a similar response in 'Dam' and 'Tower' samples. The pattern of winter decline in density has been reported by several authors (e.g. Adcock 1979) and the puzzling feature is why this should not occur in every year to so great an extent. Possibly this is an effect of water temperature directly or indirectly affecting survival. This would certainly be buffered in a large water body so that only severe weather, as occurred in 1978–79, would produce such an effect. (By contrast, the water mass studied by Adcock (*op. cit.*) is very small.)

In *A. meridianus,* the population stayed low, but note that the mean density in October 1977 of $>200\ m^{-2}$ is, in actual terms of the south arm whence nearly all the specimens came, an actual density of $>800\ m^{-2}$ in that region.

Up to June 1979, there appeared to be little or no overlap in the distribution of the two species; although the numbers of *A. meridianus* dwindled in 1978 and remained low in 1979, there was no incursion of *A. aquaticus* until July 1979, when numbers in the south arm transect changed dramatically from $<10\ m^{-2}$ to $>400\ m^{-2}$ in one month and exceeded 800 m^{-2} by late August. The initial assumption was that *A. meridianus* had been eliminated, but subsequent investigations revealed high densities further up the south arm. Thus in October 1979 a density of about 100 m^{-2} was recorded in the south arm transect, while the November series of transects in the south arm gave a mean density $>900\ m^{-2}$. Meanwhile, *A. aquaticus* increased in the south arm and currently maintains densities four to five times greater than *A. meridianus.* Interestingly, the latter species has been recorded regularly, albeit at low densities (about 10 m^{-2}) in the north arm since December 1979.

The position is puzzling, and only further monitoring will show what is the ultimate fate of the two species in the reservoir. *A. aquaticus* is generally recognised as an invasive species which it is claimed is ousting *A. meridianus* from British waters (Moon 1957; Williams 1962) and this may well be the eventual outcome in this case. Strangely, no convincing evidence of the actual mechanism of competition has yet been described but neither is there a satisfactory indication of niche segregation. Certainly both species occur in the same sample very frequently, and there is no indication of a negative correlation between the two species. We speculate that in the early stages, food supply may have limited the rate of spread and the population density. Further, there may be some form of niche segregation resulting from differing optima of physical conditions, or that some selective predator may maintain a dynamic balance between the two species. Certainly, there is evidence of long-term coexistence in some water-bodies (Moon, pers. comm.)

Gammarus pulex (Fig. 3). Although *G. pulex* was abundant in both north and south streams prior to flooding, colonisation of the reservoir floor was slow and not until November 1977 were specimens encountered regularly along all four transects; and it took until July 1978 for the population estimates on the 'Dam' and 'Tower' transects to approach the estimates for the north and south arms. Thereafter, populations have varied markedly, with a major peak in late summer 1978, followed by a decline in the following winter and a generally low population, oscillating around 100 m^{-2} thereafter.

Crangonyx pseudogracilis (Fig. 3). This is an introduced species, which we first recorded in Rutland Water in March 1978 and the population has increased markedly in both north and south arms in 1980, although the main pool, west of the dam, has yet to be successfully colonised. This species was not recorded in the feeder streams, although Moon (pers. comm.) recorded the species in the northern stream a number of years ago. In the areas where it is present the density is close to that of *G. pulex* and one may speculate whether these two species may interact. There are, however, behavioural differences between the two species and evidence of differences in their foraging behaviour (Hardwick, unpublished) which may effectively segregate the two species.

Gastropoda

Limnaea pereger (Fig. 4), *Valvata piscinalis* and *Bithynia tentaculata*. These are the only three species to appear consistently in the grab samples. *B. tentaculata* appears only at very low density and

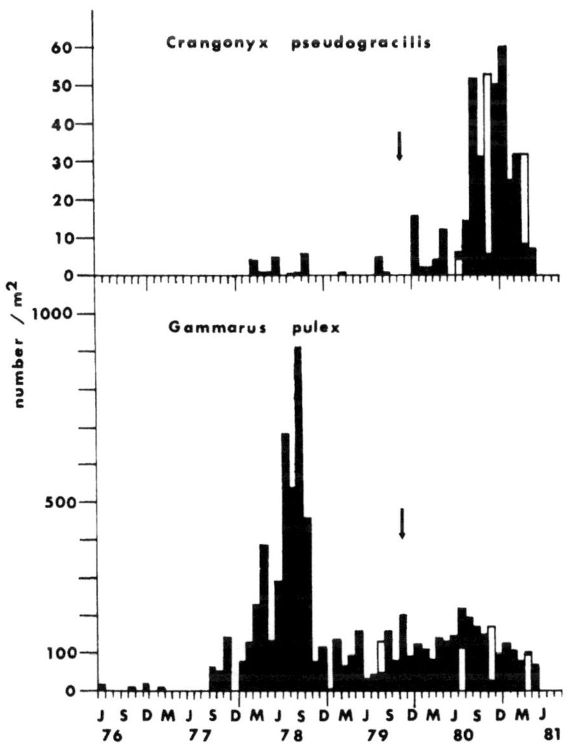

Fig. 3. Population fluctuations in *Gammarus pulex* and *Crangonyx pseudogracilis*. Further explanation in Fig. 2.

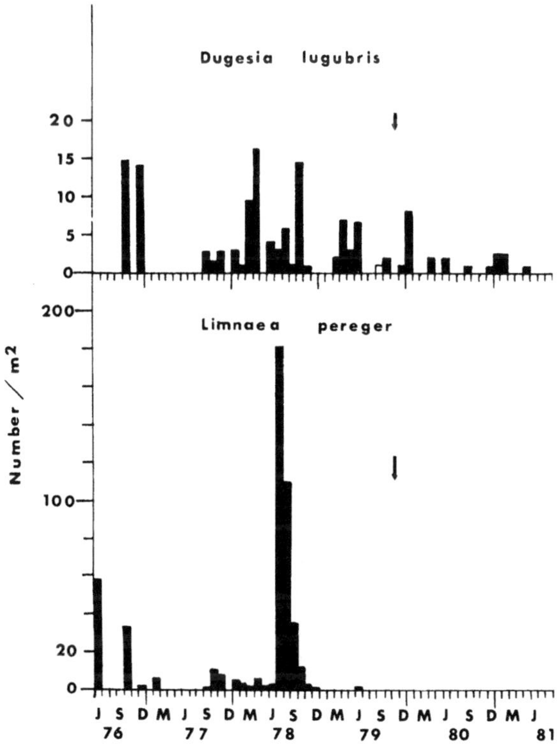

Fig. 4. Population fluctuations in *Limnaea pereger* and *Dugesia lugubris*. Further explanation in Fig. 2.

intermittently, and must be assumed to be incurring from more suitable habitats. *L. pereger* having initially been widely dispersed on all four transects declined through 1977 and early 1978, suddenly increased in July 1978, although this was exclusively along the 'Dam' transect, only to decline again and disappear by November 1978. This species has been recorded in littoral samples since then, but only at low densities. *Valvata piscinalis* has been more consistent than either of the preceding species, showing some preference for the southern arm.

In all cases there is an impression that the gastropods are predominantly strays from breeding populations elsewhere, and even on the littoral fringe, gastropod populations have declined greatly.

Other detritivores

As mentioned, the Chironomidae (Fig. 5), predatory as well as detritivorous, have been analysed by Dr. A. Brown. The density of larvae has continued to rise in our samples, with regular declines in spring to early summer when adults emerge. The larvae are very aggregated in their distribution, but there is a suggestion that the group prefers the arms to the main pool, possibly because the shallower water offers better availability of food, etc. It is also perhaps significant that peak numbers are recorded in March each year. We suggest that this is no true peak but is a reflection of either the availability of the larvae or of greater ease in sorting large larvae.

Oligochaetes (Fig. 6) remain unstudied. Initially numbers were low but with peaks in 1978 of 33 m^{-2}, in 1979 of 95 m^{-2} and of 230 m^{-2} in 1980, the density of this taxon has increased making it an important detritivore and also a food source for predators.

Hirudinea (Fig. 7).

Helobdella stagnalis was initially present at low densities, but the population increased in the first year of the study, and appears to be continuing its increase with successive peaks of 217 m^{-2} (1978), 227 m^{-2} (1979) and 327 m^{-2} (1980) in August or September of the year. *Erpobdella octoculata* has, by contrast, shown only a modest increase with a

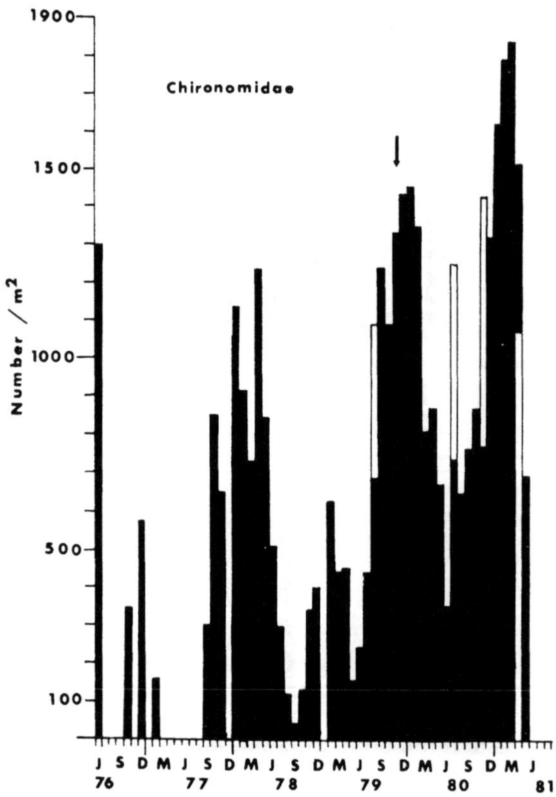

Fig. 5. Fluctuations in the density of chironomid larvae. Further explanation in Fig. 2.

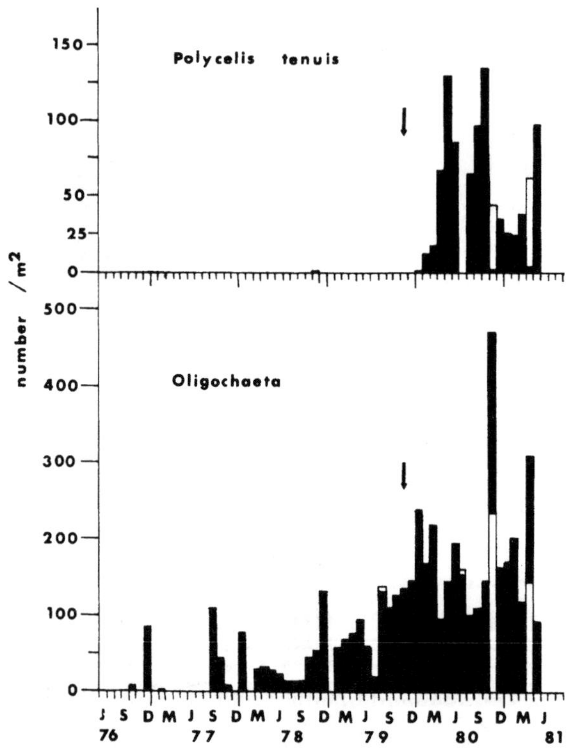

Fig. 6. Population fluctuations in Oligochaeta and *Polycelis tenuis*. Further explanation in Fig. 2.

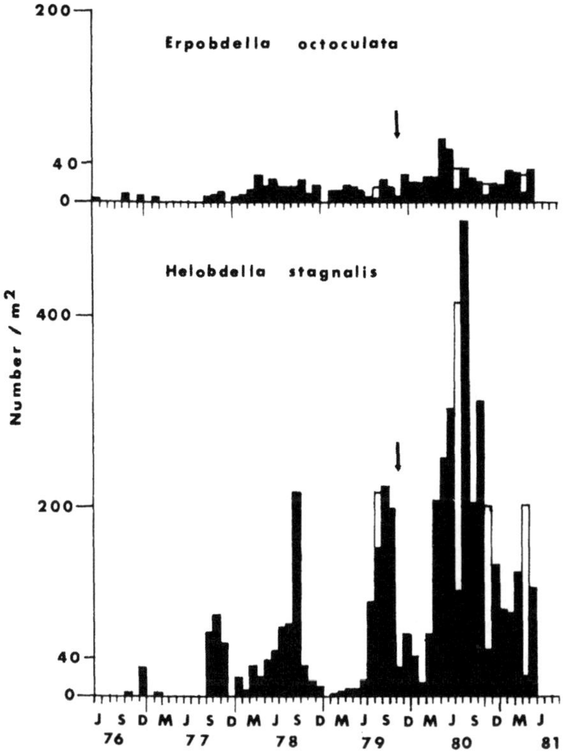

Fig. 7. Population changes in *Helobdella stagnalis* and *Erpobdella octoculata*. Further explanation in Fig. 2.

maximum density of only 69 m^{-2} in 1980 compared with peaks of 26 m^{-2} in 1978 and 22 m^{-2} in 1979.

Both species are possibly responding to the increased density of prey species, and both are very aggregated in their distribution with far greater numbers occurring in the two arms than in the main pool. *H. stagnalis* is sanguivorous, feeding on a wide range of hosts including chironomids, oligochaetes and crustaceans; *E. octoculata* is macrophagous and, while showing a wider range of prey, is likely to be unable to cope with large specimens of *Asellus* sp. This may be important in the relation between *A. aquaticus* and *A. meridianus* because the former attains a greater size than the latter and consequently must be free of predation for a longer period of time. *Glossiphonia complanata* is a sanguivore, primarily feeding on molluscs (Elliott & Mann 1979). Although the species has maintained a presence, it has remained at low densities, and never exceeding 10 m^{-2}. It seems likely that the poor supply of gastropods is a major factor in limiting density, although other taxa, primarily chironomids, can be taken as prey.

Tricladida

Dugesia lugubris (Fig. 4) has been present throughout the study although numbers have never been very high (maximum 16 m^{-2} in 1978) and there has been a marked decline since late 1978. *D. lugubris* is reported by Reynoldson & Young (1963) to prey on oligochaetes and molluscs, and it may be that the molluscan element, particularly *Limnaea pereger,* represents an essential component of the diet.

Polycelis tenuis (Fig. 6) was recorded in the feeder streams throughout 1975 and 1976, but took a long time to become established in the reservoir, the species being only recorded regularly in grab samples from January 1980, although two specimens were taken in November 1978. This species preys on oligochaetes (Reynoldson *et al., op. cit.*) and there is no apparent reason for its slow rate of colonisation. This may be purely an artifact, since it began appearing soon after our change in sampling pattern, but samples taken by Dr. A. Brown in 1978, and kindly made available to us, from further up the arms, reveal only one or two specimens.

Dendrocoelum lacteum presents a puzzle. It has been recorded rarely in grab or dredge samples, but has appeared in large numbers on artificial substrata, and also on stony stretches of the littoral zone. This species is primarily a predator of crustaceans. Macan & de Silva (1979) report it as a significant predator of *Asellus aquaticus* and we have shown that it is prepared to take *A. meridianus* in the laboratory. *D. lacteum* is unlikely to be restricted by the availability of prey and a more likely limiting factor is the lack of a stony substratum. This is not to say that there are no stony areas on Rutland Water, but they are relatively few and are the least suitable for sampling.

Detailed distribution

One of the problems encountered in the sampling programme is a lack of consistency in the samples. By this we mean that although our grab samples are taken at approximately the same spot each time, the densities recorded at a particular sample point are inconsistent between months. Detailed comparison of the individual samples over the course of two years show very little consistency in the apparent distribution of the major species. This is perhaps not so surprising when one considers that each grab collects 227 cm^2 of the substratum (in a reservoir of about 1 300 ha^{-1}) and there are certain to be inconsistencies due to chance predation and similar factors. Despite this, over the total number of samples, a fair level of consistency in overall estimates of density is obtained. Thus in August 1979, two sets of samples were taken on consecutive days, and the geometric means for the abundance of each species lie within 20% of each other. Since the animals are clearly aggregated, we consider this a satisfactory outcome. Nonetheless, we are puzzled by the microdistribution of species, particularly the detritivores, which appear to have a relatively small scale level of spatial aggregation or to have very mobile patches of high density. The latter seems the less likely character but no ready explanation for the pattern is apparent. Analysis of oxygen concentrations and of the substrates reveal no readily correlated factors. The occurrence of predators is one obvious factor, while microtopography of the bottom may also have some significance.

Discussion

The progressive development of the benthos appears to represent at least three phenomena.

First there was colonisation by a range of species, many of which have since been eliminated as conditions became unsuitable. Amongst these may be numbered *Limnaea pereger* and *Planorbis albus*, which, being pulmonates, are likely to be susceptible to the effects of oxygen stress. Macan (1975) cites evidence that above 12 °C *L. pereger* must rise to the surface to satisfy its oxygen requirements. The result is that the species is likely to become confined to the littoral zone in water bodies where such temperatures occur regularly in the summer months. Depletion of oxygen in the hypolimnion by aerobic bacteria will exacerbate the situation.

It is noteworthy that two species which are known to prey to a major extent on molluscs, *Dugesia lugubris* and *Glossiphonia complanata*, have failed to increase, although maintaining an occasional presence. Whether their failure is a response to low prey density or to some more direct factors must remain a puzzle.

A second phenomenon has been the maturation of the reservoir, with a change in available food from dead terrestrial vegetation, which supplied rotting debris, to the planktonic debris. The plentiful supply of plant debris in the early years possibly accounts for the peaks in population density in *Asellus aquaticus* and later in *Gammarus pulex*. Such oscillations in density are characteristic, however, of an immature system, and it is clear that a further influence has been the increase in density of the invertebrate predators. These latter have been slower to increase and there is clear evidence that *Helobdella stagnalis* is still increasing, and *Polycelis tenuis* has only recently appeared in large numbers. A further aspect of this process of maturation, is the marked softening of the substratum and deposition of organic debris. This factor may account for the upsurge in *Crangonyx pseudogracilis* which tends to burrow more than does *G. pulex*. This same factor could provide a possible segregation between the two species of *Asellus*, of which *A. meridianus* has been observed to be more cryptic in behaviour.

The third phenomenon affecting the benthos is the effect of reservoir management; the major factors are draw-down and fish stocking. The first will mainly affect the littoral zone providing an unstable environment only suitable to the more opportunistic and mobile species. Predation by stocked salmonid fish will maintain a continuous pressure on the benthos, as is evidenced by the studies of trout diets (Harper 1982; Warlow & Oldham 1982).

Finally, we should reiterate that our account is an overview of the benthic situation. As such we have ignored the differences caused by sampling at different depths, and in the main the differences between transects.

Acknowledgements

We are grateful to the Anglian Water Authority for permission to work on the reservoir and provision of facilities and assistance. We acknowledge the assistance of our colleagues, Messrs. M. Majeed and Y. T. Daoud in the collection and analysis of samples. All funding for this study has been provided by the University of Leicester.

Author's note: The data presented in the figures are for the period June 1976 to May 1981 except Fig. 2, which shows data from June 1976 to January 1981, together with the Dam and Tower values for April 1981.

References

Adcock, J., 1979. Energetics of a population of the isopod Asellus aquaticus: life history and production. Freshwat. Biol. 9: 343–355.
Brown, A. & Oldham, R., 1982. Bathymetric distribution of some Chironomidae (Diptera) in Rutland Water. In: Harper D. M. & Bullock, J. A. (Eds) Rutland Water – Decade of Change, pp. 159–169. Junk, The Hague.
Elliott, J. M. & Mann, K. H., 1979. A Key to the British Freshwater Leeches (2nd edn). Freshwater Biological Association Scientific Publication No. 40.
Harper, D. M., 1982. The feeding of trout during the filling phase of Rutland Water. In: Harper, D. M. & Bullock, J. A. (Eds) Rutland Water – Decade of Change, pp. 191–198. Junk, The Hague.
Macan, T. T., 1975. Freshwater Ecology (2nd edn). Longman, London.
Macan, T. T. & de Silva, P. K., 1979. On the occurrence of Dendrocoelum lacteum (Müller) and Asellus aquaticus (L.) as predator and prey in the stony substratum of Windermere. Arch. Hydrobiol. 86(1): 95–111.
Moon, H. P., 1957. The distribution of Asellus in Windermere. J. Anim. Ecol. 26: 113–123.
Rawson, D. S., 1947. An automatic-closing Ekman dredge and other equipment for use in extremely deep water. Limnological Society of American Special Publication No. 18.
Reynoldson, T. B., 1978. A Key to the British Species of Freshwater Triclads (2nd edn). Freshwater Biological Association Scientific Publication No. 23.
Reynoldson, T. B. & Young, J. O., 1963. The food for four species of lake-dwelling triclads. J. Anim. Ecol. 32: 175–191.
Warlow, A. & Oldham, R., 1982. Temporal variations in the diet of brown trout (Salmo trutta L.) and rainbow trout (Salmo gairdneri L.) in Rutland Water. In: Harper, D. M. & Bullock, J. A. (Eds) Rutland Water – Decade of Change, pp. 199–206. Junk, The Hague.
Williams, W. D., 1962. The geographical distribution of the isopods Asellus aquaticus (L.) and A. meridianus Rac. Proc. Zool. Soc. Lond. 139: 75–96.

Discussion

In introducing this joint paper, Dr. Bullock emphasised that he and his colleagues wished to stress that the results presented were intended to provide an over-view of the fluctuations in the abundance of the benthic invertebrates. This work provided the basic background to more detailed studies of the factors causing fluctuations in populations. Laboratory investigations were currently concentrated on the interrelation of leech species by Mr. M. Majeed, studies on the dynamics of the two species of *Asellus* and serological studies of predation intensity by Mr. Y. T. Daoud and, most recently, detailed studies of the amphipods.

The ultimate aims of the study were to explain the interactions and to investigate the total production of the benthos. Although such studies might be regarded as academic, he believed that they would be of considerable value in the continued management of the reservoir as a fishery. In closing, he drew attention to the co-operation which continued between Dr. Oldham and his associates at Leicester Polytechnic, who were primarily interested in chironomid larvae, and the authors' group at Leicester University.

Dr. Oldham, in opening the discussion, pointed out that he and his associates had used a similar sampling technique, albeit with a van Veen grab, which gave population values that were not significantly different from those obtained by Dr. Bullock. They too had used a transect technique, but he felt that stratified random sampling might well be more satisfactory. Detailed analysis of the chironomid data showed marked contagion with a significant correlation between the degree of contagion and the size of population. This was possibly a result of behavioural differences. He demonstrated that a total of 30 samples in one small area of 8 m^2 at a depth of 5 m gave values approaching randomness. He suggested that the dispersion pattern was determined by substrate structure, although Dr. Bullock reported no difference between substrates in relation to distribution.

Dr. Oldham demonstrated that with about 10 samples in a small area the number of individuals per sample levelled out, but that 20 samples were necessary before the number of species approached an asymptote. This strongly indicated the need for large samples.

He drew attention to the similarity between peaks in *Asellus aquaticus, Gammarus pulex* and *Limnaea pereger* recorded in his work to that reported by the authors, and to the importance of this in relation to trout diet, demonstrated by Mr. Warlow and himself in a poster paper. Dr. Oldham closed by drawing attention to the changes in the Chironomidae abundances which were again in agreement with the authors' figures, and indicated that this was due to species succession and the decline of opportunistic *Chironomus* spp. etc.

In reply, Dr. Bullock said that he agreed that other sampling patterns might well have been tried, but that the use of transects had been decided upon because (i) transect lines could be reasonably accurately discerned on succeeding occasions, (ii) they covered the range of depths in an orderly manner, and (iii) they appeared to give reasonably reproducible results overall. The problem with stratified sampling would be the location of suitable strata and deciding the essential criteria to be used.

He added that while it was true that no certain difference between substrata could be shown, this was partly due to the heterogeneity of the substrate and he had no doubt that many species were indeed affected by the nature of the substrate. For example, the transect near the dam was predominantly over clay and generally gave markedly lower values than elsewhere. The substrate was more patchy and results less predictable over the rest of the reservoir, and there was the additional complication that although the substrate might be suitable, few organisms were found.

Miss S. Eldin asked about the present situation in the littoral region. Dr. Bullock replied that he had stopped sampling the littoral zone because the zone was likely to be greatly disturbed by draw-down and he speculated that only species which showed high mobility could utilise the area.

Miss S. Eldin also enquired about the changes in the species list (numbering over 70 species) to which Dr. Bullock explained that the list applied to the streams running into the reservoir, and not to Rutland Water itself. He regretted that a full species list was not available.

The leeches of Rutland Water

M. Majeed
Department of Zoology, University of Leicester, Leicester LE1 7RH, England

Keywords: Hirudinea, production, population

Abstract

Seven species of leeches have been recorded in Rutland Water between September 1977 and October 1979 by monthly samples taken by Ekman grab.

The preliminary analysis of leeches from the four study sites of the reservoir (46 grab samples) showed that *Helobdella stagnalis* was the most abundant species with *Erpobdella octoculata* and *Glossiphonia complanata* (cf. Bullock *et al.* 1982). The life cycle of *H. stagnalis* was elucidated from these samples by dividing the animals into weight classes (2 mg live weight) and revealed that this species has two breeding seasons. The overwintering population reproduces in June. The juveniles grow rapidly and produce the next generation in August and September. This life cycle is similar to that reported in Europe and Canada.

The population density ranged from 32 m^{-2} to 90 m^{-2}, being generally higher during Summer and Autumn after breeding, but low in Winter and Spring. The population biomass increased from June until October each year because of high growth and then declined through the Winter. There are marked variations in density and biomass between the two years of the study period, which probably reflects the dramatic change in benthic abundance and distribution found in many new impoundments.

In order to calculate the energy budget, the respiration rate of *H. stagnalis* has been studied by using a Gilson respirometer operated at three temperatures 5 °C, 10 °C and 15 °C, and the respiration rate of individual *H. stagnalis* increased logarithmically with weight. The relationship between O_2 consumption and dry weight was examined using linear regression (log/log) for the three temperatures. There are significant differences between the three regressions with slopes of 0.54, 0.69, and 0.73 at 5 °C, 10 °C and 15 °C respectively. The repiratory coefficient (Q_{10}) for the range 5 °C to 15 °C was almost constant for all class weights.

The food of the leech was determined qualitatively by exposing potential prey organisms to it, using all the common zoobenthos of Rutland Water. *H. stagnalis* fed extensively on oligochaetes (Tubificidea). Food consumption was estimated at the three different test temperatures for different weight classes. Similar studies have been undertaken on *E. octoculata* and *G. complanata.*

References

Bullock, J. A., Clark, F. & Ison, S., 1982. Benthic invertebrates of Rutland Water. In: Harper, D. M. Bullock, J. A. (Eds) Rutland Water – Decade of change, pp. 147–156. Junk, The Hague.

Hydrobiologia 88, 157 (1982). 0018-8158/82/0882-0157/$00.20.

Studies on two species of *Asellus* in Rutland Water

Y. T. Daoud
Department of Zoology, University of Leicester, Leicester LE1 7 RH, England

Keywords: *Asellus,* predation, serology, population

Abstract

The study comprised a comparison of the ecology of two species of *Asellus (A. aquaticus* and *A. meridianus),* and had two main aims. First the energy budget of the population of *A. aquaticus,* which is one of the most abundant species in the reservoir was assessed. Secondly, the two species appear to be ecologically very similar and it has been assumed by many workers that they are in competition. The study aimed to investigate how these two species might be able to coexist.

The population dynamics of both species and the interaction between them is under study, as are their relations with the other benthic animals recorded in the reservoir (see Bullock *et al.* 1982). The sampling programme to cover variations in depth and substratum is also described by Bullock *et al.* 1982. *A. aquaticus* occurs in both arms while *A. meridianus* exists only in the south arm. A preliminary examination of the available data reveals that the density of *A. aquaticus* has increased in the south arm while the density of *A. meridianus* decreased drastically in February 1980. Since then low numbers of *A. meridianus* have been recorded every month even in the summer (breeding) season. It is noteworthy that simultaneously the number of predators, particularly *Helobdella stagnalis,* increased in comparison with previous years.

It was therefore decided to study the predation on *Asellus* in the laboratory as well as in the field by using a serological technique. Antisera were produced against each species of *Asellus* by injections into rabbits of precipitated proteins (antigens). Thus a positive reaction obtained from a homogenised predator when tested with the specific antiserum indicates which species of *Asellus* had been consumed and the percentage of predators which had fed on *Asellus* in every sample could be calculated. The results obtained to date reveal that considerable numbers of *Asellus* spp. are removed from the environment by the predators *(Helobdella stagnalis, Erpobdella octoculata, Polycelis tenuis. Dendrocoelum lacteum* and *Dugesia lugubris).*

Size class structures of the population of both species were constructed to study their life cycles and to estimate the population production. Comparitive respirometry of the two species was carried out at 4 °C, 10 °C and 16 °C, using a Gilson differential respirometer. This study aimed firstly to assess one of the essential parameters in the energy budget study of *A. aquaticus* and secondly, since *A. aquaticus* appears to be more active than *A. meridianus,* it would be expected that a comparison of their metabolic rate at different temperature would reveal marked differences between two species. It was found that there was no significant difference between the two species at 4 °C and 10 °C, but *A. aquaticus* had a significantly higher metabolic rate than *A. meridianus* at 16 °C.

Consumption and assimilation of decaying oak leaves and of *Cladophora glomerata* (both of which are available in the environment) by two species of *Asellus* are being measured at present in the laboratory at three different temperatures (4 °C, 10 °C and 16 °C), and will contribute to the estimation of energy budgets.

Reference

Bullock, J. A., Clark, F. & Ison, S., 1982. Benthic invertebrates of Rutland Water. In: Harper, D. M. & Bullock, J. A. (Eds) Rutland Water – Decade of Change, pp. 147–156. Junk, The Hague.

Hydrobiologia 88, 158 (1982). 0018-8158/82/0882-0158/$00.20.

Bathymetric distribution of some Chironomidae (Diptera) in Rutland Water

A. E. Brown[1] & R. S. Oldham
School of Life Sciences, Leicester Polytechnic, Leicester, England
[1] *Present address: Department of Biological Sciences, Bayero University, Kano, Nigeria*

Keywords: Chironomidae, bathymetric, distribution, Rutland Water, reservoir, SPSS

Abstract

The bathymetric distribution of the eight most abundant taxa of chironomid larvae is described from benthic grab samples collected in Rutland Water from May 1977 to April 1979. *Cricotopus, Psectrocladius* and *Endochironomus* larvae were generally restricted to water less than 9 m deep. *Procladius, Chironomus, Microtendipes, Polypedilum* and Tanytarsini larvae were all recorded at depths down to 25 m. However, the highest population densities of the latter five taxa were recorded in shallow water. The results are discussed in relation to the diet of trout and certain management practices at the reservoir. Also described is the application of SPSS (Statistical Package for the Social Sciences) to the analysis of these data.

Introduction

Benthic organisms may be distributed in response to many environmental parameters, especially those varying with water depth. Brundin (1951) stressed the importance of oxygen concentration at the mud surface in limiting the distribution of species. Mundie (1955, 1957) described the distribution of some chironomid larvae in a eutrophic reservoir, Kempton Park East, England. He suggested that other factors such as sediment depth and differing intensities of predation may also be important in determining distributions. This paper describes the bathymetric distributions of the numerically dominant chironomid taxa at Rutland Water.

Also described is the application of a computer software package, SPSS* (Statistical Package for the Social Sciences) to tha analysis of the data. Due to its ease of use and the large number of statistical computations available, SPSS is finding increasing use in other disciplines. The program example is intended to describe some of the facilities SPSS offers and its suitability for the analysis of this type of information. The paper does not contain detailed instructions for the use of SPSS; these are provided in the SPSS manuals (Klecks *et al.* 1975; Nie *et al.* 1975).

* This package is an item of proprietary software, the Licensor being SPSS Inc., 444 Michigan Avenue, Suite 3300, Chicago, IL 60611, U.S.A.

Methods

Sampling

Two transects located at the end of the north and southarm of the reservoir (Fig. 1, 2NA and 2SA) were sampled monthly using a modified van Veen grab. The grab is described and compared with other grab samplers by Brown (1981). One grab sample was taken at ten equidistant points along each transect. The first and last grab sample along each transect were each taken at approximately 0.5 m deep. The depth of intermediate grab samples was recorded at the time of sampling. The grab contents were sieved through a 500 μm sieve and the

Hydrobiologia 88, 159–169 (1982). 0018-8158/82/0882-0159/$02.20.

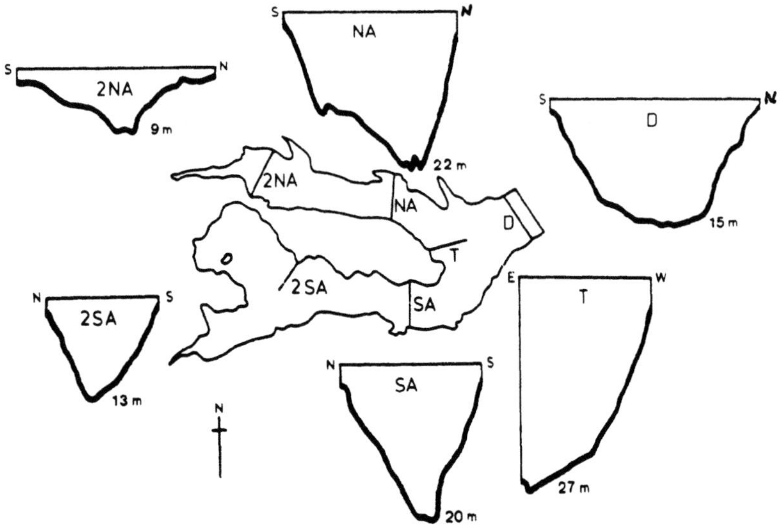

Fig. 1. Locations and profiles of transects.

```
                                      Control field          Specification field
                                      (columns 1-15)         (columns 16-80)

                                     RUN NAME
                                     DATA LIST
                   data              INPUT MEDIUM
                   definition        N OF CASES
                   cards             MISSING VALUES
control                              VAR LABELS
cards                                VALUE LABELS
                                     TASK NAME
                                     CONDESCRIPTIVE
                   procedure         OPTIONS
                   cards             STATISTICS
                                     READ INPUT DATA
data                                     31 876 1 3        3    2   132 24  6
cards                                    31 876 2 3    2   3    3    43 25  2
                                         31 876 3 5    7        4           2
control card                         FINISH
```

Fig. 2. Basic SPSS instructions required for a program run. Only examples of control fields are included.

organisms sorted live. Samples considered here were collected between May 1977 and April 1979. Data on the maximum depths at which larvae were recorded are based on Ekman grab samples collected along four transects located at the eastern end of the reservoir (Fig. 1, NA, SA, T and D).

Computing – SPSS

Arrangement of data for input to SPSS

Instructions and data are given to SPSS on computer cards or via a remote terminal. The methods are similar but computer cards were used in the present study. The cards can be divided into two groups: 'control cards' define the structure of the data and state which statistical procedures are required and 'data cards' record the actual results of the survey (Fig. 2).

Control cards

SPSS control cards are composed of two parts or fields. The 'control field' occupies columns 1–15 on the card and contains a unique control word or set of words which informs SPSS of the specification that will follow, so that the proper procedures can be called to act upon the information on the card. The 'specification field' occupies columns 16–80. Here the user supplies the system with parameters, labels and instructions required to perform the function indicated by the control field.

Control cards can be divided into two groups: 'data definition cards', which define the structure of the data to be analysed, and 'procedure cards', which request the type of statistical analysis (Fig. 2).

Data cards

In this survey a grab sample is the basic unit of analysis and is referred to as a 'data case'. Nineteen variables were recorded in the data case. Table 1 shows the columns that each variable occupies, the SPSS variable names and their descriptions and codes. The first six letters of each taxon of chironomid were used as variable names for ease of recognition.

Table 1. Example codebook defining the variables given in each data case.

Column	SPSS variable names	Variable description and codes
1– 4	SEQNOS	Sequence numbers. These identify individual cards but are excluded from programs
7–11	SADATE	Sampling data (cols 7 & 8 = day; cols 9 & 10 = month; col. 11 – year)
12	TRANNO	Transect number in reservoir, numbers 1–6 are the codes that identify each transect
13–14	GRABNO	Grab number in transect (1 to 99)
15–16	DEPTHS	Depth of water in which grab sample was taken (1 to 99)
17–20	SAMPWT	Weight of grab sample (1 to 9999)
21–23	PROCLA	No. of larvae of genus *Procladius* (max. value 999)
24–26	CRICOT	No. of larvae of genus *Cricotopus* (max. value 999)
27–29	ORTHOC	No. of larvae of genus *Orthocladius* (max. value 999)
30–32	PSECTR	No. of larvae of genus *Psectrocladius* (max. value 999)
33–35	CORYNO	No. of larvae of genus *Corynoneura* (max. value 999)
36–38	CHIRON	No. of larvae of genus *Chironomus* (max. value 999)
39–41	CRYPTO	No. of larvae of genus *Cryptochironomus* (max. value 999)
42–44	ENDOCH	No. of larvae of genus *Endochironomus* (max. value 999)
45–47	GLYPTO	No. of larvae of genus *Glyptotendipes* (max. value 999)
48–50	MICROT	No. of larvae of genus *Microtendipes* (max. value 999)
51–53	POLYPE	No. of larvae of genus *Polypedilum* (max. value 999)
54–56	TANYTA	No. of larvae of tribe *Tanytarsini* (max. value 999)
57–59	PENTAN	No. of larvae of tribe *Pentaneurini* (max. value 999)

Program steps	Control field	Specification field
1	RUN NAME	EXAMPLE TASKS FROM FILE CHIRSTUD
2	GET FILE	CHIRSTUD
3	TASK NAME	SCATTER DIAGRAM EXAMPLE
4	COMMENT	THE NUMBER OF PSECTROCLADIUS LARVAE PER GRAB IS PLOTTED AGAINST DEPTH OF WATER. THE OPTIONS CARD PRODUCES AUTOMATIC SCALING.
5	RUN SUBFILES	ALL
6	SCATTERGRAM	DEPTHS WITH PSECTRO
7	OPTIONS	7
8		

Example program

The SPSS subprogram, SCATTERGRAM, produces a plot of one variable against another, in this case the number of larvae of a particular taxon against water depth, for each sample.

Steps explained

1. These control words enable the user to include any message in the specification field which identifies the run. This message is printed at the top of each page of output generated by the run.
2. This accesses the file, named CHIRSTUD, which was created to store the data using the data definition cards in Fig. 2.
3. These control words enable the user to identify specific tasks within a program. The message in the specification field is printed below the run label on the top of each page of output to which it applies.
4. Comments may be placed anywhere appropriate in the control card deck by the use of this control word. These comments are printed with the other control cards in the output. Control cards 1, 3 and 4 act merely as labeles to identify the output.
5. All the subfile information was divided into subfiles each containing data from one transect on one sampling data. These control words allow the selection of any of those subfiles for analysis. In this case they are ALL to be analysed.
6. This is the control word used to produce the scatter diagram. Here only one plot is requested, water depth against *Psectrocladius.* Further variables may be plotted against depth by writing either PROCLA, CRICOT or PSECTRO in the specification field for those three species between those two on the data list (Table 1).
7. This OPTIONS card produces automatic scaling of the axes, i.e. the maximum and minimum values of the variables supplied in the data are the limits of the axes and the intervals are such that the whole graph can be drawn on one page of printed output.
8. No statistics were required on the plot so the statistics card was omitted. Various statistics may be requested from each SPSS subprogram in a similar manner to the OPTIONS card. All the various subprograms, options and statistics are described in the SPSS manual (Nie *et al.* 1975).

The output from the program is shown in Fig. 3. The program was modified slightly to produce the plots on which the results given in this paper are based. Firstly, only the subfiles for the transects 2NA and 2SA were used in the analysis and secondly the scales for the plots were set.

Results

Larvae of the genus *Procladius* (subfamily Tanypodinae) occurred in relatively large numbers down to 13 m (Fig. 4). Populations showed some decline in density below 14 m in the four deeper transects (Fig. 1) although larvae were still recorded at the maximum depths sampled, approximately 28 m. The maximum population density recorded at this

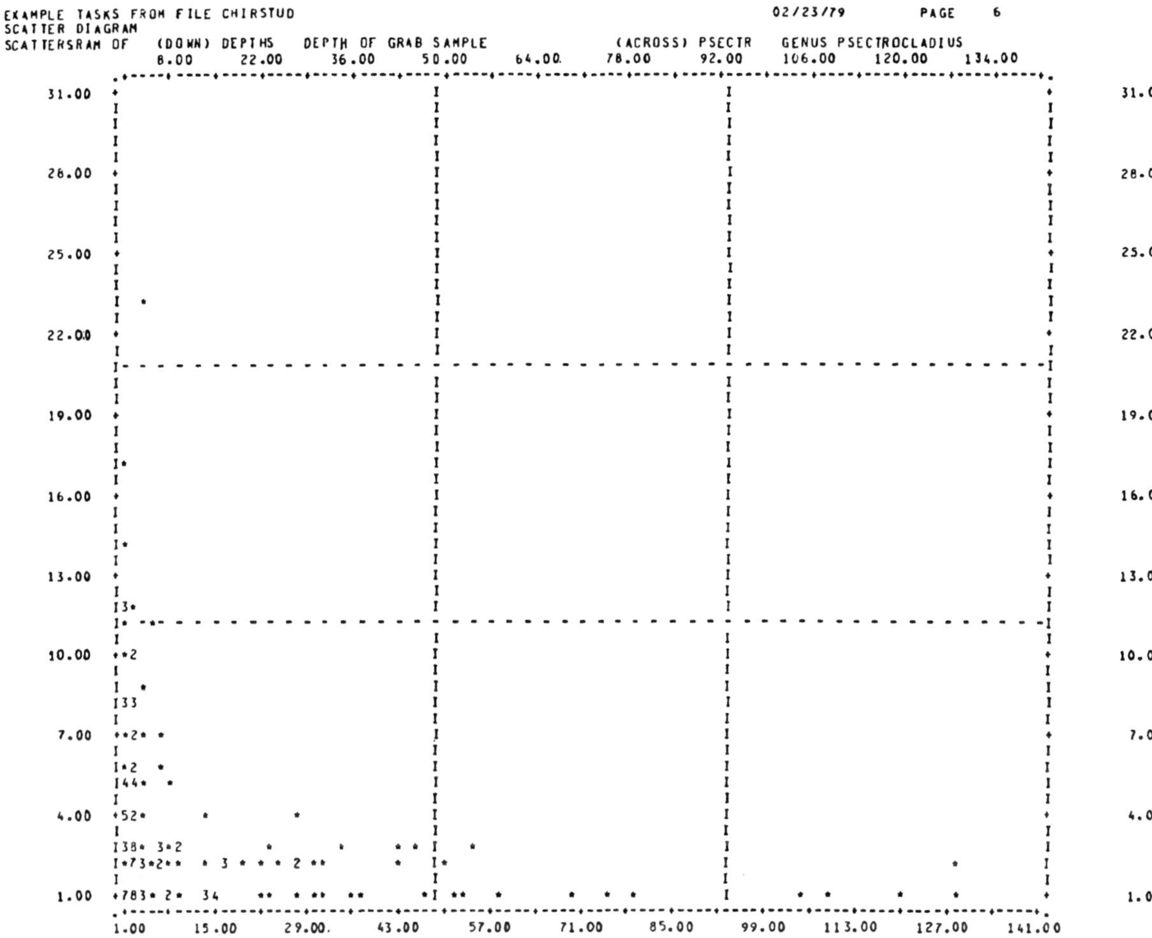

Fig. 3. Output from example program. Single data points are represented by an asterisk. If two to eight cases fall in the same position the actual number of cases is written. Nine or more cases are represented by the number 9.

depth was 400 m^{-2}. The dominant species at Rutland Water, *P. choreus* (Meigen), has been found to have a wide depth distribution in other lakes and reservoirs (Mundie 1957; Hunt & Jones 1972). Miller (1941) suggested that the wide depth distribution in the Tanypodinae is related to their predatory and free-living habit.

In general the subfamily Orthocladiinae are phytophilous, abundant in the littoral zone of lakes and reservoirs, and also occur in many lotic habitats (Thienemann 1954). Five species of *Chricotopus* were recorded at Rutland Water, *C. bicinctus* (Meigen), *C. trifascia* Edwards, *C. intersectus* (Staeger), *C. sylvestris* (Fabricius) and *C. tricinctus* (Meigen) and most larvae were found in water less than 3 m deep (Fig. 5). A few isolated individuals were recorded down to 16 m but it is likely that these were carried into deep water by water currents. *Cricotopus* larvae were found to be very numerous on mats of filamentous algae.

Two species of *Psectrocladius* (subfamily Orthocladiinae) were recorded at Rutland Water, *P. barbimanus* (Edwards) and *P. sordidellus* (Zetterstedt). *Psectrocladius* larvae showed a general decrease in density with increasing water depth (Fig. 6). Low population densities were recorded from 5 m to 16 m. *Psectrocladius* was the numerically dominant orthoclad genus found in the reservoir.

The remaining species considered here belong to the subfamily Chironominae. This group incorporates those species of larvae best adapted to withstand the low oxygen concentrations found in the hypolimnion. Three species of *Chironomus*

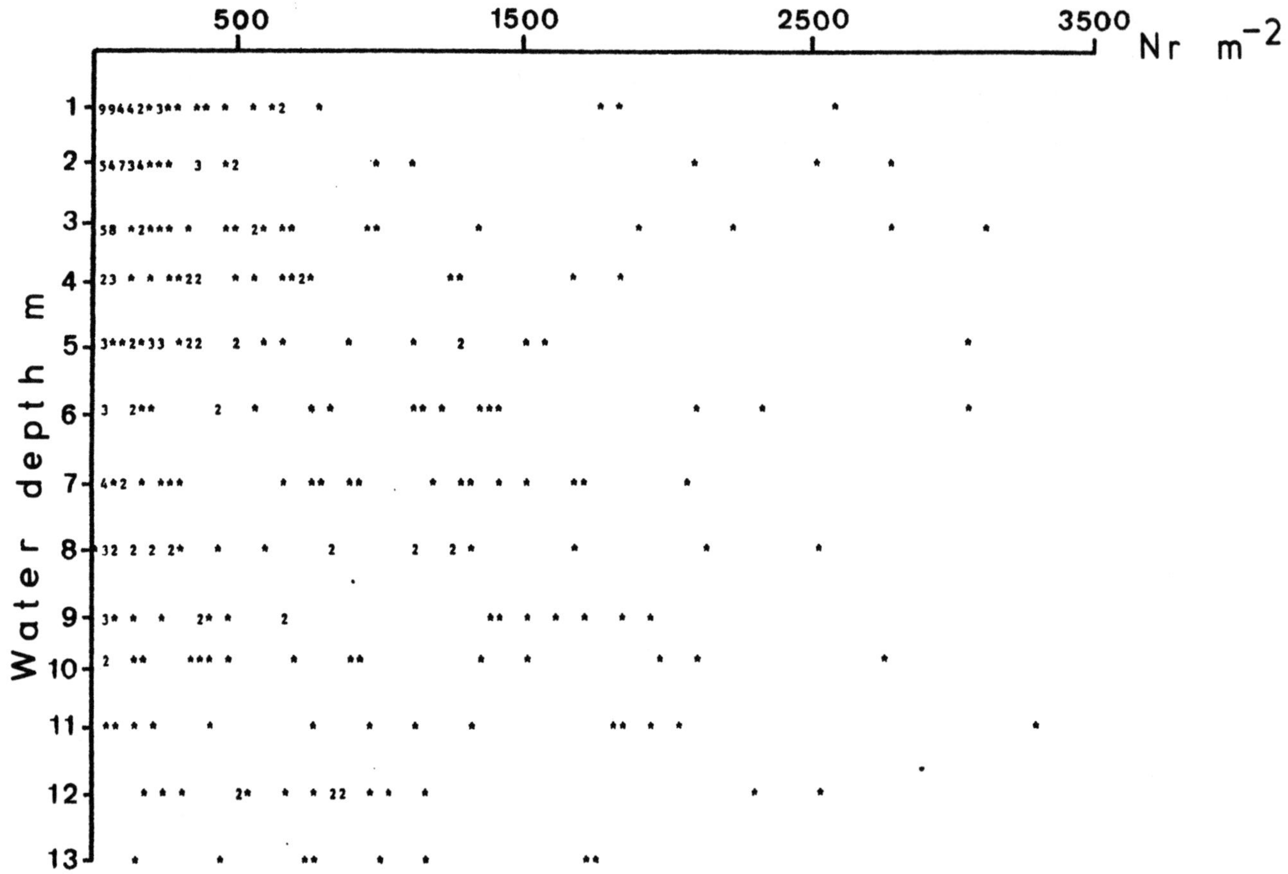

Fig. 4. Depth distribution of *Procladius* larvae.

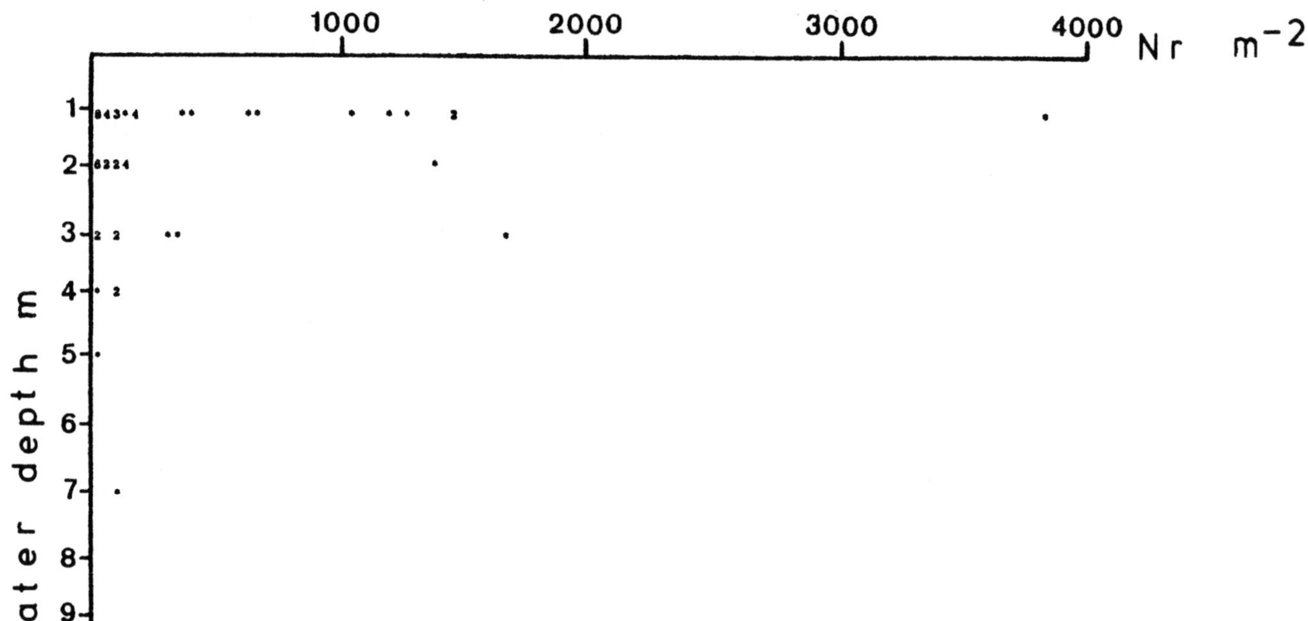

Fig. 5. Depth distribution of *Cricotopus* larvae.

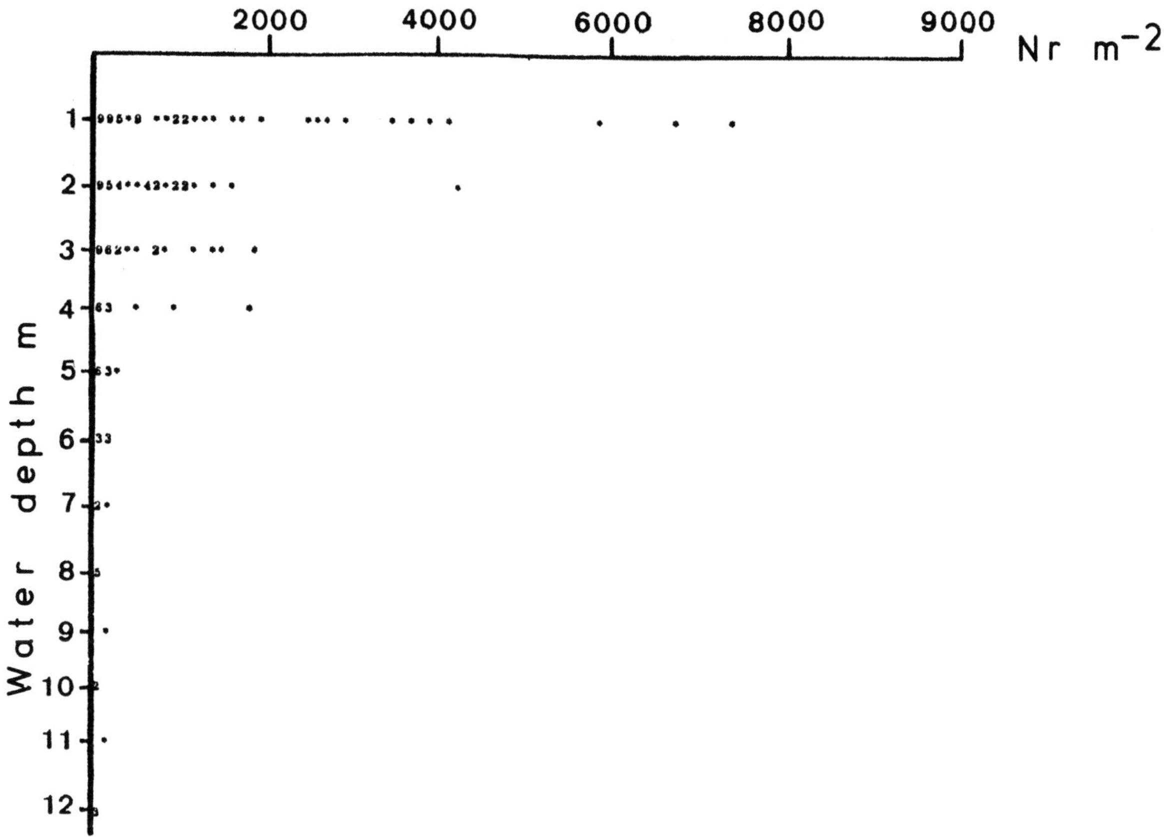

Fig. 6. Depth distribution of *Psectrocladius* larvae.

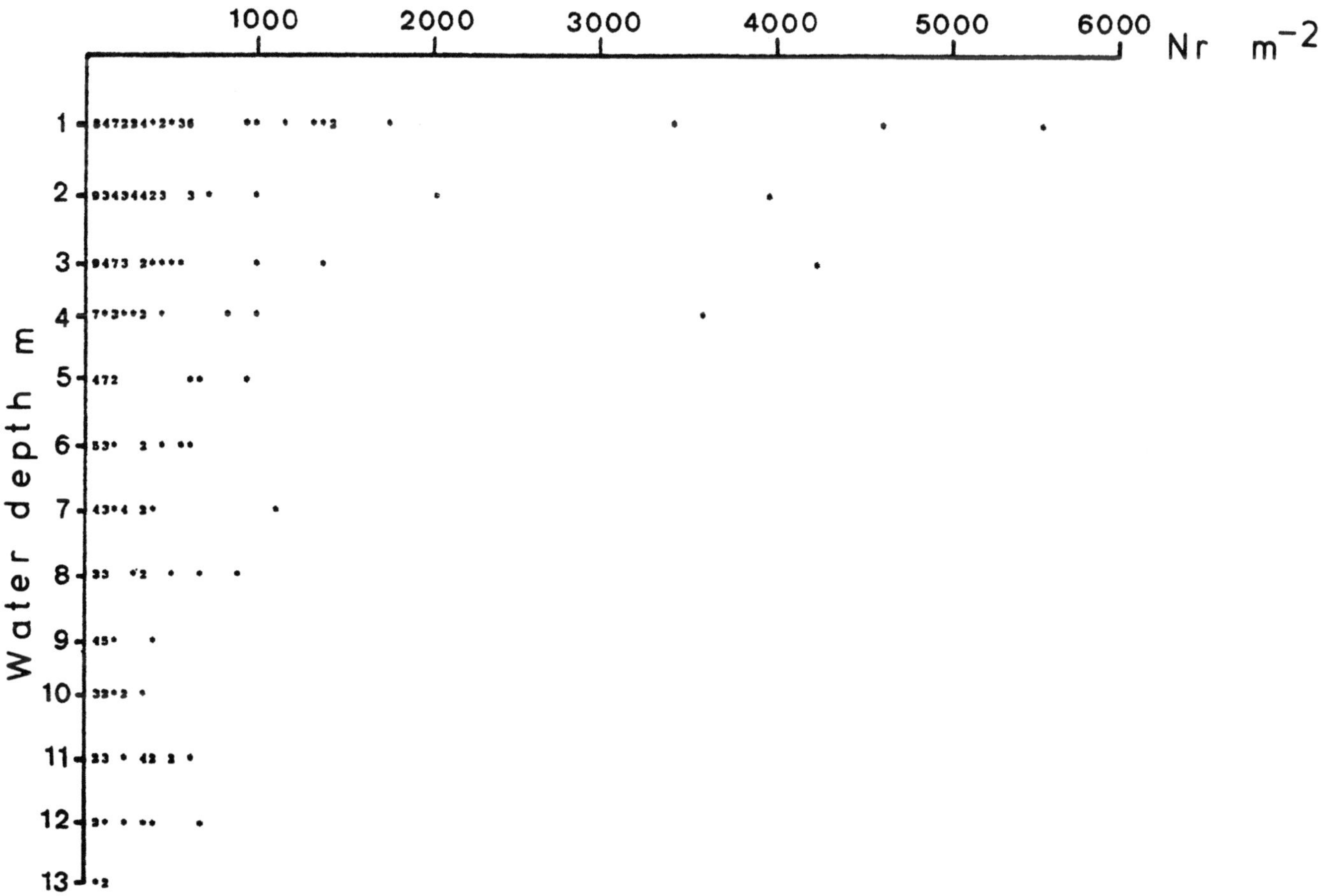

Fig. 7. Depth distribution of *Chironomus* larvae.

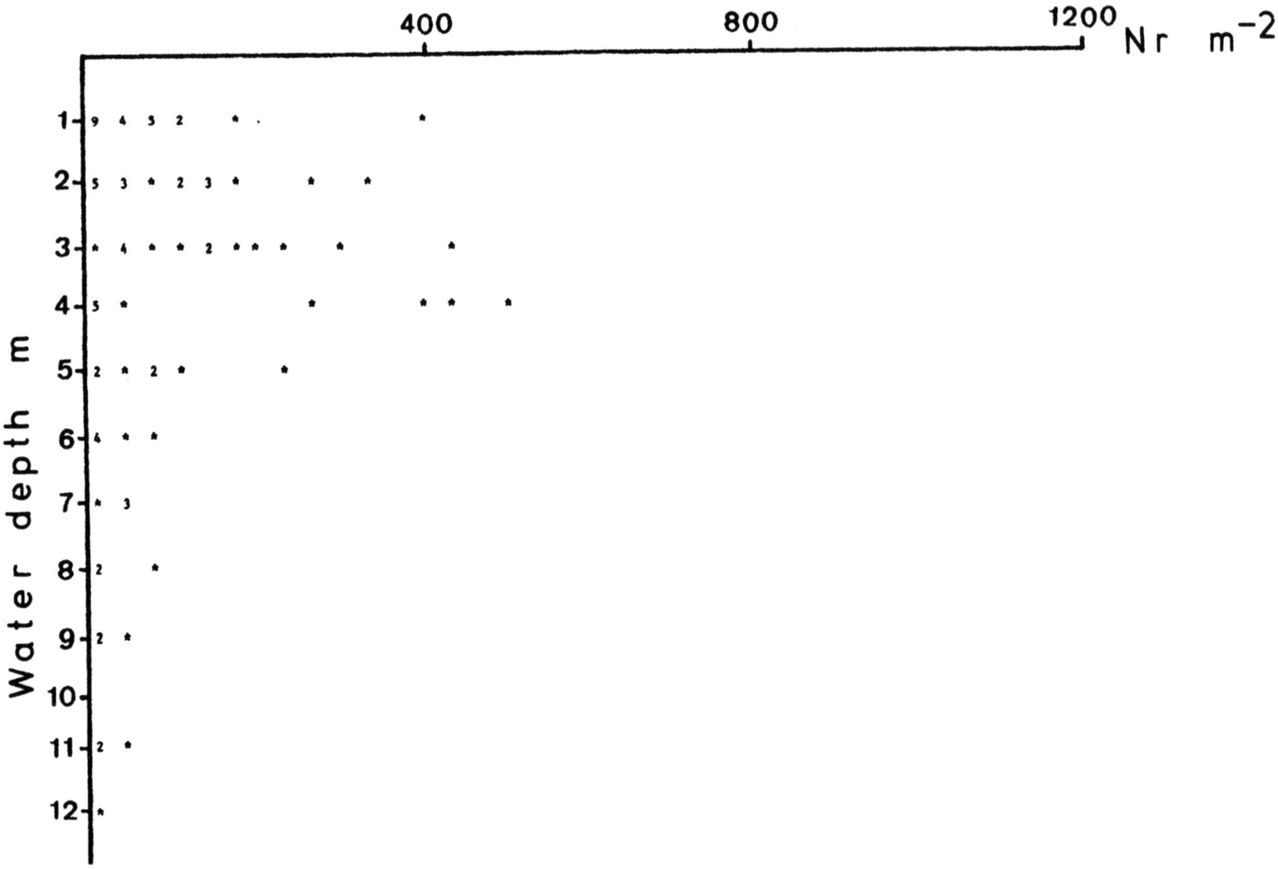

Fig. 8. Depth distribution of *Endochironomus* larvae.

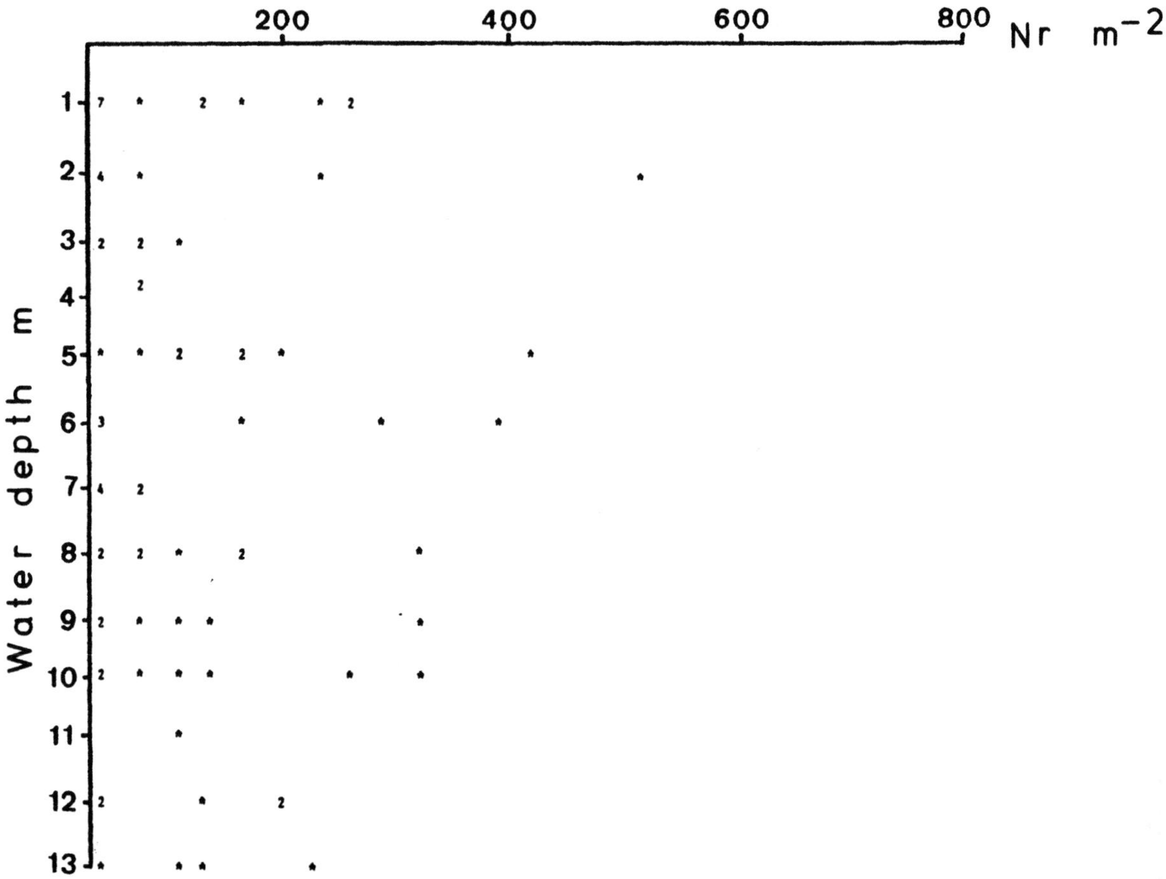

Fig. 9. Depth distribution of *Microtendipes* larvae.

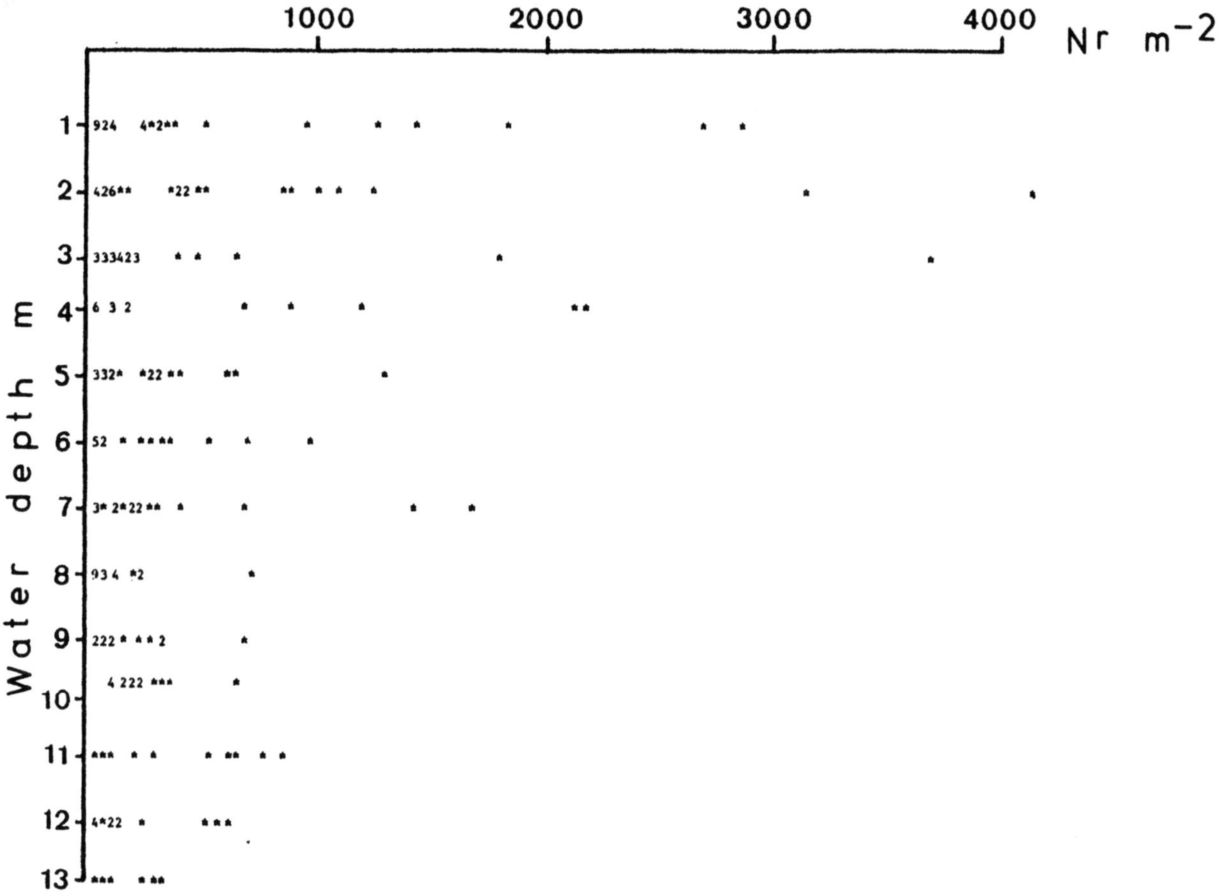

Fig. 10. Depth distribution of *Polypedilum* larvae.

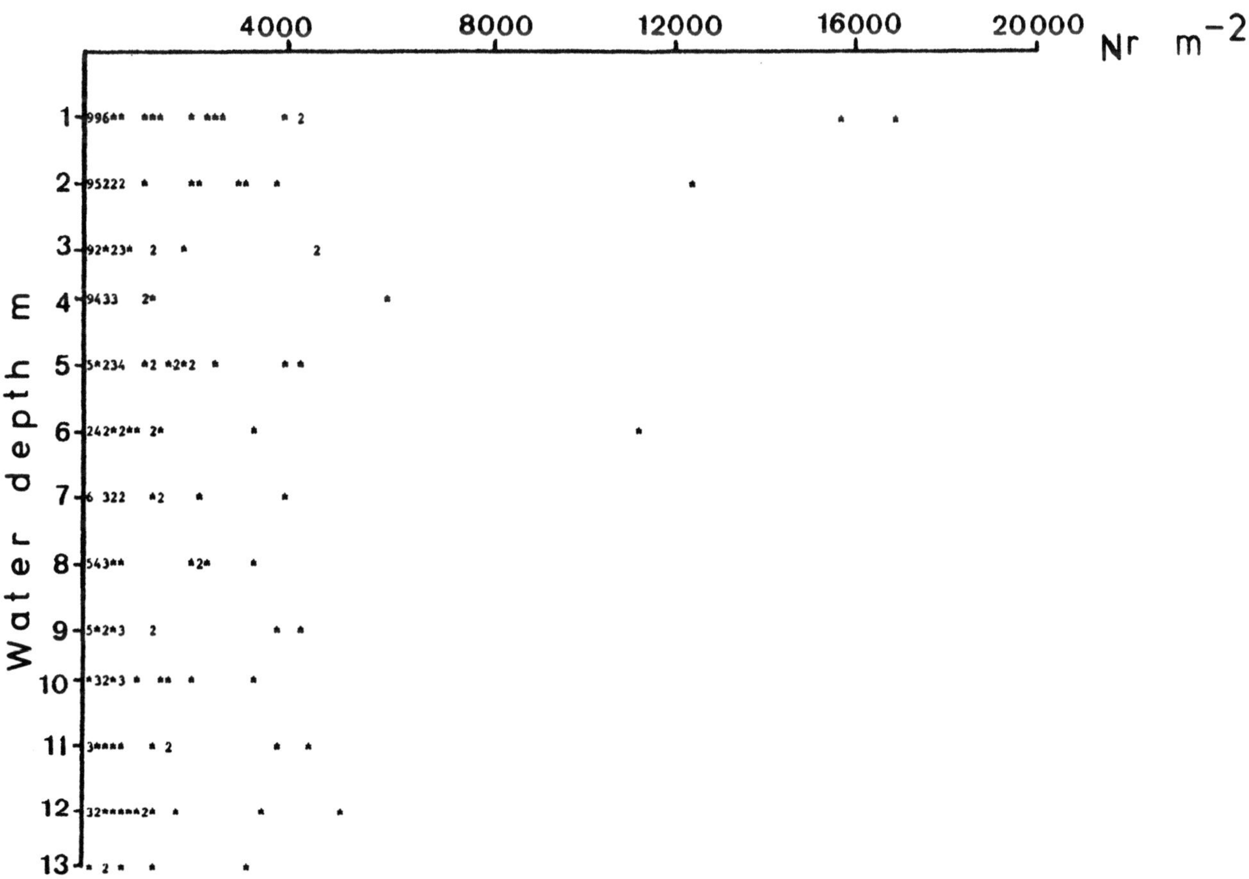

Fig. 11. Depth distribution of *Tanytarsini* larvae.

larvae were recorded at Rutland Water, *C. longistylis* Meigen, *C. plumosus* (L.) and *C. riparius* Meigen. *Chironomus* larvae were recorded throughout the depth range of the reservoir, especially between 0 and 4 m (Fig. 7). The maximum population density recorded at 28 m was 700 m^{-2}.

Endochironomus albipennis (Meigen) larvae showed a peak in maximum population density at 4 m (Fig. 8). Larvae were recorded at low densities at all depths down to 23 m. This species is known to be a typical inhabitant of algal mats in eutrophic lakes (Menche 1939) and also builds tubes on submerged macrophyte leaves (Walshe 1951).

Microtendipes chloris (Meigen) larvae were recorded in approximately uniform densities down to 23 m. The highest densities were recorded between 2 and 6 m (Fig. 9). In contrast, at Kempton Park Reservoir, Mundie (1957) did not find this species in the profundal zone.

Polypedilum nubeculosum (Meigen) is a common species in eutrophic lakes and has a wide geographical distribution (Mundie 1957). At Rutland Water *P. nubeculosum* showed a similar bathymetric distribution to *M. chloris* although the larvae occurred in larger numbers (Fig. 10).

Two genera, *Micropsectra* and *Tanytarsus*, were not routinely separated as larvae and have been treated together here as Tanytarsini. Only one species of *Micropsectra (M. lindrothi* Goetghebuer), and seven species of *Tanytarsus* were recorded; *T. bathophilus* Kieffer, *T. gracilentus* Holmgren, *T. gregarius* Kieffer, *T. holochlorus* Edwards, *T. lestagei* Goetghebuer, *T. longitarsus* (Kieffer) and *T. pallidicornis* (Walker). Tanytarsini larvae were the numerically dominant chironomids in Rutland Water from May 1977 to April 1979. The tribe as a whole are widely regarded as being characteristic species of oligotrophic lakes and favour well aerated water (Thienemann 1925; Myadi 1933; Brundin 1949). Results from Rutland Water and elsewhere (Mundie 1957; Cantrell 1975) indicate that some species of *Tanytarsus* can survive eutrophic conditions and even occur in the profundal zone. High population densities of Tanytarsini larvae were recorded in Rutland Water down to 20 m, with the largest populations in shallow water, 0–4 m (Fig. 11).

Discussion

In this investigation SPSS was used for the rapid plotting of large amounts of data. Further analysis of the data might involve examining bathymetric distributions from different transects or examining temporal changes in bathymetric distributions. This can be achieved with very little alteration of the program example described. The authors have also used SPSS for statistical analysis of other benthic invertebrate survey data. SPSS is particularly convenient for ecological work because it requires no previous computer programming experience and because there is a wide range of statistical tests, data management and data modification procedures available.

The bathymetric distributions of larvae were generally found to be in close agreement with published data. Orthocladiinae larvae, *Cricotopus, Psectrocladius* and *Endochironomus*, were restricted to water less than 4 m deep. They utilise the vegetation as a substrate on which to build tubes, and also graze epiphytic algae. Each of the three taxa is an important component of brown and rainbow trout diet in Rutland Water (Brown *et al.* 1980; Brown 1981). *Procladius* (Tanypodinae) and most of the Chironominae were recorded throughout the depth range of the reservoir but these too showed their highest population densities in shallow water.

Two management practices at Rutland Water may affect this predominantly littoral chironomid fauna, together with other littoral invertebrates. Firstly, the abstraction of water for supply may result in a rapid and large drawdown of water. Several authors have shown that water level fluctuations cause a reduction in littoral fauna and fish food organisms (e.g. Runnstrom 1955; Hynes 1961). Secondly, filamentous algae and macrophytes are mechanically removed from the reservoir margin. The benefits of this practice include a clear littoral region for wading fishermen and propellor-driven boats and reduced decomposition on exposed shores following drawdown. The main deficit is the reduction of littoral fauna which are potential food items for brown and rainbow trout. The authors are unaware of any quantitative investigation of the effects of mechanical removal of vegetation and disturbance of the littoral fauna.

Reservoir drawdown is clearly inevitable and the

immediate benefits of vegetation clearance understandably override the ill-defined costs in fish productivity. However, even within economic constraints there are probably several options open to the water authority in terms of the timing, periodicity, duration and spatial extent of their management operations. An investigation of these factorss may be rewarding in determining the most efficient overall resource utilisation.

Acknowledgements

The authors would like to thank Leicester Polytechnic computer centre for the use of their facilities and the Anglian Water Authority for permission to work on Rutland Water. The work was carried out during the tenure of an S.R.C. studentship by A.E.B.

References

Brown, A. E., Oldham, R. S. & Warlow, A., 1980. Chironomid larvae and pupae in the diet of brown trout (Salmo trutta) and rainbow trout (Salmo gairdneri) in Rutland Water, Leicestershire. In: Murray, D. A. (Ed.) Chironomidae, pp. 323–329. Pergamon Press, Oxford.

Brown, A. E., 1981. The ecology of the Chironomidae (Diptera) in a new eutrophic reservoir. Unpubl. Ph.D. thesis, Leicester Polytechnic.

Brundin, L., 1949. Chironomiden und andere Bodentiere der Sudschweden Urgebirssen. Rep. Inst. Freshwat. Res. Drottningholm 30: 1–914.

Brundin, L., 1951. The relation of O_2 microstratification of the mud surface to the ecology of the profundal bottom fauna. Rep. Inst. Freshwat. Res. Drottningholm 32: 32–44.

Cantrell, M. A., 1975. Development of the benthic fauna during the filling of a new lake. Ph.D. thesis, University of Newcastle-upon-Tyne.

Hunt, P. C. & Jones, J. W., 1972. The profundal fauna of Llyn Tegid, North Wales. J. Zool. 168: 9–49.

Hynes, H. B. N., 1961. The effect of water level fluctuations on filling of a new lake. Ph.Dlittoral fauna. Ver. int. Verein. Limnol. 14: 652–656.

Klecka, W. R., Nie, N. H. & Hadlai Hull, C., 1975. SPSS Primer. Statistical Package for the Social Sciences, Primer. McGraw-Hill, London.

Menche, A., 1939. Die Fauna im Algenbewuchs. Nach Untersuchungen im Littoral ostholsteinischer Seen. Arch. Hydrobiol. 34: 349–520.

Miller, R. B., 1941. A contribution to the ecology of the Chironomidae of Costello Lake, Algonquin Park, Ontario. Univ. Toronto Stud. Biol. 49: 1–63.

Miyadi, D., 1933. Studies on the bottom fauna of Japanese lakes: X. Regional characteristics and a system of classifying Japanese lakes based on the bottom fauna. Jap. J. Zool. 4: 417–437.

Mundie, J. H., 1955. On the distribution of Chironomidae in a storage reservoir. Verh. int. Verein. Limnol. 12: 577–581.

Mundie, J. H., 1957. The ecology of Chironomidae in storage reservoirs. Trans. R. Ent. Soc. Lond. 109: 149–232.

Nie, N. H., Hadlai Hull, C., Jenkins, J. C., Steinbrenner, K. & Bent, D. H., 1975. Statistical Package for the Social Sciences, 2nd edn. McGraw-Hill, London.

Runnestrom, S., 1955. Changes in fish production in impounded lakes. Proc. int. Ass. Limnol. 12: 176–182.

Thienemann, A., 1925. Die Binnengewasser Mitteleuropas. Eine Limnologische Einfuhrung. Binnengewasser 1: 1–255.

Thienemann, A., 1954. Chironmus Leben. Verbreitung und wirt-schaftliche Bedentung der Chironomiden. Binnengewasser 20: 1–834.

Walshe, B. M., 1951. The feeding habits of certain chironomid larvae (subfamily Tendipedinae). Proc. Zool. Soc. Lond. 121: 63–79.

The protozoan epizooites found on the gills of *Gammarus pulex*

M. H. Walker & E. Roberts
Department of Zoology, University of Leicester, Leicester LE1 7RH, England

Keywords: *Gammarus*, ciliates, epizooites, scanning electron microscopy

Abstract

Four protozoan epizooites are found attached to the gills of *Gammarus pulex* collected from Rutland Water. The suctorian, *Dendrocometes paradoxus*, and the peritrichs, *Epistylis* sp. and *Lagenophrys* sp., are relatively abundant, but the chonotrich, *Spirochona gemmipara* is only occasionally found. SEM studies of various aspects of these protozoa are described. SEM has been useful in distinguishing between the two species of *Lagenophrys* found on *Gammarus*. *L. ampulla* is found on the gills and *L. nassa* is found on the limbs. The two species are distinguished by the differences in the structure of the loricastome. The relationship between the epizooites and their host is discussed briefly.

Introduction

Many species of ciliate exist as commensals, and many of these are found on crustaceans and are often abundant. Among freshwater invertebrates, amphipods of the genus *Gammarus* frequently carry many commensal ciliates. Penard (1922) named 14 species belonging to 11 genera as commensals of *Gammarus*. We have been concerned with the epizooites found on the gills of *Gammarus pulex* collected from Rutland Water. There are four genera that we have found on the gills of *G. pulex;* the suctorian *Dendrocometes paradoxus,* the peritrichs *Epistylis* sp. and *Lagenophrys* sp., and the chonotrich *Spirochona gemmipara.* Several aspects of the biology of these epizooites are currently under investigation at Leicester. These include the seasonal and spatial distribution of the epizooites in shallow and deep water samples, the relationship between the host moult cycle and the behaviour of the epizooites, and scanning electron microscope and ultrastructural studies. This present paper is restricted to scanning electron microscope studies of these epizooites.

Materials and methods

Gammarus pulex were collected from Rutland Water, Leicestershire, by netting from shallow water and dredging from deep water. Gills were removed from decapitated *Gammarus* and placed in filtered reservoir water. Gills were then mounted on slides in reservoir water and examined under phase contrast. Gills bearing epizooites required for scanning microscopy (SEM) were collected in a 1 ml volume of reservoir water. To fix the gills 1 ml of 2% OsO_4 was added. After 10–15 minutes the gills were rinsed briefly in distilled water, dehydrated through an alcohol series and placed in amyl acetate. The gills were dried using a Samdri 780 critical point drier, mounted on aluminium stubs, and coated with gold using a Polaron Sputter coater.

Whole *Gammarus* were prepared for SEM by narcotization in increasing alcohol concentrations. After dehydration they were placed in amyl acetate, dried and coated as described above. All specimens were examined using an ISI mini SEM or an ISI 60 scanning electron microscope.

Hydrobiologia 88, 171–176 (1982). 0018-8158/82/0882-0171/$01.20.

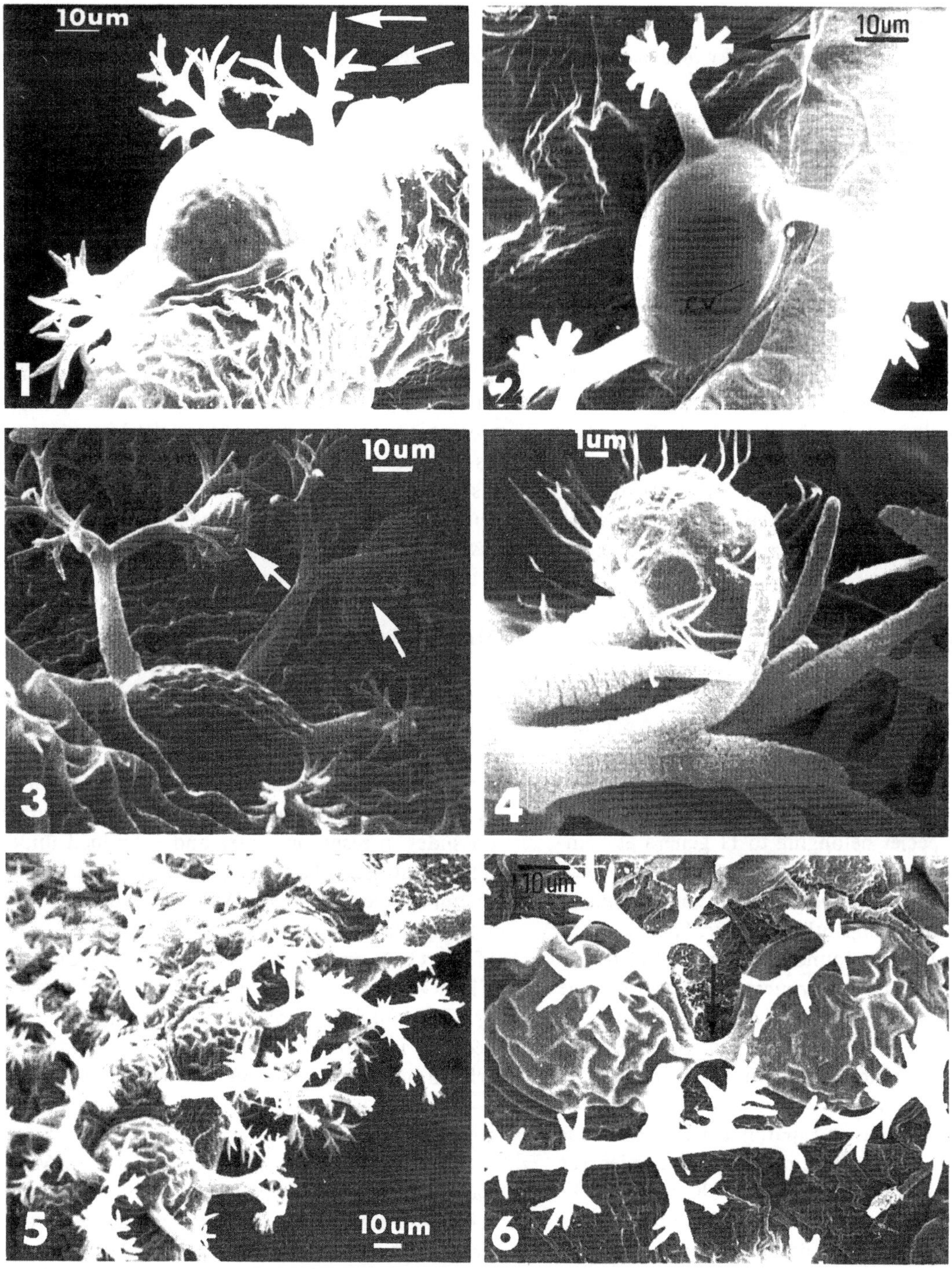

Fig. 1. Dendrocometes paradoxus attached to the periphery of a *Gammarus* gill. Note the plano-convex body and the branched arms with pointed tentacule tips (arrowed).

Fig. 2. Dendrocometes showing stumpy tentacules (arrowed). The opening of the contractile vacuole (cv) is seen near the base of the body.

Fig. 3. Dendrocometes feeding. Note two ciliates attached to the tentacules (arrowed).

Fig. 4. Tentacule of *Dendrocometes* embedded (arrowed) in ciliate prey.

Fig. 5. Part of a *Gammarus* gill bearing many *Dendrocometes.*

Fig. 6. Two *Dendrocometes* linked together by a conjugation process (arrowed).

Observations

Dendrocometes paradoxus

The suctorian *Dendrocometes paradoxus* is found on a high percentage of gills examined. *Dendrocometes* has a planoconvex body from which extend a variable number of tentacular arms (Fig. 1). The opening of the contractile vacuole can be seen towards the base of the body (Fig. 2). *Dendrocometes* are attached to the gill by the base of the body (Figs. 1 and 2). Variations are noted in the tentacule tips of different individuals. In some individuals the tentacle tips are pointed (Fig. 1) and in others they are stumpy (Fig. 2). In feeding *Dendrocometes* prey are trapped on the pointed tentacule tips. Several prey may be captured at once (Fig. 3). The tentacle tip enters the prey (Fig. 4) and the prey diminishes in size as the prey cytoplasm is passed down the arms to the body of the *Dendrocometes*. Usually *Dendrocometes* are situated around the periphery of the gill, but where there are large numbers of *Dendrocometes* present, they may cover a large percentage of the gill surface (Fig. 5). On gills which bear large numbers of *Dendrocometes*, pairs of animals can be observed linked together by a conjugation process across which nuclei are exchanged (Fig. 6). The conjugation processes are usually found towards the bases of the bodies, although it cannot be seen if there is any attachment of the conjugation process to the gill surface.

Epistylis sp.

The colonial peritrich *Epistylis* is also found on a high percentage of *Gammarus* gills. Sometimes single specimens are found, but more often they are in pairs or groups of up to 10 individuals. The stalks of *Epistylis* are short and the body of the animal is held close to the gill (Figs. 7–9). Frequently threads are seen attached to the stalk (Figs. 7 and 8). These threads have been shown to be algal stands (Dearden, unpublished transmission EM observations). *Epistylis* vary in their position on the gill and may be situated on the periphery of the gill (Fig. 7), usually towards the stalk, or they may be found on the gill faces (Figs. 8 and 9). In animals fixed with the ciliary crown protruded, the spiral nature of the crown can be clearly seen (Fig. 9). The body surface is covered with circular ridges that point towards the ciliary crown (Fig. 9). A clearly marked groove can be seen below the ciliary crown in animals where the crown is protruded (Fig. 9). In animals where the crown is withdrawn, this groove is not seen (Fig. 8).

Lagenophrys sp.

Lagenophrys is a loricate peritrich and the lorica is usually hemispherical and attached to the gill surface. The opening of the lorica, the loricastome, is situated at one end of the lorica. The lips of the loricastome are smooth and in some preparations the ciliary crown can be seen protruding through the loricastome (Fig. 10). *Lagenophrys* are usually found on the gill faces and never around the periphery of the gill. Where there are *Lagenophrys* on any area of the gill surface, *Dendrocometes* are never found on that area of the gill face. In one *Gammarus* that was about to moult a group of empty loricas were observed. This demonstrates the orientation of the loricas. Where there are several *Lagenophrys* on the same gill they are always orientated with the loricastomes facing (Fig. 11) in the same direction.

Whole Gammarus preparations

Preparation of whole *Gammarus* for SEM allows the visualisation of the epizooites *in situ* on the gills (Fig. 12). These preparations also allowed the examination of other parts of the *Gammarus* body surface. On the limbs of *Gammarus, Lagenophrys* were sometimes found. These *Lagenophrys* show the same orientation as those on the gills, i.e. all *Lagenophrys* on any particular structure are all orientated so that their loricastomes point in the same direction. However, SEM shows very clearly that the loricastome differs in the *Lagenophrys* found on the limbs from the loricastome of the *Lagenophrys* on the gills. In the *Lagenophrys* on the limbs, the lips of the loricastome bear very regular crenulations.

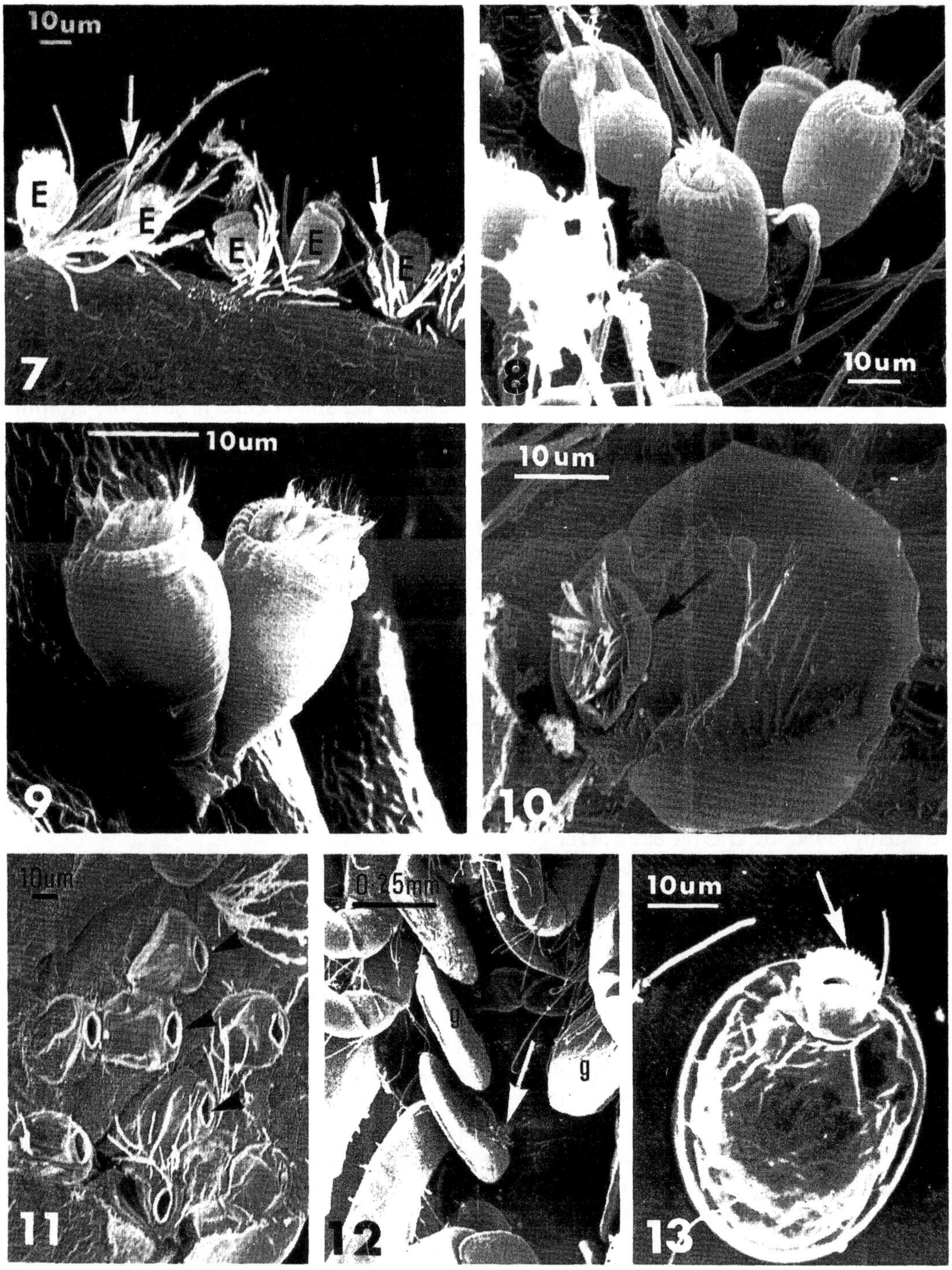

Fig. 7. Epistylis (E) situated on the edge of a *Gammarus* gill. Note the algal threads (arrowed).

Fig. 8. A cluster of *Epistylis* on the gill face with their cilia in varying stages of protrusion.

Fig. 9. A pair of *Epistylis* showing the spiral arrangement of the ciliary crown.

Fig. 10. Lagenophrys ampulla on the gill face of *Gammarus.* Note the smooth edge to the loricastome (arrowed) through which the cilia are protruded.

Fig. 11. A group of empty *L. ampulla* loricae on the gill face of a *Gammarus* that is about to moult. All the loricastomes (arrowed) are facing the same direction.

Fig. 12. Part of the ventral surface of *Gammarus* showing the gills (g) in situ. A *Dendrocometes* (arrowed) is seen on one of the gills.

Fig. 13. Lagenophrys nassa attached to the cuticle of a *Gammarus* limb. Note the regular crenulations on the loricastome (arrowed).

Discussion

SEM provides a useful method for the examination of the surface features of the epizooites found on the gills of *Gammarus pulex*. Instantaneous fixation followed by critical point drying enables the cilia to be displayed while alternative methods of fixation preclude the observation of the cilia. The observations presented here concerning the tentacules and feeding in *Dendrocometes* complement the transmission observations of Bardele (1972). At the tips of the extended tentacules there are haptocysts by which prey are caught. The membranes of the prey and predator fuse and the tip of the tentacule penetrates into the prey while the prey cytoplasm is passed down the arms into the body of the *Dendrocometes*. In the non-feeding tentacules, the tip of the tentacule is withdrawn, rather like the finger of a glove, giving the stumpy appearance. The withdrawl of the tip also causes the withdrawal of the haptocysts and the *Dendrocometes* are therefore unable to catch prey.

Observations of the sexual conjugation process in *Dendrocometes* have been made at the light microscope level by Hickson (1902). He describes the simultaneous protrusion of short processes from two *Dendrocometes* sitting adjacent to one another on the gill of a *Gammarus*, their eventual fusion and the passage of the nuclei across this process. It was suggested by Butschli (1877) that the conjugation processes are modified arms. This suggestion was disputed by Hickson (1902) and from the SEM observations presented here the conjugation process does not look like modified arms, although ultrastructural studies are required to confirm this observation.

In *Epistylis* SEM shows very clearly the spiral organization of the ciliary crown. This feature is not always readily visible in animals studied under phase contrast, especially animals which are situated on the face of the gill surface and not on the periphery. In particular SEM has been very useful in distinguishing between the two different species of *Lagenophrys* found on *G. pulex*. *Lagenophrys ampulla*, which is found on the gills has a smooth loricastome, whereas *L. nassa* which is found on the limbs has regular crenulations around the edge of the loricastome (Penard 1922; Plate 1886). These features are difficult to distinguish by phase contrast microscopy.

Although *Spirochona gemmipara* was named in the introduction as one of the four epizooites found on the gills of *G. pulex* taken from Rutland Water, to date no material has been available for SEM preparation. Of all the four epizooites, *Spirochona* is the most rare in its occurrence. When it is present it is always situated at the distal edge of the gills and always located on the periphery of the gill. The other epizooites described here are usually common but from a monthly sampling programme, in which 30 *Gammarus* were examined each month, Borowski (personal communication) has shown that on average *Spirochona* was only present on one *Gammarus* in each sample and usually there was only a single individual although very occasionally there may be up to eight *Spirochona* on a single gill.

The spatial distribution of the epizooites on the gills is fairly constant for the different species. *Spirochona* is always found at the tips of the gills and always on the gills' periphery. *Epistylis* and *Dendrocometes* are found either around the gill periphery or on the gill faces, and *Lagenophrys* is always found on the gill face. Sleigh (1973) suggests that this distribution may be related to the water currents set up by *Gammarus* which provide an abundant food supply for the protozoa, and may carry different food types to the different parts of the gill normally occupied by the different species.

The presence of large numbers of ciliates on the cuticle of the gills of *Gammarus* invites the question of what happens to these animals when *Gammarus* moults? Is there a relationship between the moult cycle of *Gammarus* and the behaviour of the epizooites? Chatton & Lwoff (1935) have suggested that among parasitic ciliates, sexual activities and ecdysis of the host frequently determine when the parasites divide and encyst. *Lagenophrys* appear to sense the onset of cuticular moulting by the arthropod host to which they are attached, as indicated by the empty loricae shown here (Fig. 11) and Corliss (1979). These ciliates free themselves from their loricae and then resettle on the appropriate surfaces of the freshly emerging host organisms and form new loricae. In the case of *Dendrocometes* and *Spirochona*, escape from the moulting host requires the formation of a ciliated motile swarmer which can be freed from the host and settle on an appropriate surface before metamorphosing into the normal sessile individual.

The occurrence of epizooites on *Gammarus* gills can be very sporadic. Some *Gammarus* may carry large numbers of epizooites, whereas others may carry very few or none. This may be due to a short range in the motile stages which reassociate with the host from which they have detached. The spread of the epizooites may be reliant on contact between *Gammarus* at copulation. Willis (1942, 1948) indicates that the swarmers of *L. tattersalli* are transferred from the female host, *G. marinus,* to the embryos in the brood pouch. It is hoped that further studies will help to elucidate the mechanism of transfer of these epizooites found on the gills of *Gammarus* and the relationship between the behaviour of the epizooites and the moulting cycle of the host.

Acknowledgements

We would like to thank George McTurk, Leicester University SEM Unit, and Stefan Hyman, for technical assistance.

References

Bardele, C. F., 1972. A microtubule model for ingestion and transport in the suctorian tentacle. Z. Zellforsch. 126: 116–134.

Bütschli, O., 1877. Ueber den Dendrocometes paradoxus. Z. Zool. XXVIII: 49.

Chatton, E. & Lwoff, A., 1935. Les Ciliés apostomes. I. Aperçu historique et général: étude monographique des genres et des espèces. Arch. zool. exp. gén. 77: 1–453.

Corliss, J. O., 1979. The Ciliated Protozoa, 2nd edn. Pergamon Press, Oxford.

Hickson, S. J., 1902. Dendrocometes paradoxus. Part I. Conjugation. Q. J. microsc. Sci. 45: 325–362.

Penard, E., 1922. Etudes sur les Infusoires d'Eau Douce. Georg et Cie., Geneva.

Plate, L, 1886. Untersuchungen einiger an den Kiemenblättern des Gammarus pulex lebenden Ectoparasiten. Z. Wiss. Zool. 43: 175–241.

Sleigh, M., 1973. The Biology of Protozoa. Arnold, London.

Willis, A. G., 1942. Studies on L. tattersalli sp. nov. (Ciliata, Peritricha, Vorticellinae). I. Structure, asexual reproduction and metamorphosis. Q. J. microsc. Sci. 83: 171–196.

Willis, A. G., 1948. Studies on L. tattersalli sp. nov. (Ciliata, Peritricha, Vorticellinae). II. Observations on the bionomics, conjugation and apparent endomixis. Q. J. microsc. Sci. 89: 385–400.

The control of a fly nuisance problem in a woodland area adjacent to Rutland Water

M. J. Knight & R. S. Oldham
School of Life Sciences, Leicester Polytechnic, Leicester, England

Keywords: fly control, *Hydrotaea irritans*, amenity, Rutland Water

Abstract

The proposed development of a caravan park in Burley Wood was postponed in 1979 due to the presence of *Hydrotaea irritans* (Fall.). This species causes intense irritation to man by feeding upon skin secretions and is also of veterinary importance. The fly population was monitored for 10 weeks using a number of trapping methods. The efficiency of one of these, the malaise trap, was found to be greatly increased by the controlled release of carbon dioxide. Used together with habitat modification and locally applied insecticides and repellants this is considered potentially valuable in ameliorating the problem. It is likely that the development of Rutland Water will increase the local range of the fly.

Introduction

Burley Wood, situated close to the northern arm of Rutland Water, is an area (160 ha) of mixed deciduous and coniferous forest. During the period June to September the area is subject to a high population density of *Hydrotaea irritans* (Fall. Diptera Muscidae). *H. irritans* causes intense irritation by swarming around people. In 1979 the development of a caravan park within the woodland was postponed due to the fly. This short research project was conducted during the summer of that year; its aims were to investigate the nature of the problem and to suggest control measures.

Biology of the fly

H. irritans is an olive green species similar in size and appearance to the common housefly. The female lays its eggs in small groups during August, the favoured site for oviposition being soil beneath pasture or long grass (Robinson & Luff 1976). The second instar larva emerges from the egg and feeds saprophagously. The third instar, however, is an active predator preying upon the larvae of other Diptera (Tarry & Kirkwood 1976). Pupation occurs by mid-May and in early June the first adults usually appear. The male fly is short-lived, present only until mid-August, and is not troublesome. After mating the female fly requires a protein meal for ovarian development to proceed (Kirkwood 1976). This requirement accounts for the close association of the fly with man and his livestock. Blood and sweat have been shown to be suitable sources of protein (Kirkwood *op. cit.*). *H. irritans* is of veterinary significance (Nielsen *et al.* 1971; Tarry & Kirkwood 1974; Hunter 1975) for inducing self inflicted wounds in sheep and as a vector of disease amongst cattle.

Monitoring scheme

The level of fly activity was monitored for the duration of the project using a hand net. A standardized sweeping procedure occupying 3 min was adopted. Several other methods were tested but

Hydrobiologia 88, 177-178 (1982). 0018-8158/82/0882-0177/$00.40.

were unsuccessful, with the exception of the malaise trap. The controlled release of carbon dioxide was found to greatly increase the number of *Hydrotaea* caught by the trap due to the involvement of carbon dioxide in the host-seeking activity of *Hydrotaea*.

Control measures

Control measures must be compatible with the intended amenity use of Burley Wood. Three approaches are considered feasible: trapping, habitat management and local use of repellants and insecticides.

The carbon dioxide baited malaise traps were so successful that they are considered potentially useful in controlling the population.

Habitat management should be approached in two ways. Firstly, the removal of understory and herbaceous vegetation would increase air flow through the site and remove the resting places used by *H. irritans*. The caravan site is advantageously placed in this respect, in relation to the prevailing wind. Secondly, the replacement of grass on the rides with gravel or tarmacadam would remove a potential breeding site.

The use of contact insecticides on the trees within the caravan site and on the caravans themselves may be advantageous. Insect repellants such as 'Crotoxyphos' (Robert Young & Co. Ltd., Glasgow) have been shown to have a lasting repellant effect upon *Hydrotaea* (Berlyn 1978).

Although clearly desirable, attempts to treat the larval habitat with insecticide would be very expensive due to the low density of the larvae in pasture (1.93 larvae m^{-2}; Robinson & Luff 1976).

It is apparent that *H. irritans* is associated with areas which contain both open woodland and moist pasture. The development of Rutland Water has increased the area of moist pasture and as a result of associated tree planting schemes there will eventually be an increase in woodland area. Thus, with the areas of suitable habitat increasing, it is likely that during the next few years the fly nuisance problem will extend to other amenity sites around the reservoir.

Acknowledgements

We should like to thank Burley Estate Limited for financial support. Thanks are also due the staff of Leicestershire Museum for help in identifying specimens, and to A. D. Berlyn for advice.

References

Berlyn, A. D., 1978. Factors affecting the sheep headfly Hydrotaea irritans (Fallen) (Diptera: Muscidae), with a note on the evaluation of repellants. Bull. Ent. Res. 68: 583–588.

Hunter, A. R., 1975. Sheep headfly disease in Britain. Vet. Rec. 97: 95–96.

Kirkwood, A. C., 1976. Ovarian and larval development of the sheep headfly. Hydrotaea irritans (Fallen) (Diptera, Muscidae). Bull. Ent. Res. 66: 757–763.

Nielsen, B. O., Nielsen, B. M. & Christensen, O., 1971. Bidray Eil plantagcfluens, Hydrotaca irritans Fall., Biologi (Diptera, Muscidae). Ent. Meddr. 39: 30–44.

Robinson, J. & Luff, M. L., 1976. The sheep headfly, Hydrotaea irritans (Fall.) (Diptera, Muscidae): larval habitat and immature stages. Bull. Ent. Res. 65: 579–586.

Tarry, D. W. & Kirkwood, A. C., 1974. Hydrotaea irritans: the sheep headfly in Britain. Br. Vet. J. 130: 180–188.

Tarry, D. W. & Kirkwood, A. C., 1976. Biology and development of the sheep headfly Hydrotaea irritans (Fall.) (Diptera: Muscidae). Bull. Ent. Res. 65: 587–594.

Establishing and maintaining the trout fishery at Rutland Water

D. E. Moore
Anglian Water Authority, North Street, Oundle, Peterborough, England

Keywords: reservoir, trout, angling, impoundment

Abstract

Rutland Water was stocked with trout in 1975 after removing the existing fish populations. Growth and survival of rainbow and brown trout are compared with the reservoir's filling regime, and re-colonisation by other fish species is described. The use of angler success as an indicator of fish stock density and its management is discussed.

Introduction

In 1972 the Welland & Nene River Authority resolved that the then-named 'Empingham Reservoir' should be developed as a trout fishery. This resolution was influenced by the success of other lowland reservoirs in the East Midlands, notably Draycote, Eyebrook, Pitsford and most recently, Grafham Water. This would create the largest fishery in Europe to be entirely dependent on farm-reared trout.

Having considered results at other reservoirs the following strategy was adopted:

1) Removal of existing fish populations.
2) Construction of trout hatchery to produce 400 000 fry.
3) Initial stocking with 400 000 brown trout (*Salmo trutta* L.) rainbow trout (*Salmo gairdneri* R.) two years before angling commenced.
4) Construction of a rearing farm for an annual stocking of 80 000 yearling trout to the reservoir fishery.

Removal of existing fish populations

The reservoir's perimeter included ponds, ditches, streams and two main tributaries of the River Gwash. Most of these contained fish and the success of introducing trout fry required a minimum of predation and competition during their early growth phase.

Re-colonisation by these fish was anticipated but the cost of preventing their transfer from the rivers Nene and Welland by pumping was prohibitive. Fish removal was instigated with the objective of reducing the population to a minimum. The contributory streams were electrofished during 1974 and the Burley Fish Ponds located in the north-west extremity were drained, allowing fish to be released elsewhere. Final removal was by poisoning the streams with 5% rotenone solution to eliminate those fish which had avoided capture.

Nineteen species of fish were recorded and are listed in Table 1.

Electrofishing was also used to reduce the pike (*Esox lucius* L.) population in the vicinity of the reservoir's river intakes where screens allowed ova and fish of less than 50 mm to pass.

Numerous ditches which contained sticklebacks (*Gasterosteus aculeatus* L.) were not poisoned as they were too numerous and the return of the species was inevitable. Sticklebacks have been an important trout food in newly flooded reservoirs (Fleming-Jones 1974).

Hydrobiologia 88, 179-189 (1982). 0018-8158/82/0882-0179/$02.20.

Table 1. Fish species recorded in the site for Rutland Water prior to impoundment in 1974 and year of subsequent recurrence.

	Species	Reservoir	Tributaries
Bleak	*Alburnus alburnus* L.	–	–
Brown trout	*Salmo trutta* L.	1975	1975
Bullhead	*Cottus gobio* L.	–	1975
Chub	*Leuciscus cephalus* L.	1977	1977
Common bream	*Abramis brama* L.	1979	–
Common carp	*Cyprinus carpio* L.	1977	1975
Dace	*Leuciscus leuciscus* L.	1980	1977
Eel	*Anguilla anguilla* L.	1978	1975
Grayling	*Thymallus thymallus* L.	–	–
Gudgeon	*Gobio gobio* L.	1976	1975
Minnow	*Phoxinus phoxinus* L.	1975	1975
Perch	*Perca fluviatilis* L.	1979	1979
Pike	*Esox lucius* L.	1977	1976
Roach	*Rutilus ritulus* L.	1977	1976
Rudd	*Scardinius erythrophthalmus* L.	1978	–
Silver bream	*Blicca bjoerkna* L.	–	–
Stickleback	*Gasterosteus aculeatus* L.	1975	1975
Stone loach	*Neomacheilus barbatulus* L.	–	1975
Tench	*Tinca tinca* L.	1976	–

Initial introductions of trout

Releasing juvenile fish into newly impounded reservoirs and allowing them to grow before permitting fishing has been accepted practice in fisheries management. The advantages provided by abundant food and a lack of predators are well documented (Campbell 1957; Fleming-Jones 1974; Frost 1956; Holloway 1947).

Filling the reservoir was planned to take several years and since pumping would not begin until the 1975/76 winter, only a relatively small volume of water would be available for the first year following impoundment in December 1974. With fishing scheduled to begin in May 1977 it was important to release trout at the first opportunity to benefit from two summers growth in the reservoir. Timing the first release was critical, however, to avoid the possibility of mortalities arising from high water temperatures and de-oxygenation in shallow water containing decomposing terrestrial vegetation. An incident of this nature eliminated the first trout introduced at Eyebrook Reservoir in 1941 (Oliver 1968).

In March 1975 approximately 120 ha of water had accumulated from natural impoundment and the first trout were released. These fish and subsequent stocks are shown in Table 2.

Two species of trout were introduced. Both are commonly stocked to trout fisheries in the southern U.K., although rainbow trout are forming an increasing proportion. Brown trout have the advantage of longevity and resistance to eyefluke infection whereas rainbow trout grow more rapidly, cost less to produce and are considered easier to catch by anglers. Both species feed on a similar invertebrate diet (Frost 1974) and larger fish are often piscivorous in U.K. reservoirs. It is a common belief that brown trout generally feed at greater depths than rainbows but both species take aerial insects and bottom fauna. Little work has been published on the feeding of brown and rainbow trout in the same environment but that of Hunt & Jones (1972) and Hunt & O'Hara (1973) support these statements.

The first trout introduced were 1+ brown trout ranging between 5 g and 45 g but the remaining fish released in 1975 were all 0+ fish weighing less than 20 g. 256000 rainbow trout and 193000 brown trout were stocked to the reservoir in 1975.

Trout released in 1976 were mostly those which could not be held in the low river flows experienced at the hatchery that year and were additional to the planned introduction of 400000 trout.

By the summer of 1976, 272000 brown and 360000 rainbow trout weighing 3.4 tonnes had been released into the reservoir.

Table 2. Early trout introductions to Rutland Water.

Date	Species	Number	Age	Total weight (kg)	Mean weight (g)
1975					
March	Brown	54000	1+	983	18.2
May	Brown	24000	0+	24	1.0
May	Rainbow	36600	0+	73	2.0
June	Rainbow	170200	0+	426	2.5
June	Brown	40000	0+	12	0.3
June	Brown	8500	0+	17	2.0
August	Rainbow	50100	0+	777	15.5
September / October	Brown	66200	0+	629	9.5
1976					
March8	Brown	7740	1+	310	40
April	Brown	14000	0+	7	0.5
May	Brown	58300	0+	49	0.8
May	Rainbow	102700	0+	131	1.3
Total	Browns	272740		2031	
	Rainbows	359600		1407	
		631340		3438	

The performance of trout introduced prior to 1977

Growth

Both growth and feeding of trout were monitored by regular samples obtained by seine netting commencing two months after their introduction and at approximately three-monthly intervals thereafter. Growth in weight is shown in Fig. 1, data after May 1977 being based on anglers catch returns.

Initial growth was rapid with fish feeding on zooplankton in the summer followed by chironomids, corixids, *Limnaea pereger* and *Gammarus pulex* during the autumn (Harper 1982).

In spring 1976, growth accelerated with increasing water temperatures but the minimal increase in water volume following a relatively dry winter did not release the large quantities of terrestrial food which had been anticipated. Growth from June 1976 to September was poor at a time when optimum growth rates should have been achieved. er temperatures in the body of the 200 ha reservoir were Watbetween 18 °C and 24 °C for 60 consecutive days between June and August; temperatures near the surface and in the shallows must have been considerably higher.

Brown trout growth ceases at 18 °C, even given an unrestricted diet (Elliott 1970), and rainbow trout have only a slightly greater tolerance of high temperatures. Adequate food organisms were available at that time and a small sample of trout were all found to have been feeding (Harper *op. cit.*).

From September onwards the reservoir's area increased continuously and an additional 600 ha was inundated within six months. Trout growth during this period of relatively low water temperatures was rapid and most fish examined were feeding almost exclusively on earthworms. Individual brown trout weighed 1.5 kg by November.

Some loss of weight was apparent by February and many mature female trout had shed their ova. In April 1977, the mean weight of rainbows was 644 g (±48) and for browns 720 g (±107).

These overall growth rates are good but not exceptional compared with trout in U.K. waters. Poor growth during the 1976 summer undoubtedly prevented fish attaining the 3 kg recorded on the opening day's angling at Grafham Water in 1966 (Fleming-Jones 1974). Growth during 1977 was monitored from fish caught by angling and both rainbow and brown trout from the original stocks

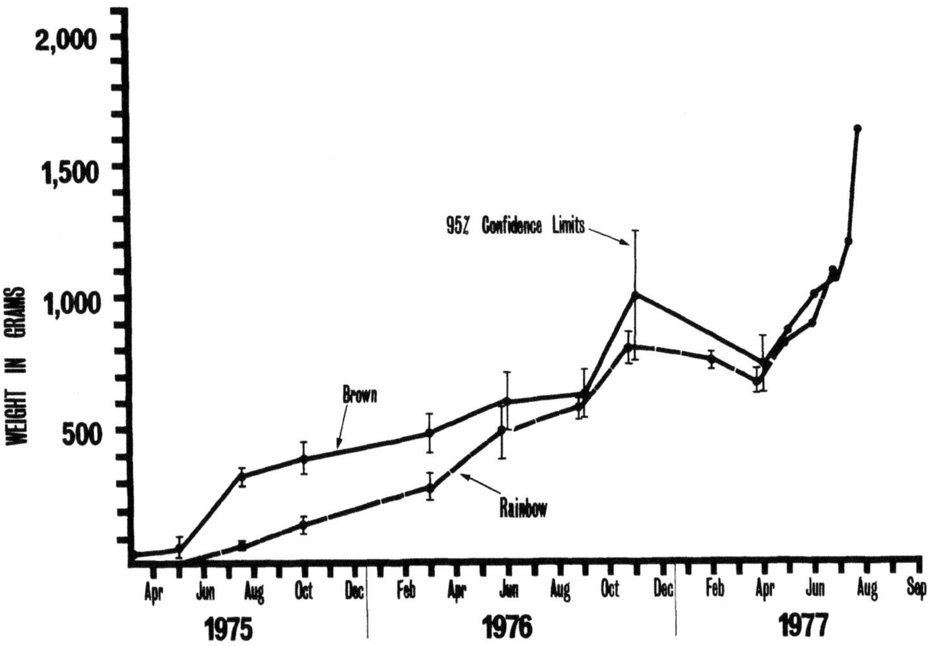

Fig. 1. Rutland Water trout fishery - growth in weight of trout stocked in 1975.

weighing approximately 4 kg were caught by 1979 and 1980 respectively.

Survival of brown and rainbow trout

Estimation of survival rates from trout introduced in 1975 and 1976 has only been made using the results of angling. Catches by seine netting were inconsistent and no conclusions could be drawn from catch/unit effort data.

On the first day of angling it was apparent that brown trout dominated anglers' catches; 54% of trout caught in the first eight weeks were brown. Samples taken by seine netting during the previous year contained only 36% brown trout and angling is generally considered to select rainbows rather than browns in lowland reservoirs (Crisp & Mann 1977).

Analysis of the monthly weight distribution of trout caught in 1977 allowed minimum estimates of survival rates from earlier trout introductions to be made. No brown trout were released in 1977 and although rainbows were released after the ninth week of the season, they weighed less than 400 g and were easily distinguished from the larger original stocks.

Rainbow and brown trout released in 1976 and browns released after March 1975 all contributed less than 1% of the 1977 angling catch and few were caught by test-nettings after this release.

The number of brown and rainbow trout available at the commencement of the 1977 angling season was estimated by fitting a regression of catch per angler against cumulative catch (Fig. 2). The predicted cumulative catch when catch per angler reaches zero was taken as the minimum estimate of fish available at the season's beginning in May.

By this method it was estimated that 31% of those brown trout introduced in May 1975 and 5.4% of rainbows introduced between May and August 1975 had survived as potentially catchable fish to May 1977.

As stated previously, the recovery of rainbow trout was disappointing but the survival of brown trout released in March 1975 was better then expected and compares favourably with releases of larger trout in established fisheries. Templeton (1971) recorded less than a 13% catch by anglers of yearling fish after one winter and Frost & Brown (1967) concluded that brown trout released in Wise Een Tarn were insignificant in catches more than one

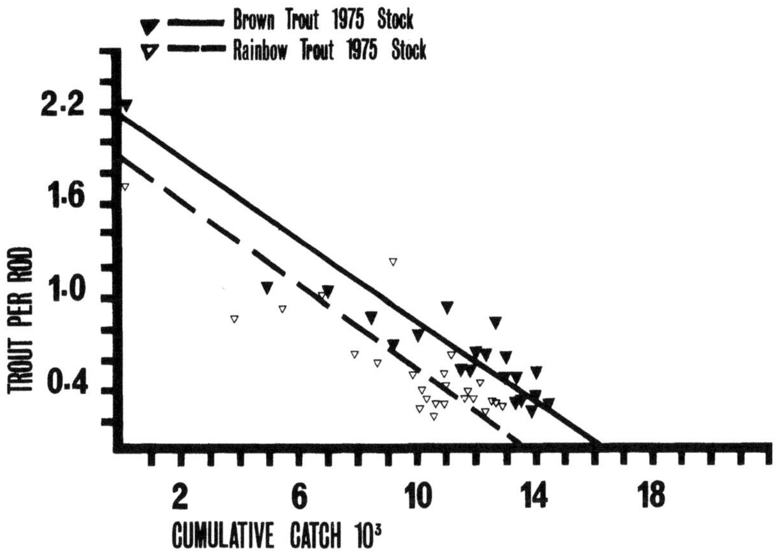

Fig. 2. Rutland Water trout fishery 1977 – trout per rod against cumulative catch.

year after their introduction.

Anglers catches of brown trout are less than 50% of the number stocked each year in most lowland reservoirs (Crisp & Mann 1977) and subsequent seasons at Rutland Water have produced similar results.

Post spawning mortality

Pronounced spawning migrations of both species were observed in the 1976/1977 winter. Rainbow trout migrated in January and February and estimates by mark-release-recapture near inflowing streams indicated several thousand fish to be concentrated in small areas and an adequate population within the reservoir. Spent fish were found in February but no corpses were observed.

The poor overwinter recovery of rainbow trout in U.K. waters is usually believed to be associated with failure to release ova but Hunt & O'Hara (1973) suggested that rainbows adopted a bottom feeding behaviour following autumn stocking and were therefore more difficult to catch. Cragg-Hine (1975) recorded low angling catch of autumn stocked immature rainbow trout in Northern Ireland lakes and concluded poor survival was related to the presence of other trout although there was no evidence of cannibalism.

Eyefluke infections

Diplostomum spathaceum and *Tylodelphys clavatum* were both common in cyprinid fish and *Lymnaea pereger* prior to impoundment of the river Gwash.

Low numbers of flukes were first found in rainbow trout during June 1976 and heavily infected specimens were recorded in September samples. Rainbows examined during 1977 showed lower infections but still 50 to 60 per lens in some larger fish (Moody & Gaten 1982). Although metacercaria counts of both species decrease during the winter (Burrough 1978), rainbow trout are known to die following large infections (Betterton 1974) and it is possible that the lower counts recorded in rainbow trout lenses in 1977 were at least partly caused by mortalities of heavily infected fish. Possibly of more importance, trout whose sight is impaired are less likely to be caught by angling. Brown trout are relatively resistant to these parasites (Betterton 1974) and only small numbers have been recorded in these fish in Rutland Water.

These two parasites are believed to have been the main cause of a relatively low rainbow catch by anglers in 1977 and a similar problem occurred following the impoundment of Grafham Water and Bewl Bridge reservoir.

Angler success and trout stocking

All anglers fishing the reservoir are required to complete a catch return incorporated into their permits. No register of anglers is maintained and so an estimate of visits from season permit holders is required to assess the total 'rod visits' on the water. Season permit holders' fishing frequency varies from almost zero to over 100 visits during the 27 week season but it is assumed that the average is equal to the relative cost of purchasing day and season permits, 1:35 in 1980.

Table 3 shows the number of returns received in each of the first four angling seasons at Rutland Water and it can be seen that these returns account for between 40% and 60% of the annual stocking. Information is analysed weekly throughout the season as an indicator of angler success. This and the effects of restocking during the 1977 season are shown in Fig. 3.

Catches were high in the first week, followed by a steady decline for rainbow and brown trout as the fish population was depleted. Restocking with relatively small numbers of rainbow trout in weeks 9 to 13 moderated the decline but an introduction of 10000 rainbows between weeks 13 and 16 induced a marked improvement in anglers' results with that species. No brown trout were released during the season and a gradual decline in catch per angler was evident throughout the season.

Table 3. Number of angler visits and recorded catches compared with trout stocked.

		Fishing season			
		1977	1978	1979	1980
(1)	Number of rod visits (estimated)	30000	42000	47000	43000
(2)	Number of returns	17812	26010	26910	23691
(3)	% (2) Returns / (1) Rod Visits	59	62	57	55
(4)	% Rainbow catch / (2) Returns	–	42[a]	57	55
(5)	% Brown catch / (2) Returns	–	42	40	67[b]

[a] Value probably lowered by heavy autumn stocking.
[b] Value probably raised by high proportion of previous year's stock.

The objective at Rutland Water is to maintain an average weekly catch of 2.5 trout per angler return throughout the season since four seasons experience has shown this meets the criteria of (1) satisfying the anglers, (2) an acceptable cost of stock fish and (3) good natural growth ensuring a high average weight of trout caught.

The expression 'trout per return' must be used

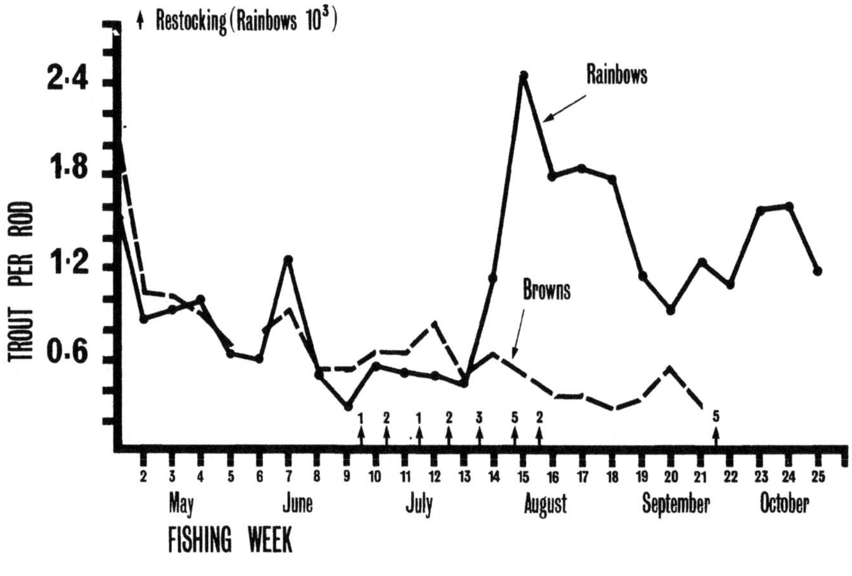

Fig. 3. Rutland Water trout fishery 1977 – weekly angling success.

cautiously. It does not equate to 'trout per anglers' since there is evidence that those anglers making a return are more successful than those anglers who do not. Comparison of the proportion of anglers making a return is similar to the proportion of stock recorded as caught and extrapolation of this to all anglers fishing could suggest that virtually all and sometimes more than the annual stock are caught. These situations are both theoretically possible since natural mortality may be low in such a new environment but nettings after the season ends always indicate that substantial stocks of trout remain uncaught. Succesful anglers are therefore more likely to make a catch return.

At Eyebrook Reservoir all anglers record their catches and 75% of the stocked fish were caught during 1962–1971 (Crisp & Mann 1977). This reservoir receives more fish and is fished by more anglers relative to its size than is Rutland Water.

An investigation into catches by anglers not making returns at Ffestiniog Reservoir (Cane 1980) concluded that returns were accurate but that they over-estimated the true average catch by approximately 35% since anglers with poor or nil catches were less likely to make a return. This problem is critical at Rutland Water since the relationship between stock density and angler success provides the predictive model for restocking requirements. Fishing mortaility and natural mortality of trout stocks must be accurately quantified for this purpose and Fig. 4 demonstrates the effect of assuming natural mortality to be zero during the fishing season and that all unreported anglers catches are zero.

Since the trout population during the season on the residual population available for the following season are both overestimated, there appears to be an anomaly in correlation of anglers' success with stock density.

The initial population size is given by the number of trout introduced immediately prior to the season with an addition for overwintered fish from the previous season. Overwintering has been evaluated by tagging fish released in November 1977 and recording those caught in the following two seasons. Results from this indicated 25% of rainbow trout to be available for capture in the spring following their release. This is high compared with other published results (Cragg-Hine 1975).

Preliminary assessment of the combined effects of natural mortality during the season and non-reported fishing mortality has been made by applying various levels of this unknown mortality rate to the correlation of angler success and estimated stock density.

A value of angler success is calculated each week

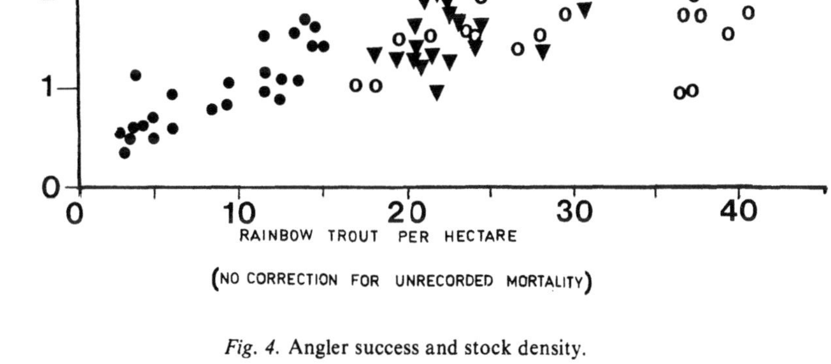

Fig. 4. Angler success and stock density.

of the fishing season using catch returns and the recorded catch is deducted from the trout population to give the population size for the following week. If the recorded catch is increased by a level of unknown mortality rate, lower populations estimates are obtained. Fixing this mortality rate at 60% of the recorded catch gave the best correlation of catch return^{-1} with stock density for rainbow trout in 1979 described by the equation:

$\log_n y = -2.3 + 1.37 \log_n x$

$(r^2 = 0.85)$ (see Fig. 5)

where x = trout ha^{-1} and y = trout return^{-1}.

Several years more data will be required before a satisfactory model for rainbow and brown trout stocking is produced since a variety of climatic and biological factors affect anglers catches. An analysis of possible influences on angler success at Eyebrook Reservoir (Taylor 1978) showed stock density to be the most important factor affecting rainbow trout catches. Water temperature and phytoplankton densities were of low importance to rainbow trout catches but brown trout were more sensitive to these factors.

At Grafham Water, total rainbow trout catches have been closely correlated with annual stocking rates and brown trout to a lesser extent (Fleming-Jones & Stent 1975).

The target of 2.5 trout per return at Rutland Water requires a stock density of 27.5 trout ha^{-1} from the equation calculated for rainbow trout. No relationship has been established for brown trout but there is evidence of lower natural mortality and similar fishing mortality rates for this species in Rutland Water.

Annual stocking requirements are based on the maintenance of 27.5 trout ha^{-1} throughout the season but with higher density in the first two weeks when angling pressure is high. The equivalent standing stock is approximately 30 000 trout over a water surface of 1 100 ha allowing for draw-down and inaccessible areas. The average weight of trout caught in the first four seasons has been 820 g and assuming this to reflect the population it gives a standing stock of 24.6 tonnes of 22.4 kg ha^{-1} at 1 100 ha. Densities in similar environments but with a varied fish population usually range from 200 to 600 kg ha^{-1} (Schaperclaus 1933). Maximum brown trout biomasses in Shelligan Burn, Scotland, were

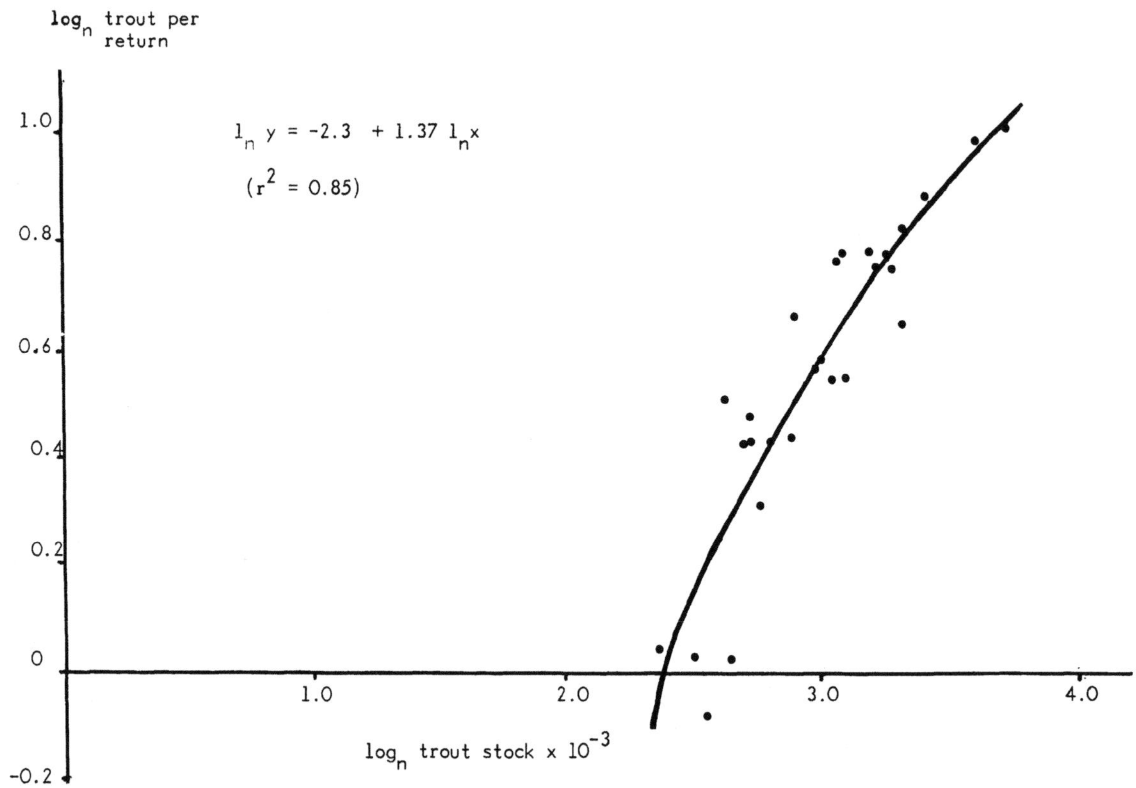

Fig. 5. Angler success and stock density of rainbow trout 1979.

86 kg and 125 kg ha^{-1} (Egglishaw 1970) and in the Horokiwi Stream, New Zealand were between 100 kg and 350 kg ha^{-1} (Allen 1952).

Little comparable data is available for trout in large lakes but Loch Leven, a eutrophic, lowland water of similar size to Rutland Water was estimated to hold between 52000 and 126000 adult brown trout in 1968–1971 (Thorpe 1974). No biomass data is given but the trout were of a similar size to those in Rutland Water and the stocks are equivalent to 39 and 95 trout ha^{-1}.

Re-colonisation by most of the fish species present before impoundment has now been recorded. Pike and roach have spawned successfully since 1978 and perch are also increasing rapidly. Total fish biomass will increase and trout will lose their dominance in the near future but this need not be reflected in slower growth. More likely consequences are increased predation by pike which have previously been too small to feed extensively on the relatively large trout and increased predation on the fry of cyprinid fish by the trout. Most lowland trout fisheries support large coarse fish populations without serious detriment apart from their 'nuisance value' to anglers who catch them.

Conclusions

Over a period of eight years the fishery has been developed and established in line with the original policy for the reservoir. Rutland Water has become a major centre for a variety of recreational pursuits (Langford 1978) with the fishery attracting more anglers and yielding a larger catch than any other U.K. trout water. In its first four seasons 162000 angler visits have caught at least 230000 trout totalling 187 tonnes.

Many problems still exist, among them the factors which limit overwinter survival of rainbow trout and the poor condition of these fish in April, which have serious financial implications. The development of early spawning strains of rainbow trout, 'sex-reversed' females, hybrids between brown and rainbows and the autumn spawning brook trout (*Salvelinus fontinalis*), are all being considered as solutions which would extend the fishing season and the financial viability of the fishery.

Anglers travelled an average of 100 km for each day's sport at Rutland Water in 1980 and this cost is becoming high in relation to that of the fishing permit. Clearly transport costs will have a direct effect on the number of anglers fishing in the future and as demonstrated above, a standing stock of 30000 trout is required for satisfactory angling results irrespective of how few anglers visit the fishery.

There has been a massive increase in the number of trout anglers since the development of large lowland reservoirs made the sport available to all and it is hoped that this trend will continue to take advantage of the high standard of fishing they can provide.

Acknowledgements

The author would like to thank Mr. A. W. Davies, Director of Scientific Services, Anglian Water Authority, for his permission to publish this paper.

Mr. P. R. Langford, Mr. R. E. Field, Dr. A. Ferguson, Mr. P. Noble and Mr. N. F. Low have all provided useful comments on the manuscript. A large project such as this cannot succeed without the help of many people and I would like to thank all of the fisheries staff who have assisted in collecting and collating data.

Miss J. Moody of Leicester University has provided information in respect of the eyefluke study and Dr. D. Harper of Leicester University and Mr. A. Warlow of Leicester Polytechnic in respect of trout feeding.

My thanks also to Miss S. Watson for assisting with the figures and Mrs. D. Thurlby for typing the manuscript.

References

Allen, K. R., 1972. A New Zealand trout stream, some facts and figures. New Zealand Marine Dept. Fish. Bull., 10A: 70 pp.

Betterton, C., 1974. Studies on host specificity of the eyefluke Diplostomum spathaceum in brown and rainbow trout. Parasitology 69: 11–29.

Burrough, R. J., 1978. The population biology of two species of eyefluke Diplostomum spathaceum and Tylodelphys clavatum in roach and rudd. J. Fish Biol. 13: 19–32.

Campbell, R. N., 1957. The effect of flooding on the growth rate of brown trout in Loch Tummel. Freshwat. Salmon Fisheries Research 14, D.A.F.S.

Cane, A., 1980. The use of anglers' returns in the estimation of fishing success. Fish. Mgmt 11: 145–156.

Cragg-Hine, D., 1975. Studies on overwintering mortality of autumn stocked rainbow trout in some lakes in Northern Ireland. Fish. Mgmt 6: 1–7.

Crisp, D. T. & Mann, R. H. K., 1977. A desk study of the performance of trout fisheries in a selection of british reservoirs. Fish. Mgmt 8: 101–109.

Egglishaw, H. J., 1970. Production of salmon and trout in a stream in Scotland. J. Fish Biol. 2: 117–136.

Elliott, J. M., 1976. The energetics of feeding, metabolism and growth of brown trout (Salmo trutta L.) in relation to body weight, water temperature and ration size. J. Anim. Ecol. 45: 923–948.

Fleming-Jones, D., 1974. Development of the trout fishery at Grafham Water. Fish. Mgmt 5: 1–9.

Fleming-Jones, D. & Stent, R. F., 1975. Factors affecting the trout fishery at Grafham Water. Fish. Mgmt 6: 55–63.

Frost, W. E., 1956. The growth of brown trout in Haweswater before and after raising of the lake. Salm. Trout Mag.: 267–275.

Frost, W. E., 1974. A survey of the rainbow trout in Britain and Ireland. Salmon & Trout Association. 36 pp.

Frost, W. E. & Brown, M. E., 1967. The Trout. New Naturalist Special, Vol. 21. Collins, London. 286 pp.

Harper, D. M., 1982. The feeding of trout during the filling phase of Rutland Water. In: Harper, D. M. & Bullock, J. A. (Eds) Rutland Water – Decade of Change, pp. 191–198. Junk, The Hague.

Holloway, R. C., 1947. Lake management studies in Oregon. Trans. Am. Fish. Soc. 77: 105–113.

Hunt, P. C. & Jones, J. W., 1972. The food of brown trout in Llyn Alaw, Anglesey, North Wales. J. Fish Biol. 4: 395–408.

Hunt, P. C. & O'Hara, K., 1973. Overwinter feeding in rainbow trout. J. Fish Biol. 5: 277–280.

Jenkins, R. M., 1979. Fish production in large U.S.A. reservoirs. Proc. 1st British Fresh wtr. Fisheries Conf.: 3–10.

Langford, P. R., 1978. Leisure and sports facilities at Rutland Water. Proc. Public Works Congress 1978 F(1)1-27.

Moody, J. & Gaten, E., 1982. The population dynamics of eyeflukes in rainbow and brown trout in Rutland Water 1974–1978. In: Harper, D. M. & Bullock, J. A. (Eds) Rutland Water – Decade of Change, pp. 207–209. Junk, The Hague.

Oliver, C. C. S., 1968. The development of a trout fishery in a lowland reservoir (Eyebrook). J. Inst. Wat. Engrs 22: 129–136.

Schaperclaus, W., 1933. Textbook of pond culture. U.S. Dept. Int. Fish Leaflet No. 311. 261 pp.

Taylor, A. H., 1978. An analysis of the trout fishing at Eyebrook – a eutrophic reservoir. J. Anim. Ecol. 47: 407–424.

Templeton, R. G., 1971. An investigation of the advantages of autumn and spring stocking with brown trout (Salmo trutta L.) in a Yorkshire river. J. Fish Biol. 3: 303–324.

Thorpe, J. E., 1974. Estimation of brown trout in Loch Leven, Kinross, Scotland. J. Fish Biol. 6: 135–152.

Discussion

Mr. Moore outlined the initial pressure to maintain the reservoir as a coarse fishing area, the decision to stock with trout, and the clearance of coarse fish from the Gwash prior to the reservoir filling. He pointed out that initially the reservoir was stocked with trout of unusually small size ($\leqslant$10 cm) which were then allowed 2 years growth before fishing commenced. In this early stage, the trout relied on chironomid larvae and other small invertebrates for food, being too small to take the large earthworms. As they grew and particularly as the reservoir increased in size in the autumn of 1976, they exploited earthworms vigorously and their growth rate increased markedly.

He drew attention to the low catches of *Salmo gairdneri* in the first season compared to those of *S. trutta* despite an initial 2:1 stocking ratio. He attributed this to heavy infestations of eye-fluke which had since declined, probably as a result of the declining abundance of the intermediate host, *Limnaea pereger*.

He then demonstrated the value of monitoring the stock density and the necessity of having an informed re-stocking policy to maintain a constant level of angler success.

Dr. R. Huggins said that he found the paper most stimulating and that he would appreciate Mr. Moore's opinion on whether the use of predictive mathematical models, no matter how simple, was essential in put-and-take fisheries. He further enquired as to the profit or loss per trout caught. Dr. Huggins noted that seine netting had give variable results and he wondered whether alternative methods of estimating fish densities had been considered. Further, as returns from anglers were unsatisfactory, he suggested that these might be improved by the institution of lotteries. Finally, bearing in mind the initial interest in coarse fishing and the recovery of various coarse fish populations, should Rutland Water be used for the culture of species other than trout.

In replying, Mr. Moore said that the use of

models to predict re-stocking rates was essential in a large water body although in small bodies removal rates might be accurately monitored. He did not consider that more elaborate models would necessarily serve any useful purpose. Regarding economics, the cost of running the fishery, including all salaries, was more or less balanced by income from fishermen.

Mr. Moore explained that seine netting was generally satisfactory except for samples taken just before the first fishing season (1977) when shoaling by gravid *S. gairdneri* caused problems. Nonetheless, alternative methods, such as sonar, would be useful in large water bodies. The problem of inadequate returns from anglers was common to all areas where an anglers' register was not maintained. A lottery was already operated with a prize of a full year's season ticket (value £168). The potential for culturing other fishes existed, although only eels (*Anguilla anguilla*) have been considered seriously. These pose problems by becoming trapped in 75–100 mm outlet pipes.

Mr. Kekwick asked whether there was a relation between eye-flukes, trout and black-headed gulls (*Larus ridibundus*). Was there any possibility that these were transmitting the fluke in winter? He also asked if there was any sign of ulcerative dermal necrosis. Mr. Moore said there was no sign of ulcerative dermal necrosis or other dermal diseases. The initial outbreak of flukes was sparked by the bloom of *Limnaea pereger*, and had clearly been present before the reservoir filled.

Miss J. Moody said that at the peak, *L. pereger* carried very low infestations. Now, with the disappearance of the snail, black-headed gulls would be releasing large quantities of fluke eggs which would be 'wasted' for lack of the intermediate host.

Mr. A. Warlow asked if, in view of the violent oscillations in the populations of *Gammarus pulex*, etc., indicated by Dr. Bullock, the system might not be more fragile than had previously been thought.

Mr. Moore replied that there was evidence that trout were showing an increased diversity in their stomach contents, indicating that they were no longer concentrating on taxa that were abundant as had occurred previously. He speculated that the trout might later turn to roach (*Rutilis rutilis*) fry.

Mr. Guiver asked if there were any plans to run Rutland Water as a mixed fishery. Mr. Moore replied that the heavy growth of weed in the shallows was not suitable for coarse fishing. There was a possibility of sanctioning fishing from the dam while boat fishing had obvious possibilities. In general, he considered that coarse fishing should only be sanctioned to reduce numbers of predatory fish, such as pike (*Esox lucius*) and perch (*Perca fluviatilis*), which were increasing in Rutland Water.

The feeding of trout during the filling phase of Rutland Water

D. M. Harper
Anglian Water Authority, Welland & Nene River Division. Oundle, Peterborough, England[1]
[1] *Present address: Department of Adult Education, University of Leicester, Leicester, England*

Keywords: trout, littoral fauna, impoundment, flooding, reservoir, Rutland Water

Abstract

Fingerling *Salmo trutta* L. and *S. gairdneri* R. were stocked in Rutland Water in early 1975. Initial rapid growth was associated with feeding on littoral and planktonic Crustacea during this year. Rapid water level changes occurred over the winter months, interspersed with longer periods of stability during the summer. These changes caused changes in trout diet. Terrestrial food (e.g. earthworms), released from flooded land was important during winter months; aquatic items which increased in abundance during periods of stable water levels (e.g. *Gammarus pulex, Limnaea pereger,* Chironomidae larvae and pupae) were important during spring, summer and autumn.

It is suggested that the rapid growth rates recorded in the first two years of the reservoir are the result of the ready availability of different food sources.

Introduction

Since the first stocking of fingerling (6–8 cm) brown trout (*Salmo trutta* L.) and rainbow trout (*S. gairdneri* R.) in 1975, seine-net samples have been regularly taken to monitor growth and for stomach analysis (Moore 1982). Seventy trout stomachs from these samples were analysed prior to the opening of the reservoir for sport fishing in early May 1977. After this time, a more extensive study was carried out using stomachs from angler-caught fish (Warlow & Oldham 1982).

This paper compares the changes which have occurred in trout diets between May 1975 and February 1977 in relation to the availability of food sources as the reservoir increased in size, and to the growth of the stocked trout.

Methods

Trout stomachs

Stomachs were removed from seine-net caught trout and preserved in 4% formalin in the field. In the laboratory they were slit longitudinally and the contents of the stomach (but not intestine) removed. Contents were sorted into their major taxonomic groups and counted. The dry weight of each group was then measured after 24 hours drying at 105 °C.

Littoral invertebrates

Littoral and not benthic invertebrates were sampled as the trout were netted from the littoral of the reservoir and, it was assumed, had been feeding there. The reservoir was sampled, usually at five sites and at monthly intervals using a hand net swept through an area of approximately 1 m^2 along the shoreline at a depth of 0.5 m. Whilst this was not strictly quantitative, it was consistent so that results

Hydrobiologia 88, 191–198 (1982). 0018-8158/82/0882-0191/$01.60.

from different months may be compared. All data from each sample date were combined. The samples were sorted in the laboratory, identified to species wherever possible, and counted.

Expression of results

The results of the stomach analyses are expressed as the percentage occurrence of organisms in each sample of stomachs and the percentage dry weight of each organism. This latter is considered to be a more accurate measure of the importance of different food items in the diet. The percentage occurrence of aquatic food items in the stomachs is compared with the percentage occurrence of the same taxa in littoral samples at the same time as a 'Forage Ratio' (Frost & Brown 1967).

Results

The first trout stocked in Rutland Water in early 1975 were around 8 cm in length and under 100 g weight. Although there were subsequent stockings of larger fish, the mean size of the population showed a steady increase as these fish, initially 0+ and 1+, matured (Fig. 1) (Moore 1982).

The increase in the dry weight of stomach contents (Fig. 2) generally followed the increase in wet weight of fish during 1975–1977. Exceptions occurred during the periods of highest and lowest water temperatures, when the trout evidently ceased feeding; there was a maximum water temperature of 23 °C recorded in June 1976 and a minimum of 2 °C during the winter of 1976/77.

Rapid water level changes occurred at certain times, mainly during the winter months, with stable water levels during the summer months. During the periods of rapid changes in water levels and flooding of new land, aquatic invertebrates were scarce in littoral samples; during periods of stability they became more common. Figure 3 shows how the number of taxa taken in littoral hand net samples was related to the water level changes. (Table 1 shows the taxa recorded during the period 1975–1977 in the littoral.)

The fluctuations in numerical abundance of the main groups of aquatic invertebrates during 1975–1977 are shown in Fig. 4. There was a dramatic increase in littoral Crustacea (*Daphnia pulex, Simocephalus vetulus, Eurycercus lamellatus*) and larvae of Chironomidae in 1975 during the first summer after impoundment. In subsequent summers this was repeated on a smaller scale. These

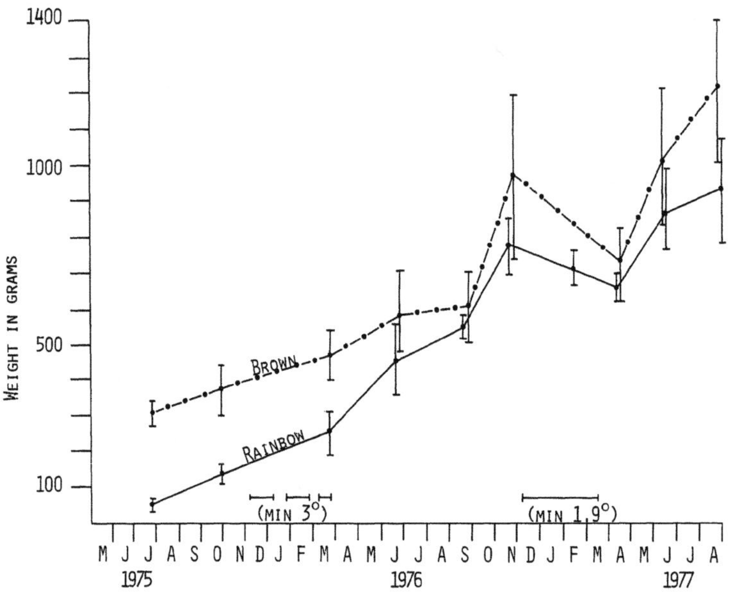

Fig. 1. The increase in weight of trout in Rutland Water 1975–1977. (Horizontal bars show periods of water temperature below 5 °C.)

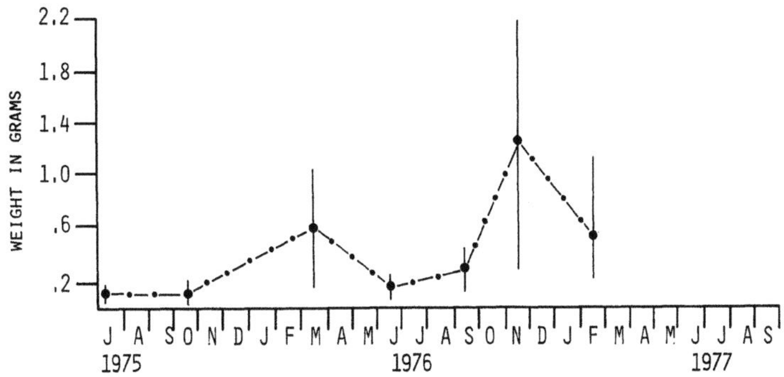

Fig. 2. The dry weight of the stomach contents of trout (both species combined) collected from Rutland Water 1975–1977.

taxa and Corixidae were present all the year round in net samples, although very much more common in summer and autumn than in winter. By contrast, *Gammarus pulex* and *Limnaea pereger* were not taken at all in net samples during periods of rapid water level change but became very much more common after periods of water level stability. The extent to which the trout diets reflected these changes in aquatic invertebrates is shown in Figs. 5 and 6. These only show the important food items; the full list of those found in trout stomachs during the study is shown in Table 2.

Table 1. Invertebrate taxa recorded from the littoral region of Rutland Water between May 1975 and January 1977 in hand net samples.

- Platyhelminthes
  - *Dugesia lugubris* (O. Schmidt)
- Annelida
  - Oligochaeta
    - Lumbricidae
    - Lumbriculidae
    - Tubificidae
    - Naididae
    - Enchytraeidae
  - Hirudinea
    - *Erpobdella octoculata* L.
    - *Helobdella stagnalis* L.
    - *Glossiphonia complanata* L.
    - *G. heteroclita* L.
    - *Piscicola geometra* L.
- Crustacea
  - *Daphnia pulex* (De Geer)
  - *Eurycercus lamellatus* (O. F. Muller)
  - *Chydorus sphaericus* (O. F. Muller)
  - *Simocephalus vetulus* (O. F. Muller)
  - *Cyclops* sp.
  - Ostracoda
  - *Gammarus pulex* L.
  - *Asellus aquaticus* L.
  - *A. meridianus* Racovitza
- Insecta
  - Coleoptera
    - Dytiscidae
  - Corixidae
    - *Sigara dorsalis* Leach
    - *S. falleni* (Fieb.)
    - *S. nigrolineata* (Fieb.)
    - *S. limitata* (Fieb.)
    - *S. lateralis* (Leach)
    - *S. concinna* (Fieb.)
    - *S. selecta* (Fieb.)
    - *Arctocorisa germari* (Fieb.)
    - *Callicorixa praeusta* (Fieb.)
    - *Hesperocorixa castanea* (Thoms.)
    - *Corixa* sp.
  - Diptera
    - Chironomidae
    - Ceratopogonidae
    - Tipulidae
    - *Limnophora* sp.
    - Syrphidae
  - Ephemeroptera
    - *Caenis moesta* Bengtsson
    - *Centroptilum luteolum* (Müller)
    - *Chloeon dipterum* (L.)
  - Trichoptera
    - Leptoceridae
- Arachnida
  - Hydrachnellidae
- Mollusca
  - Gastropoda
    - *Limnaea pereger* (Muller)
    - *Planorbis albus* (Muller)
    - *P. laevis* Alder
    - *P. leucostoma* Miller
    - *P. vortex* (L.)
    - *P. carinatus* Muller
    - *P. planorbis* (L.)
    - *Valvata piscinalis* (Müller)
    - *Potamopyrgus jenkinsi* (Smith)
  - Bivalvia
    - *Pisidium* sp.

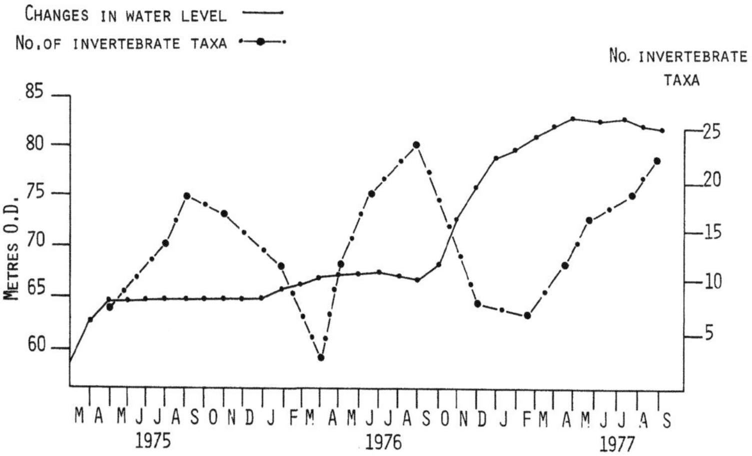

Fig. 3. The numbers of invertebrate taxa recorded in hand net samples shown against the rise in water level of Rutland Water 1975–1977.

The percentage occurrences of the major food items are shown in Fig. 5. Littoral Crustacea were widespread in stomachs, mainly *D. pulex* in 1975 and *E. lamellatus* with Ostracoda in 1976. Corixidae and larvae of Chironomidae occurred consistently, but pupae of Chironomidae were only important during their emergence in the summer months. *Limnaea pereger* and *Gammarus pulex* were important as each was abundant in the reservoir.

In March 1976 (the first sample taken after the winter increase in reservoir volume) and from November 1976, two months after the second period of rapid inundation (Fig. 3), earthworms were prominent in the diets, together with other terrestrial arthropods and plant debris such as grass stalks.

The stomach contents expressed as percentage dry weight are shown in Fig. 6. This emphasises more quantitatively the importance of these different dietary items. Littoral Crustacea became progressively less important as the trout increased in size and the range of potential food organisms increased. Chironomidae became dominant in each summer during emergence. Earthworms were dominant in March 1976 and again in November 1976 and February 1977, together with substantial quantities of plant detritus.

Table 3 shows the percentage abundance of

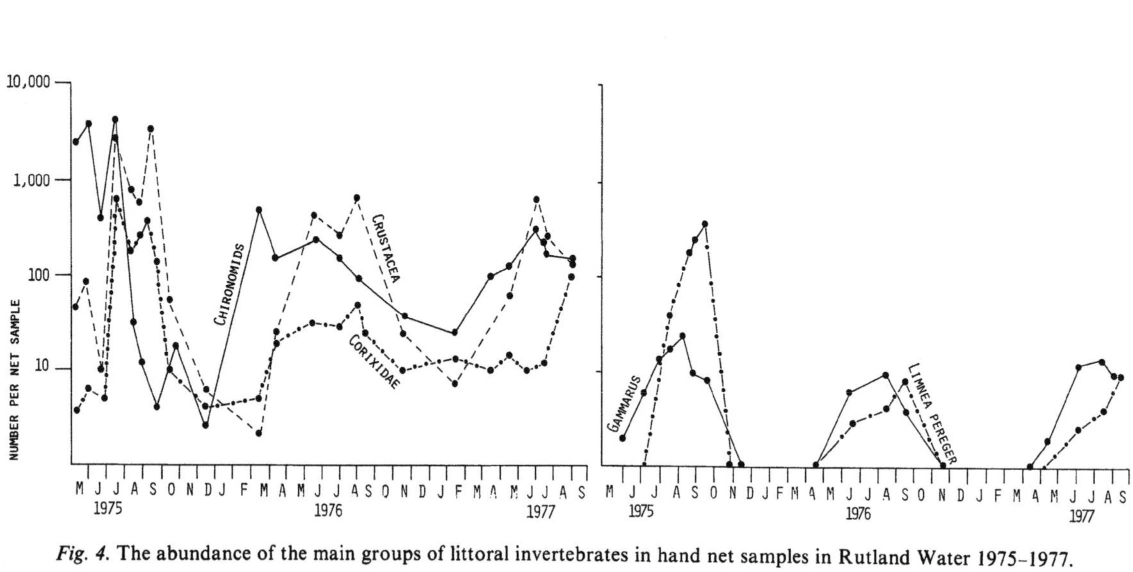

Fig. 4. The abundance of the main groups of littoral invertebrates in hand net samples in Rutland Water 1975–1977.

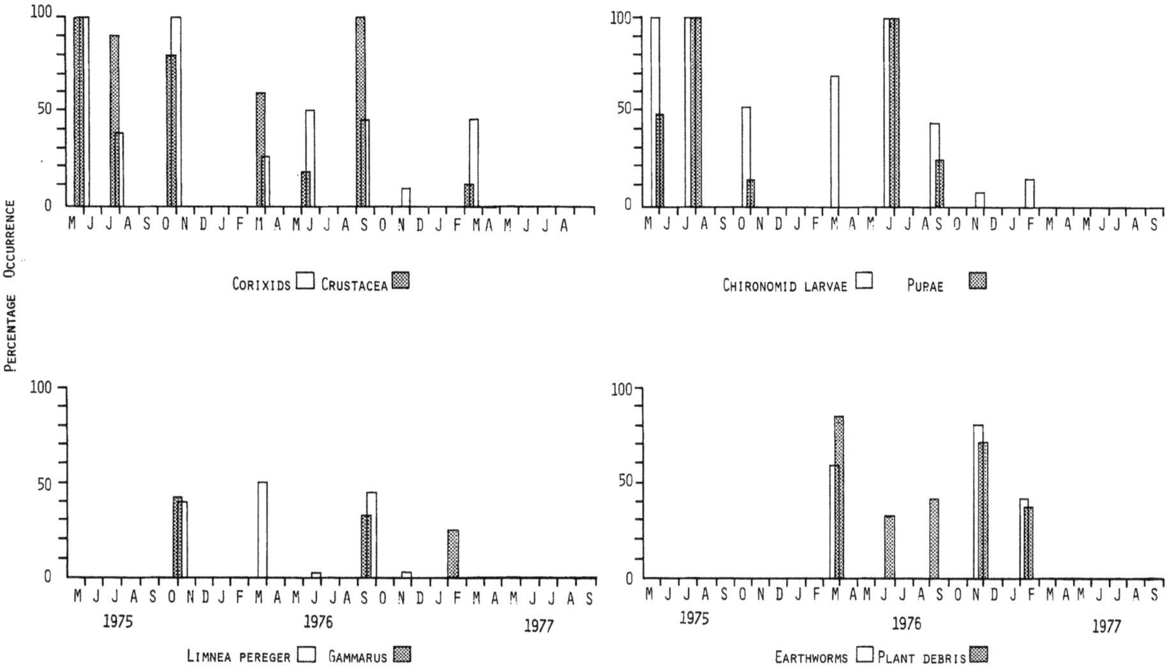

Fig. 5. The percentage occurrence of the major food items in trout diets in Rutland Water 1975–1977.

organisms in trout stomachs compared with the percentage abundance of organisms in littoral samples [the 'Forage Ratio' (Frost & Brown 1967)].

Where this Forage Ratio exceeds 1, it is considered that trout are selecting the particular food organism (Frost & Brown 1967). Table 3 indicates that Crustacea, *Gammarus pulex* and *Limnaea pereger,* were selected; larvae of Chironomidae and Corixidae were not selected; Hirudinea, Oligochaeta and Hydracarina were hardly taken at all.

The maximum percentage dry weight of each organism found in any single trout is shown in

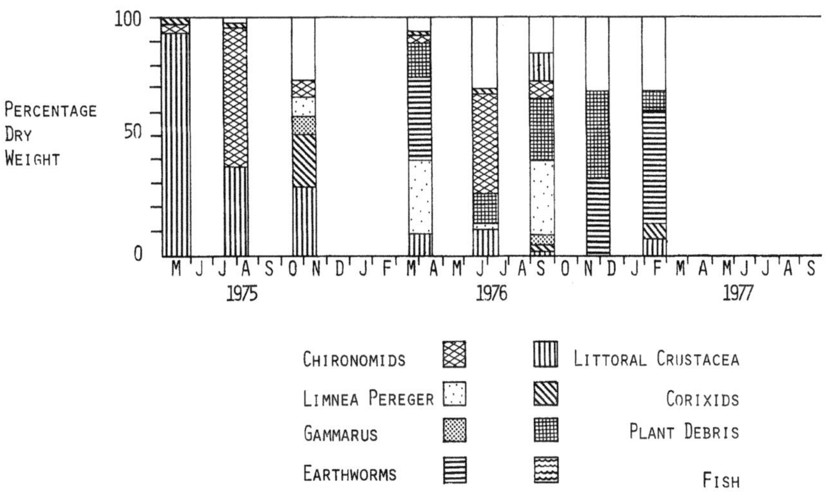

Fig. 6. The percentage importance of food items by weight in Rutland Water 1975–1977. (Uncoded sections represent food not identifiable and items of minor importance.)

Table 2. Items recorded in trout stomachs from Rutland Water between May 1975 and February 1977.

Aquatic:
- *Daphnia* sp.
- *Cyclops* sp.
- *Eurycerus lamellatus*
- *Simocephalus vetulus*
- Ostracoda
- Corixidae
- *Gammarus pulex*
- *Limnaea pereger*
- Chironomidae larvae
- Chironomidae pupae

Aquatic (rare: found in <10% of all samples):
- Dytiscidae
- *Caenis* sp.
- *Planorbis* spp.
- *Asellus aquaticus*
- Zygoptera (nymphs)
- Hirudinea
- Oligochaeta
- Hydracarina

Terrestrial:
- Oligochaeta
- Arthropoda (chiefly Arachnida and Coleoptera)
- Plant debris

Terrestrial (rare: found in <10% of all samples):
- Mollusca
- Tipulidae larvae
- Isopoda
- *Sorex araneus* (one)

Table 4. These data confirm the results of the Forage Ratios in that food organisms with a Forage Ratio >1 were found in quantities of over 80% in individual stomachs. Organisms with Forage Ratios <1 were found to make up a smaller maximum percentage (64% and 59%). Earthworms, plant debris and Chironomidae pupae, for which Forage Ratios could not be calculated, were also selected by individual trout in high quantities.

Table 3. Mean Forage Ratios for aquatic invertebrate food items in Rutland Water.

Littoral Crustacea	1.4
Gastropoda	6.0
G. pulex	1.62
A. aquaticus	0.4
Corixidae	0.37
Chironomidae (larvae)	0.5
Hirudinea	0
Oligochaeta	0
Hydracarina	0

Table 4. Maximum percentage dry weight of food items in single trout stomachs.

Food item	Date of sample	% dry weight	'Stomach fullness' scale 1-5	'Forage Ratio'
Daphnia pulex	20.5.75	94	4	1.4
Ostracoda	20.5.76	93	3	1.4
Gammarus pulex	24.10.75	83	4	2.0
Limnaea pereger	20.9.76	96	3	1.7
Sigara sp.	24.10.75	64	4	0.8
Chironomid larvae	20.5.75	59	4	0.5
Chironomid pupae	20.5.75	69	3	–
Plant detritus	17.11.76	91	3	–
Oligochaeta	17.11.76	89	4	–

Discussion

The results of this study may be compared with limited data from other lowland lakes and artificial impoundments, and more extensive data from natural trout lakes which are usually upland, oligotrophic lakes. The only large lowland eutrophic lakes in Britain with natural trout populations are Loch Leven and Lough Neagh, and only from the former are data available.

Few data are available on trout populations of new reservoirs during their initial stages after impoundment, or on artificially-regulated lakes, and those data are primarily for upland waters in Wales (Llyn Tegid), the Lake District (Haweswater), Scotland (Loch Tummel) or Sweden. Data are available for trout diets in a wide variety of natural lakes in Britain, Europe and the United States.

The development of the populations of aquatic invertebrates in Rutland Water was similar in some respects to changes recorded in other newly-impounded lakes. Runnstrom (1955) and Nilsson (1964) have reported increased production of littoral Crustacea following impoundment of Swedish Lakes; these authors and Hynes (1961), working on Llyn Tegid, noted that Amphipoda and Mollusca, together with insect larvae, were the groups least able to adjust to water level fluctuations. Comparisons of trout diet in regulated lakes in Sweden compared to unregulated ones (Nilsson 1964) showed that fish fry, the littoral crustacean *Eurycercus lamellatus* and terrestrial insects became more important in regulated lakes, whilst Trichoptera, *Gammarus pulex* and *Limnaea pereger* were more important in unregulated lakes.

Where impoundment has raised the water level of existing lakes, flooding grassland and marshes, the additional food made available to the trout was heavily utilised. Frost (1956) found that 41% of the diet of trout in Haweswater in 1941, the year following the flooding, was of terrestrial origin, including earthworms and slugs. Campbell (1957) found that in Loch Tummel trout stomachs contained mainly this new bottom fauna after flooding and estimated that earthworms could be present for up to 18 months after flooding. Kennedy & Fitzmaurice (1971) reported earthworms and terrestrial tipulid larvae in trout stomachs of newly-flooded Irish lakes or reservoirs. Runnstrom (1955) reported earthworms and littoral Crustacea in the diet of trout in newly-flooded Swedish lakes. All these studies found increased growth rate of trout in the years immediately following impoundment, declining after 5 to 10 years.

Comparison of the trout diet at Rutland Water with other lowland trout waters shows some similarities but many differences. The main similarity is the heavy use of Chironomidae pupae during the periods of emergence. This is similar to Chew and Blagdon reservoirs (Wilson 1971), Loch Leven (Thorpe 1974) and Grafham Water (Church 1974). However, in Chew and Blagdon, Hydracarina, Oligochaeta and Corixidae were important at certain times of the year (Wilson 1971); in Grafham Water, *Daphnia* and fish fry were important food items (Church 1974), and fish fry were also important in Eyebrook Reservoir (Oliver 1968) and Loch Leven (Thorpe 1974). They began to appear in late 1977 in Rutland Water (Warlow & Oldham 1982) as stocks of coarse fish became established in the reservoir. There are major differences between the diet of trout in Rutland Water and in natural upland lakes. In these latter lakes [e.g. Windermere (Allen 1941), Malham Tarn (Holmes 1960), Llyn Tegid (Dunn 1954; Ball 1961)], Ephemeroptera, Trichoptera and/or Plecoptera were important dietary components, with *Gammarus* and *Asellus* becoming significant over winter. In addition, food items of terrestrial origin, usually adult insects, blown on to the water surface, were taken heavily in summer. These aquatic insect larvae are not present in any number in Rutland Water, and terrestrial insects do not appear to be a significant summer food source, perhaps because of adequate aquatic alternatives.

Conclusions

The initial impoundment and filling of the reservoir between 1975 and 1977 resulted in a population explosion of littoral Crustacea in the summer of 1975, repeated to a lesser extent in the following two summers. During the course of each summer, following periods of water level stability, a diverse invertebrate fauna, which included Amphipoda and Mollusca, built up. The food of aquatic origin taken by trout reflected these changes in fauna.

During each winter period when water level increased rapidly the littoral Crustacea, chironomid larvae and Corixidae were observed to be able to respond to the changes by migrating into the new littoral regions, whereas Amphipoda and Mollusca were much less able to do so.

During each winter and following spring, after new land was flooded, large quantities of terrestrial organisms, principally earthworms, became available and were taken by trout.

The rapid growth rates of trout stocked in 1975 (Moore 1982) is most likely the result of two nutritional factors. The initial population explosion of Crustacea and Chironomidae provided an abundant food supply for the fingerling trout, whilst terrestrial food was available during those periods when aquatic food was scarce.

There was an unexpectedly high incidence of plant debris found in stomachs during periods of water level change, and this appears to have been due to active selection of these items by some trout. It seems unlikely that these items were particularly nutritious but they were possibly taken by mistake, because of their abundance in shallow water in the same places as the drowned terrestrial invertebrates.

Acknowledgements

The author wishes to thank the Anglian Water Authority for permission to publish this paper, and its employees, particularly fisheries staff, for assistance in the field. David More provided the data on trout growth, and I am grateful to him, together with A. Warlow and R. Hancock for valuable discussion of the results. The mistakes remain the responsibility of the author, and the opinions expressed are his own.

References

Allen, K. R., 1941. Some observations of the biology of the trout (Salmo Trutta) in Windermere. J. Anim. Ecol. 7: 333–349.
Ball, J. N., 1961. On the food of the brown trout in Llyn Tegid. Proc. Zool. Soc. Lond. 137: 599–622.
Campbell, R. N., 1957. The effects of flooding on the growth rate of brown trout in Loch Tummel. Freshw. Salmon Fisheries Research 14. Department of Agriculture and Fisheries for Scotland, Edinburgh.
Church, V., 1974. An investigation into the diet of rainbow trout in Grafham Water during the months of May to September 1974. Unpublished MSS.
Dunn, D. R., 1954. The feeding habits of some of the fishes and some members of the bottom fauna of Llyn Tegid. J. Anim. Ecol. 23: 224–233.
Frost, W. E., 1956. The growth of brown trout in Haweswater before and after the raising of the lake. Salm. Trout Mag., 267–275.
Frost, W. E. & Brown, M. E., 1967. The Trout. Fontana New Naturalist.
Holmes, P. F., 1960. The brown trout of Malham Tarn. Salm. Trout Mag. 159: 127–145.
Hynes, H. B. N., 1961. The effect of water level fluctuations on the littoral fauna. Verh. int. Verein. Limnol. 14: 652–656.
Kennedy, M. & Fitzmaurice, P., 1971. Growth and food of brown trout (S. trutta) in Irish waters. Proc. R. Irish Acad. 7: 269–350.
Moore, D. E., 1982. Establishing and maintaining the trout fishery at Rutland Water. In: Harper, D. M. & Bullock, J. A. (Eds) Rutland Water – Decade of Change, pp. 179–189. Junk, The Hague.
Nilsson, N. A., 1964. Effects of impoundment on the feeding habits of brown trout and char in lake Ransareen (Swedish Lapland). Verh. int. Verein. Limnol. 15: 444–452.
Oliver, G. S., 1968. The development of a trout fishery in a lowland reservoir. J. Inst. Wat. Engrs 22: 129–137.
Runnstrom, S., 1955. Changes in fish production in impounded lakes. Verh. int. Verein. Limnol. 12: 176–182.
Thorpe, J. E., 1974. Trout and perch populations at Loch Leven, Kinross. Proc. R. Soc. Edinb. B, 74: 295–313.
Warlow, A. & Oldham, R., 1982. Temporal variations in the diet of brown trout (Salmo trutta L.) and rainbow trout (Salmo gairdneri R.) in Rutland Water. In: Harper, D. M. & Bullock, J. A. (Eds) Rutland Water – Decade of Change. pp. 199–206. Junk, The Hague.
Wilson, R. S., 1971. The limnology of Chew and Blagdon lakes. Interim Report, University of Bristol.

Temporal variations in the diet of brown trout (*Salmo trutta* L.) and rainbow trout (*S. gairdneri* R.) in Rutland Water

A. D. Warlow & R. S. Oldham
School of Life Sciences, Leicester Polytechnic, Leicester, England

Keywords: trout, *Salmo trutta, Salmo gairdneri,* diet, Rutland Water, reservoir

Abstract

The diet of brown trout (*Salmo trutta* L.) and rainbow trout (*S. gairdneri* Richardson) in Rutland Water were compared during the first two fishing seasons (April–October 1977 and 1978).

Fortnightly samples of approximately forty stomachs were obtained from boat and bank, rod-and-line caught trout giving a total of 1 046 stomachs over the two seasons.

During 1977 seasonal changes in the diet were divided into two phases; the first being a period of abundant drowned terrestrial food until June. This was followed by a period of more stable water level from July onwards when chironomid larvae and pupae were consistently the most important food items and the diversity of food also increased.

In 1978 the proportion of chironomid pupae and larvae declined and they were replaced in the diet by *Gammarus* and *Asellus.*

Introduction

The food of brown trout in Great Britain has been well studied in streams (Maitland 1965; Egglishaw 1967; Woolland 1972) and in natural and impounded lakes (Kennedy & Fitzmaurice 1971; Hunt 1970; Thorpe 1974; Wilson *et al.* 1974) but few studies have been made on the diet of brown and rainbow trout populations artificially maintained for game fisheries. Work on the effect of impoundment on the growth rate and diet of trout is restricted to upland reservoirs where large rivers are impounded (Campbell 1957; Crisp *et al.* 1978) or existing water levels raised (Frost 1956; Campbell 1963). Such studies indicate an improved growth rate immediately following impoundment which appears to be linked to the abundant supply of terrestrial food made available after flooding, confirmed for Rutland by Harper (1982). The stocking policy at Rutland Water was designed to utilise food resources which were anticipated by stocking large numbers of small trout at the start of inundation (Moore 1982). This paper reports on the diet of the trout in the 2 years after inundation, when terrestrial food disappeared and a high biomass of trout was forced to find alternative food resources in this new environment.

Methods

Sampling began in April 1977 at the start of the first fishing season. A total of approximately forty stomachs was removed from rod-and-line caught trout at fortnightly intervals throughout the 1977 and 1978 fishing seasons (April–October). The ratio of brown trout to rainbow trout in a sample was dependent upon the ratio of the two species caught on that day. Stomachs were obtained from bank caught fish at the site of capture and from boat caught fish when anglers returned to the fishing lodge. Trout were sampled in January 1978

Hydrobiologia 88, 199–206 (1982). 0018-8158/82/0882-0199/$01.60.

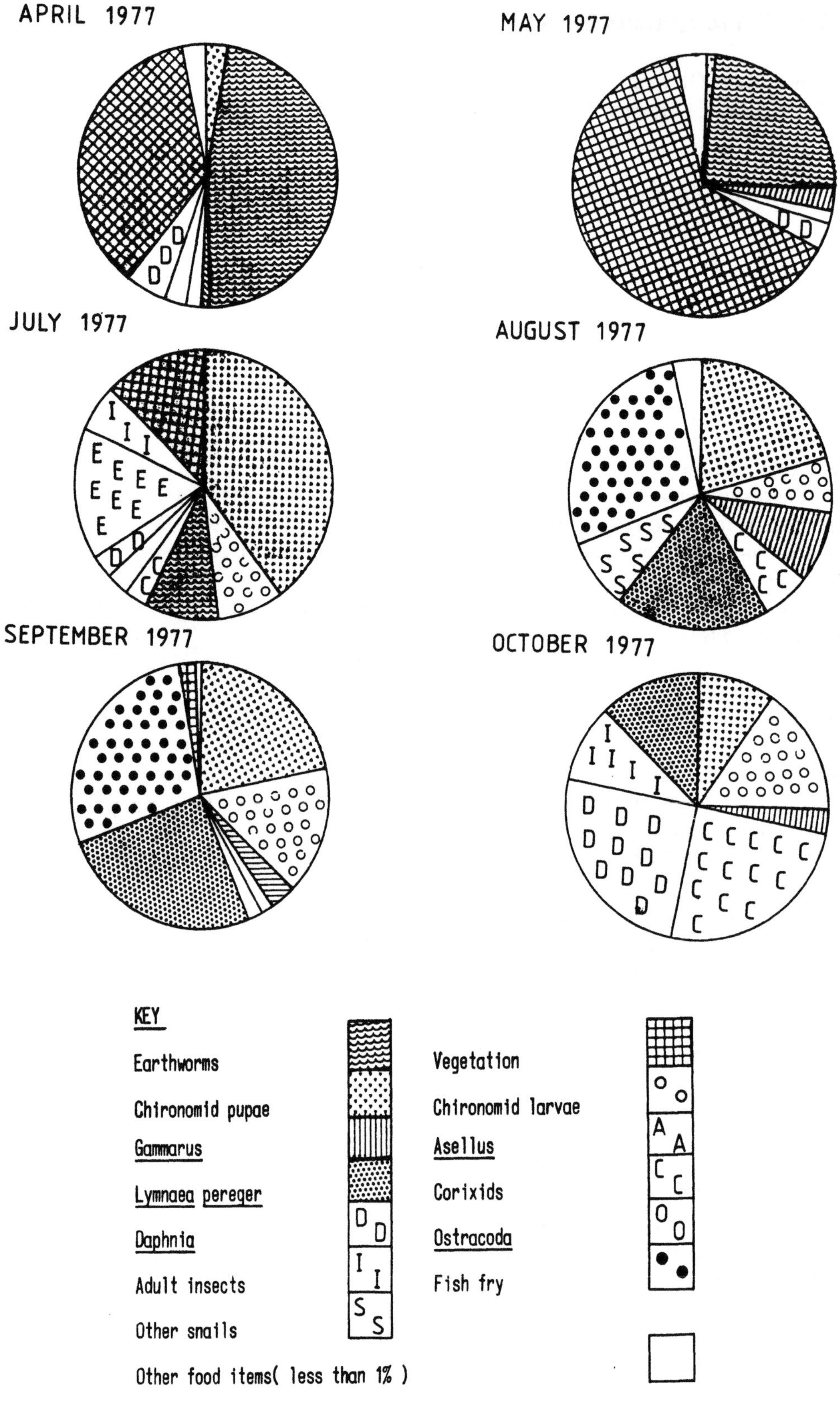

Fig. 1. Seasonal changes in the diet of brown trout and rainbow trout in Rutland Water during 1977.

using a 150 m beach seine net of 1 cm mesh. In all cases, stomachs were removed from the fish in the field and placed in 4% formalin. In the laboratory the 'standard stomach' (Ball 1961) was opened and the contents transferred to 70% alcohol. A total of 1 046 stomachs was examined in 1977 and 1978. The stomach contents were sorted into major taxonomic groups and counted. Weights were obtained for each group after drying at 105 °C for 24 hours.

Results

The results of dietary analysis are expressed in terms of 'percentage occurrence', 'percentage number' and 'percentage dry weight'. Percentage occurrence (Dunn 1954) expresses the number of stomachs containing one or more individuals of each category as a percentage of all those containing food. Percentage number (Crisp, Mann and McCormack 1978) expresses the number of individuals in each food category, recorded for all stomachs, as a percentage of the total individuals in all food categories. Percentage dry weight (Crisp *et al.* 1978) expresses the dry weight of each food category, recorded for all stomachs, as a percentage of the total dry weight of all food categories.

Statistical analysis of the results was facilitated by computer using the 'Statistical Packages for the Social Sciences' (Nie *et al.* 1975).

During 1977 the diet of the trout was divided into two main phases, the first being the end of the period of rapid flooding from September 1976 to March 1977 when up to 250 ha of pasture were inundated monthly. During this period trout were utilising the abundant terrestrial food comprising mainly earthworms, but also slugs, spiders and terrestrial tipulid larvae (Fig. 1). Associated with the earthwors were large amounts of terrestrial vegetation – grass, sticks and plant stems. A comparison of the proportion of terrestrial invertebrate and aquatic invertebrate food (Fig. 2) shows that 80% of the dry weight of food consumed in April and May 1977 was terrestrial compared with less than 1% in August. In addition, microcrustaceans were common in the stomachs in April and May, reaching a mean of 282 per stomach (Table 1). Although surface water temperatures

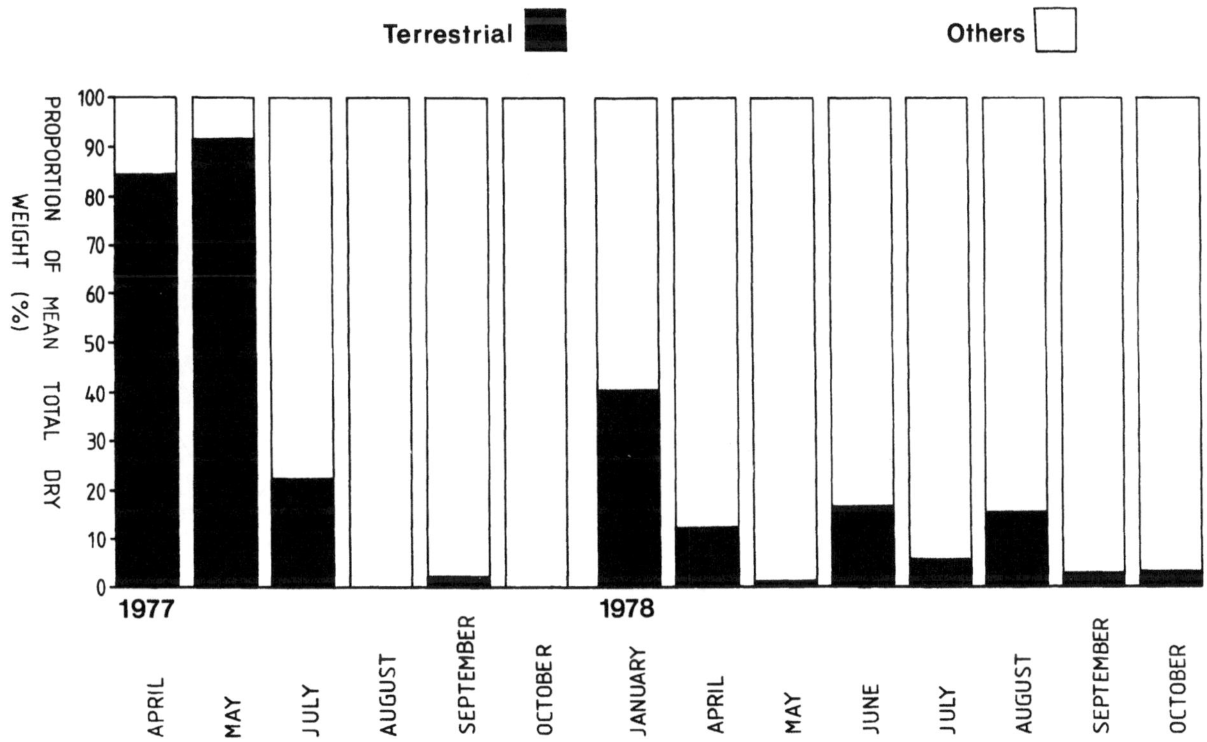

Fig. 2. The importance of terrestrial food compared with the remaining food items in the diet of brown and rainbow trout.

Table 1. The occurrence and mean number of micro-crustaceans, mainly Cladocera, in the stomachs of brown and rainbow trout.

Year and month		Number of stomachs examined	Number containing Cladocera	% occurrence in non-empty stomachs	Mean number per stomach (non-empty)
1977	April	69	20	30	282
	May	73	14	20	113
	July	75	20	28	91
	August	79	3	4	11
	September	70	3	4	23
	October	40	17	52	107
1978	January	21	12	71	51
	April	173	23	14	95
	May	83	1	1	1
	June	39	4	11	58
	July	80	14	20	33
	August	81	16	23	143
	September	80	6	8	18
	October	54	15	31	156
1977		406	77	19	104
1978		611	91	15	69

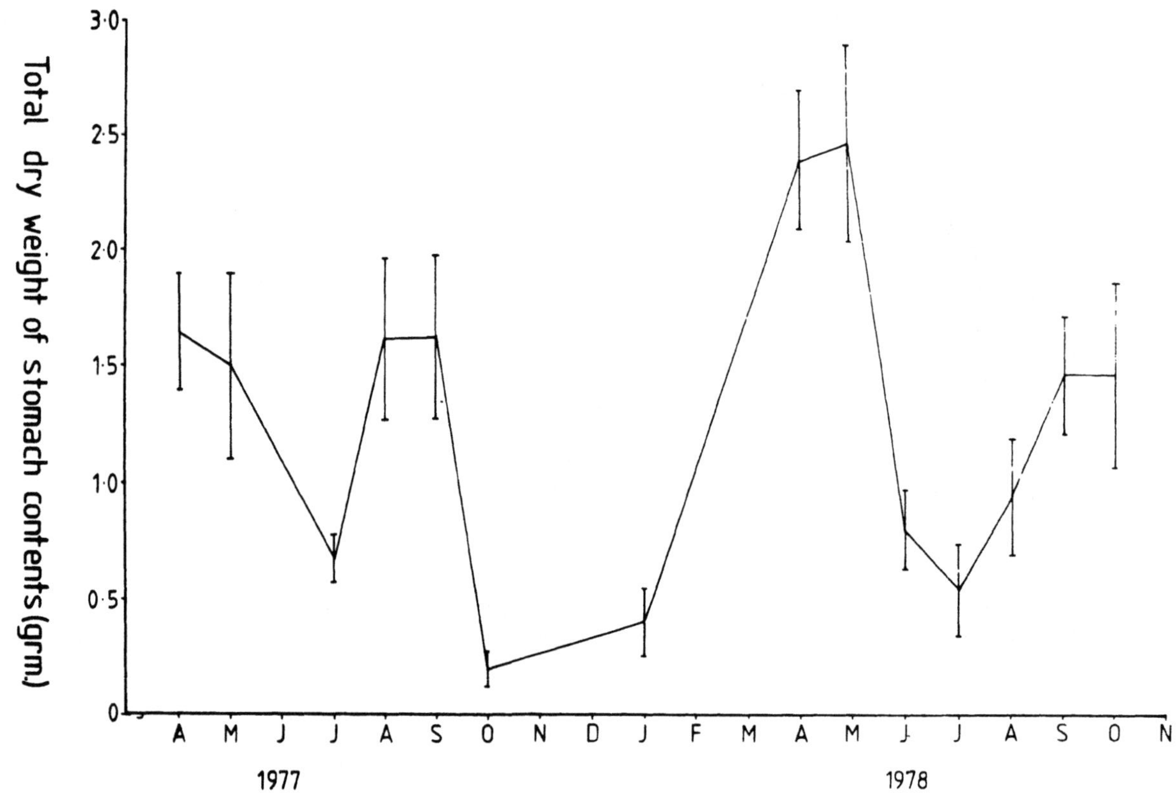

Fig. 3. Seasonal variation in the mean total dry weight of stomach contents (± s.e.m.) of brown and rainbow trout during 1977 and 1978.

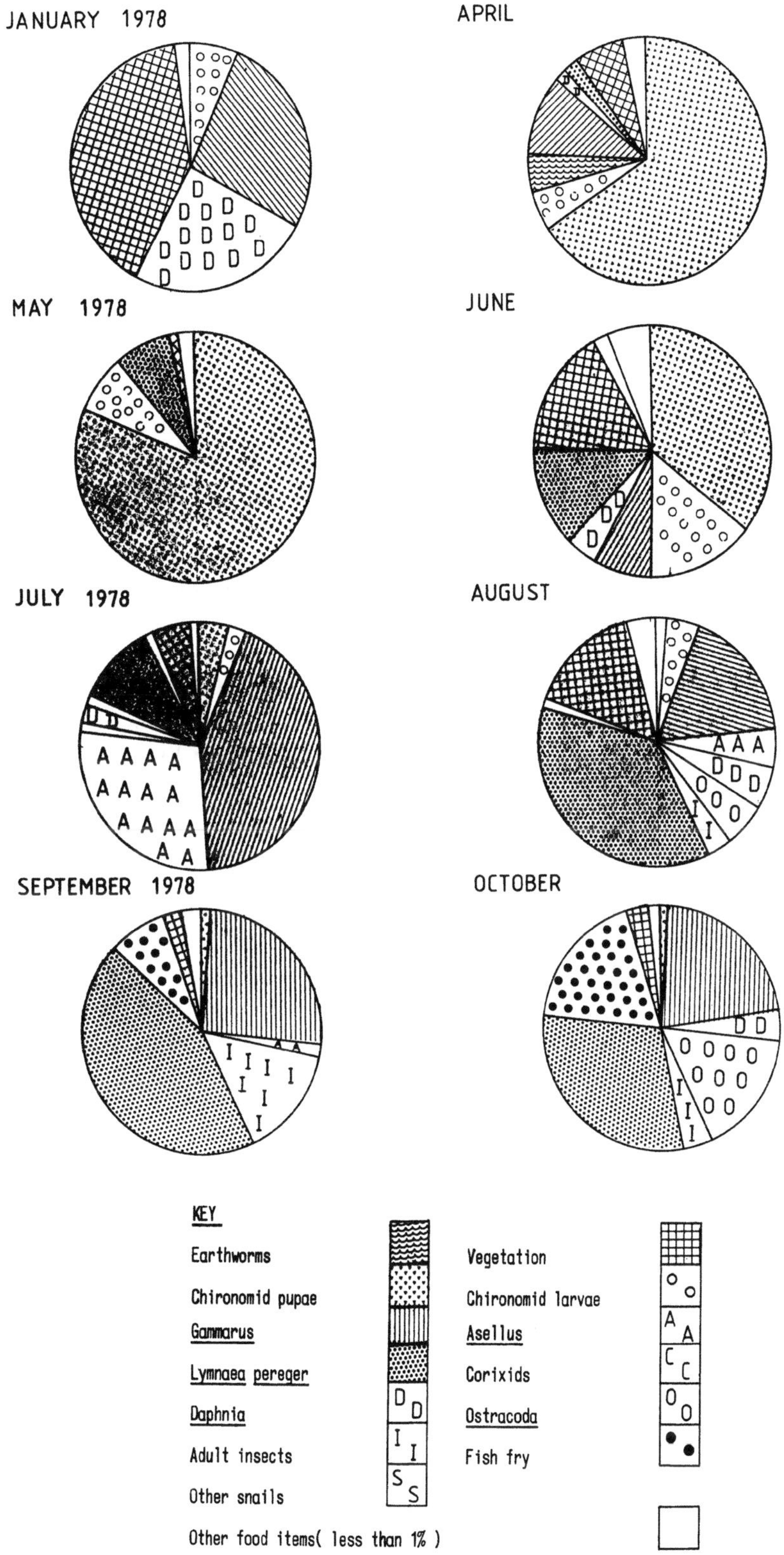

Fig. 4. Seasonal changes in the diet of brown and rainbow trout during 1978.

were low at this time, 8–12 °C, the mean total dry weight of stomach contents was high (Fig. 3).

The second phase, of more stable water level, occurred from July onwards, when a greater variety of food organisms was present with the most important food items, chironomid pupae and larvae (Fig. 1), accounting for over one quarter of the dry weight of stomach contents from July to October, and 80% of the diet by number. Maximum numbers were reached in September 1977 when the mean number of pupae and larvae per stomach rose to 94.0 and 106.0 respectively. Although numbers of chironomid pupae were high, few adult chironomids were found in the stomachs and numbers of all adult insects were always insignificant. The remaining food items in this year fell into two groups – those present in small numbers throughout the year in a few fish, e.g. *Gammarus pulex* and corixids, and those occurring for part of the year in many fish, e.g. *Limnaea pereger* and fish fry. Gastropods and fish fry appeared in the diet in August 1977 and during August and September each accounted for one quarter of the dry weight of stomach contents (Fig. 1).

In contrast to 1977, the water level rose only slightly during the winter 1977/78 and was comparatively stable during the 1978 fishing season (April–October). In January 1978 a low feeding rate was probably due to the low surface water temperature (2 °C), with *Gammarus pulex* the most important food item. In April and May 1978 the trout were feeding heavily and the mean total dry weight of stomach contents rose to 0.8 g; chironomid larvae and pupae accounted for over 75% of the diet, with up to 250 pupae and 35 larvae per stomach (Fig. 4). During June the diet showed a greater diversity of food items including *Gammarus pulex, Daphnia* and *Limnaea pereger,* although the mean total weight of stomach contents fell to 0.28 g and continued to fall to 0.19 g in July (Fig. 3) when the water temperature was at its summer peak. During July the change in the diet continued as numbers of chironomids in the stomachs fell dramatically and numbers of *Gammarus* and *Asellus* increased, contributing 70% of the dry weight of the diet. The fall in the numbers of chironomid larvae and pupae continued into October. High numbers of microcrustacea (mainly *Cypris* sp.) were also present in the diet in August and October (Table 1). From August to October the mean total dry weight of stomach contents rose (Fig. 3), due mainly to the increased consumption of *Limnaea pereger* in August and fish fry as well during September and October. The importance of adult insects in the diet rose throughout the summer to almost 50% by number in September when large numbers of small Bibionidae were found in the stomachs.

Discussion

The diet of the trout at Rutland Water during 1976 (Harper 1982) and early 1977 was similar to that found by other workers on newly flooded water bodies in that terrestrial food formed the bulk of the diet (Campbell 1957, 1963; Frost & Brown 1967, Kennedy & Fitzmaurice 1971; Crisp *et al.* 1978). In common with the findings of Crisp *et al.* (1978) the terrestrial food comprised eathworms and larval Tipulidae and Coleoptera. Campbell (1963) and Crisp *et al.* (1978) also noted the presence of terrestrial vegetation, mainly grass, in stomachs containing earthworms, a feature of the trout diet at Rutland Water. However, whereas these workers found the terrestrial food to be available to the trout for several years, at Rutland Water the rapid flooding during the winters of 1975/6 and 1976/7 released a super-abundance of terrestrial food over a very much shorter period. During much of this time, low winter temperatures restrained growth rates but from April to August 1977 the growth rate was high (Moore 1982). By August 1977 the terrestrial food had disappeared, only three months after the end of flooding. Laboratory work by Roots (1956) showed that earthworms could survive immersion for a maximum of 50 weeks but it seems unlikely that any survived longer than 12 weeks at Rutland Water. If flooding had taken place over several years it is likely that the available food would have been used more efficiently and growth may have been prolonged. Once the terrestrial food had disappeared the trout switched to chironomid larvae and pupae as the most important food item, a pattern established in the first two summers (Harper 1982) and elsewhere (Crisp *et al.* 1978), and the high rate of growth was maintained (Moore 1982).

Chironomids were also important dietary components in early 1978 but during the later part of the year their importance declined. Large populations of adult chironomids were present in 1977, so it is not clear whether egg laying and larval recruitment were poor in 1978, or whether there was a behavioural shift in the trout feeding pattern (Macan 1966; Ware 1972). The change to *Gammarus pulex* and *Asellus* was accompanied by a decline in the dry weight of stomach contents from May through June and July. *Limnaea pereger* then became important during August. Hunt & Jones (1972) found that in nutrient rich water bodies trout consume *Gammarus pulex* or *Asellus meridianus* rather than the Trichopteran larvae found to be important in less nutrient rich waters (Southern 1935; Allen 1938; Frost & Smyly 1952; Ball 1957; Pedley & Jones 1978). Trout in richer waters appear less likely to consume aerial insects although when large numbers of insects were available to the trout at Rutland Water (September 1978) they were readily consumed. As many insects emerge under still-water conditions, usually in the early morning or late evening, sampling of trout stomachs during the day may not cover these periods when adult insects are available to the fish.

A study of the relationship between chironomids in the trout diet and those in the reservoir substrate (Brown *et al.* 1979) showed differences in the diet of the two species of trout pointing towards a spatial separation of the two species within the reservoir, brown trout feeding in deeper water and rainbow trout in surface waters. Further analysis of the diet of the two species is continuing, to determine the extent of this separation.

Acknowledgements

We thank Anglian Water Authority and Rutland Water anglers for their co-operation throughout the project and B. A. Long for his help in the field. The work was completed during the tenure of an S.R.C. studentship to A.D.W.

References

Allen, K. R., 1938. Some observations on the biology of the trout in Windermere. J. Anim. Ecol. 7: 333–349.

Ball, J. N., 1957. The biology of the brown trout of Llyn Tegid. Ph.D. thesis, Liverpool University.

Ball, J. N., 1961. On the food of the brown trout of Llyn Tegid. Proc. Zool. Soc. Lond. A. 110: 81–96.

Brown, A. E., Oldham, R. S. & Warlow, A., 1980. Chironomid larvae and pupae in the diet of brown trout (Salmo trutta) and rainbow trout (Salmo gairdneri) in Rutland Water, Leicestershire. In: Murray, D. A. (ed.) Chironomidae, pp. 323–329. Pergamon Press, Oxford.

Campbell, R. N., 1957. The effect of flooding on the growth rate of brown trout in Loch Tummel. Sci. Invest, Freshw. Fish. Scot. 14: 1–7.

Campbell, R. N., 1963. Some effects of impoundment on the environment and growth of brown trout in Loch Garry (Inverness-shire). Sci. Invest. Freshw. Fish. Scot. 30: 1–27.

Crisp, D. T., Mann, R. H. K. & McCormack, J. C., 1978. The effects of impoundment and regulation upon the stomach contents of fish at Cow Green, Upper Teesdale. J. Fish. Biol. 12: 287–302.

Dunn, D. R., 1954. The feeding habits of some of the fishes and some members of the bottom fauna of Llyn Tegid. J. Anim. Ecol. 23: 2.

Egglishaw, H. J., 1967. The food, growth and population structure of salmon and trout in two streams in the Scottish Highlands. Freshwat. Salm. Fish. Res. 38: 32 pp.

Frost, W. E. & Smyly, W. J., 1952. Brown trout of a moorland fishpond. J. Anim. Ecol. 21: 62–86.

Frost, W. E., 1956. The growth of brown trout in Haweswater before and after raising the lake. Salm. Trout Mag. 267–275.

Frost, W. E. & Brown, M. E., 1967. The Trout. New Naturalist, Collins, London.

Harper, D. M., 1982. The feeding of trout during the filling phase of Rutland Water. In: Harper, D. M. & Bullock, J. A (eds) Rutland Water – Decade of Change pp. 191–198. Junk, The Hague.

Hunt, P. C., 1970. Biological investigations in regulated reservoirs. Ph.D. thesis, Liverpool University.

Hunt, P. C. & Jones, J. W., 1972. The profundal fauna of Llyn Tegid, North Wales. J. Zool. 168: 9–49.

Kennedy, M. & Fitzmaurice, P., 1971. Growth and food of brown trout Salmo trutta (L) in Irish Waters. Proc. R. Ir. Acad. 71B: 270–348.

Macan, T., 1966. Predation by Salmo trutta in a moorland fishpond. Verh. Int. Ver. Theor. Angew. Limnol. 16: 1081–1087.

Maitland, P. S., 1965. The feeding relationships of salmon, trout, minnows, stoneloach and three-spined stickleback in the River Endrick, Scotland. J. Anim. Ecol. 34: 109–133.

Moore, D. E., 1982. Establishing and maintaining the trout fishery. In: Harper, D. M. & Bullock, J. A. (eds) Rutland Water – Decade of Change, pp. 179–189. Junk, The Hague.

Nie, N. H., Hadlai Hull, C., Jenkins, J. C., Steinbrenner, K. & Bent, D. H., 1975. Statistical Package for the Social Sciences (2nd ed). McGraw-Hill, London.

Pedley, R. B. & Jones, J. W., 1978. The comparative feeding behaviour of brown trout Salmo trutta L. and Atlantic salmon, Salmo salar L. in Llyn Dwythwch, Wales. J. Fish. Biol. 12: 239–256.

Roots, B. I., 1956. The water relation of earthworms. II. Resistance to desiccation and immersion, and behaviour when submerged and when allowed a choice of environment. J. Exp. Biol. 33: 29–44.

Southern, R., 1935. Reports from the limnological laboratory. III. The food and growth of brown trout from Loch Derg and the river Shannon. Proc. R. Ir. Acad. 42B: 87–172.
Thorpe, J., 1974. Estimation of brown trout in Lock Leven, Kinross, Scotland, J. Fish. Biol. 6: 135–152.
Ware, D. M., 1972. Predation by Rainbow trout (Salmo gairdneri): the influence of hunger, prey density and prey size. J. Fish. Res. Bd Can. 29: 1193–1201.
Wilson, R. S., Maxwell, T. R. A., Mance, G., Sleigh, M. A. & Milne, R. A., 1974. Biological aspects of Chew Valley and Blagdon lakes, England. Freshwat. Biol. 5(4): 379–394.
Woolland, J. V., 1972. Studies on salmonid fishes in Llyn Tegid and the Welsh Dee. Ph.D. thesis, Liverpool University.

The population dynamics of eyeflukes *Diplostomum spathaceum* and *Tylodelphys clavata* (Digenea: Diplostomatidae) in rainbow and brown trout in Rutland Water: 1974–1978

J. Moody & E. Gaten
Department of Zoology, University of Leicester, Leicester LE1 7RH, England

Keywords: eyeflukes, trout, fisheries, fish parasites, Diplostomatidae

Abstract

The development of populations of eyeflukes *Diplostomum spathaceum* and *Tylodelphys clavata* in trout introduced into Rutland Water is related to the availability of infected *Limnaea pereger*, the first intermediate host, and the differential resistance to eyeflukes in *Salmo gairdneri* and *S. trutta*. The numbers of snail and fish intermediate hosts were monitored from impoundment of the Gwash in 1974 to completion of the reservoir in 1978. High infection rates in coarse fish in the feeder streams in 1974 are related to high infection rates in *Limnaea pereger*. Large eyefluke populations in introduced rainbow trout but not in brown trout confirm their differential resistance. Differences in levels of infection in seine-netted and rod-caught rainbow trout suggest that heavily infected fish could not respond to the fishermens' lures. Reduced numbers of eyeflukes in rainbow trout from 1979 onwards are related to the vast increase in volume of the reservoir in 1977 and decrease in abundance of *Limnaea pereger*.

Introduction

Heavy eyefluke infections in rainbow trout were recorded at Grafham Water, a lowland reservoir in Cambridgeshire, a few years after the introduction of trout. It was predicted that similar infections would occur at Rutland Water. Infected *Limnaea pereger* (Mull.), the first intermediate host of the two eyeflukes, *Diplostomum spathaceum* (Rudolphi) and *Tylodelphys clavata* (Nordmann), were present in the River Gwash system: infected coarse fish, second intermediate hosts, were also present. Appropriate definitive hosts, gulls and grebes, were also available.

Materials and methods

A large number of coarse fish removed from the feeder streams by the Anglian Water Authority in November and December 1974 were examined for eyeflukes. Metacercariae of *Diplostomum spathaceum* from the lens and *Tylodelphys clavata* from the vitreous humour were counted.

Eyes of rainbow trout (*Salmo gairdneri*) and brown trout (*S. trutta*) netted by the Anglian Water Authority between June 1976 and April 1977 were examined. After angling began in May 1977, eyes of rod-caught trout were examined: these included a number of lenses removed from stock fish introduced to the reservoir after July 1977.

Collections of *Limnaea pereger* were made from October 1974 by handnet from the feeder streams. These were measured, crushed and examined for larval stages of *Diplostomum spathaceum* and *Tylodelphys clavata*.

From May 1976, regular monthly samples of snails were collected from the body of the reservoir by dredging using a Dory 11 dinghy and hand-netting in the littoral region. Between May 1976 and August 1978, 7 868 *Limnaea pereger* were collected: 3 548 were incubated and the emergent

Hydrobiologia 88, 207–209 (1982). 0018-8158/82/0882-0207/$00.60.

cercariae recorded. After incubation snails were crushed and examined for larval eyefluke stages.

Results

1. Fish infection levels

Coarse fish: River Gwash 1974

The incidence of infection and mean number of flukes per fish is shown in Tables 1 and 2.

Trout: Rutland Water 1976–1978

The incidence and mean number of *D. spathaceum* and *T. clavata* found in brown and rainbow trout is shown in Tables 3 and 4. The maximum infection recorded was 324 metacercariae, a mixed infection of *D. spathaceum* and *T. clavata* in a netted rainbow trout.

2. Snail infection levels

River Gwash: October to December 1974

174 *Limnaea pereger* collected from the reservoir feeder streams, 119 from Hambleton on the River Gwash, and 54 from Fox Bridge on the Oakham tributary, were examined. 15 snails, 8.67% of those examined, were infected with furcocercariae.

Table 1. Infection of coarse fish by *Diplostomum spathaceum.*

Host species	No. of fish examined	% of fish infected	Mean no. of flukes per fish
Dace (*Leuciscus leuciscus*)	7	85.7	97.4
Roach (*Rutilus rutilus*)	39	97.4	40.5
Gudgeon (*Gobio gobio*)	9	55.6	7.1
Chub (*Leuciscus cephalus*)	32	81.2	5.5
Pike (*Esox lucius*)	1	100.0	2.0
Eel (*Anguilla anguilla*)	3	66.7	0.7
Bullhead (*Cottus gobio*)	21	33.4	0.4
Stickleback (*Gasterosteus aculeatus*)	75	14.7	0.3
Stone loach (*Noemacheilus barbatulus*)	16	6.2	0.1

Table 2. Infection of coarse fish by *Tylodelphys clavata.*

Host species	No. of fish examined	% of fish infected	Mean no. of flukes per fish
Dace (*Leuciscus leuciscus*)	7	18.6	14.3
Roach (*Rutilus rutilus*)	39	17.9	0.5
Chub (*Leuciscus cephalus*)	32	9.4	0.2

Table 3. Infection of trout by *Diplostomum spathaceum.*

Host species		No. of fish examined	% of fish infected	Mean no. of flukes per fish
Brown trout	– net caught	8	0	0
	– rod caught	22	13.64	0.14
Rainbow trout	– net caught	24	100	172.55
	– rod caught	25	100	45.57
	– stock fish	32	–	5.69

Table 4. Infection of trout by *Tylodelphys clavata.*

Host species		No. of fish examined	% of fish infected	Mean no. of flukes per fish
Brown trout	– net caught	8	0	0
	– rod caught	22	0	0
Rainbow trout	– net caught	24	100	53.79
	– rod caught	25	94.4	17.22
	– stock fish*	–	–	–

* Lenses only: vitreous humour not examined.

Rutland Water: May 1976 to August 1978

Results obtained from the body of the reservoir are summarised in Fig. 1.

Discussion

As predicted, rainbow trout introduced into the reservoir in 1975–1976 acquired heavy eyefluke infections: brown trout, which are known to show higher resistance to eyefluke infection (Betterton 1974), showed a low incidence of *Diplostomum spathaceum* and no *Tylodelphys clavata*. Heavy infections in rainbow trout occurred when the first intermediate host, *Limnaea pereger*, was abundant in the littoral region of the partly filled reservoir. The snails had a comparatively high infection rate during this period. Analysis of trout stomach contents by Harper (1982) indicates that *L. pereger* was a major item in the diet at this time: close contact between the two intermediate hosts, snail and trout, ensured high infection rates of the latter. Fall in incidence of infection of snails in the first half of 1977 followed the period of maximum filling of the reservoir between October 1976 and March 1977 when the volume of water rose from 17 to 115 million cubic metres. The resulting fall in concentration of fluke eggs and miracidia probably led to the low infection rates after this time. No infected snails have been collected since June 1977. Samples of rainbow trout taken in 1979 showed low eyefluke incidence: a mean of 18 metacercariae as opposed to 180 in 1976–1977. This can be related not only to absence of infected snails but also to the overall reduction in the population density of *L. pereger* shown by Bullock *et al.* (1982).

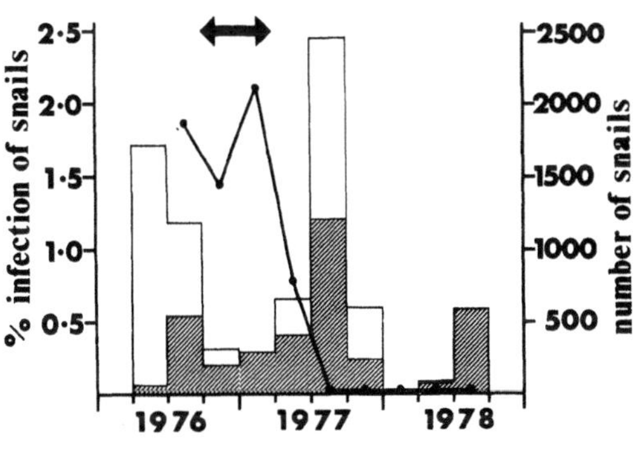

Fig. 1. The % infection of *Limnaea pereger* from Rutland Water by larval eyeflukes (●—●): number of *L. pereger* collected by dredging (plain histogram): number of *L. pereger* examined for larval eyeflukes (hatched histogram): period of maximum filling of the reservoir (⟷).

Rainbow trout caught by anglers between May and September 1977 had relatively low numbers of metacercariae whereas those seine-netted from the reservoir were heavily infected. This apparent inability of the heavily infected trout to respond to the anglers' lures was probably the main cause of the low rainbow catch reported in 1977 by Moore (1982). Crowden & Broome (1980) demonstrated that in dace (*Leuciscus leuciscus*) which were infected with *Diplostomum spathaceum*, the efficiency with which the fish fed on *Gammarus* declined as the number of parasites present in the eye increased.

The infection of snails and fish in the bunds of the nature reserve of Rutland Water has not been investigated: vast flocks of gulls and other definitive hosts suggest the possibility of high infection rates if *Limnaea pereger* is present in sufficient numbers.

Acknowledgements

We are grateful to D. Colman for permission to use his snail and coarse fish infection data of 1974. We are also grateful to the Anglian Water Authority, Mr. D. Moore and Mr. A. Warlow, for supplies of fish heads and eyes.

References

Betterton, C., 1974. Studies on host specificity of the eyefluke, Diplostomum spathaceum in brown and rainbow trout. Parasitology 69: 11–20.

Bullock, J. A., Clark, F. & Ison, S., 1982. Benthic invertebrates of Rutland Water. In: Harper, D. M. & Bullock, J. A. (Eds.) Rutland Water – Decade of Change, pp. 147–156. Junk, The Hague.

Crowden, A. E. & Broom, D. M., 1980. Effects of the eyefluke, Diplostomum spathaceum, on the behaviour of dace (Leuciscus leuciscus). Anim. Behav. 28: 287–294.

Harper, D. M., 1982. The feeding of trout during the filling phase of Rutland Water. In: Harper, D. M. & Bullock, J. A. (Eds.) Rutland Water – Decade of Change, pp. 191–198. Junk, The Hague.

Moore, D. E., 1982. Establishing and maintaining the trout fishery at Rutland Water. In: Harper, D. M. & Bullock, J. A. (Eds.) Rutland Water – Decade of Change, pp. 179–189. Junk, The Hague.

Rutland Water Nature Reserve: concept, design and management

T. P. Appleton
Reserve Warden, Leicestershire and Rutland Trust for Nature Conservation, Fishponds Cottage, Stamford Road, Oakham, Rutland, Leics., England

Keywords: nature reserve, reservoir, planning, design, management, wildfowl

Abstract

This paper illustrates how practical conservation management has helped to create a nature reserve of international importance for wildfowl. The development of existing and introduced habitats has produced a wide diversity of fauna and flora without causing an unnatural ecological balance and close liaison with other reservoir users has established zoning areas free from disturbance resulting in both man and birds benefiting from such agreements.

Introduction

Rutland Water Nature Reserve, previously known as Empingham Reservoir Nature Reserve, is unique in that it was declared a nature reserve before it existed. Within a decade the reserve has been transformed from plans on the drawing board to the reality of a nationally important wildfowl refuge.

For the first time in this country, the wildlife potential of a man-made reservoir has been developed to the full through practical conservation management. Old habitats have been preserved and new ones have been created to provide a diversity of fauna and flora rarely encountered within reservoir margins.

Concept

The wildlife potential of the proposed reservoir was identified during the late sixties. Three prominent naturalists, the late James Fisher, Ron Hickling and John Thompson, surveyed the site in 1969 and formulated some of the earliest recommendations for the establishment of a reserve. The Countryside Commission and Nature Conservancy Council were both involved in discussions with the Welland and Nene River Authority at an early stage and played an important part in laying the foundations for a wildlife conservation area at the reservoir.

A Recreation and Amenity Facilities working party was set up for the reservoir in September 1970 and topics discussed included reserve boundaries and the possibility of the construction of lagoons and islands. The committee recognised that with careful planning the reservoir could at some future date become a site of international importance for wildfowl. In 1971 a working committee, formed under the auspices of the River Authority, reported on the proposed recreation and amenity facilities for the reservoir. A more detailed report on the conservation requirements was submitted by the Nature Conservancy Council in consultation with the Leicestershire and Rutland Trust for Nature Conservation (Nature Conservancy Council 1972). In this report the importance of islands, shelter belts, lagoons and bunds was emphasised, as was the provision of adequate educational facilities for both the general public and serious naturalists. The need for a warden to manage the reserve was also noted.

Hydrobiologia 88, 211–224 (1982). 0018-8158/82/0882-0211/$2.80.

In June 1972 the River Authority invited the conservation bodies to set up an advisory committee and the Empingham Nature Reserve Committee was formed.

During the same period the River Authority agreed in principle to enter into a formal agreement with the Leicestershire and Rutland Trust for Nature Conservation, under which the Trust would accept responsibility for the day to day management of the nature reserve.

The nature conservation movement received a further boost when the 1973 Water Act, Section 22, charged the Water Authorities with the responsibility of 'having regard for' nature Conservation and amenity, in addition to their responsibility in respect of recreation defined in Section 20.

In 1974 re-organisation of the Water Authority took place and the Anglian Water Authority, Welland and Nene River Division became the new promoters of the reservoir project.

In 1975 the Leicestershire and Rutland Trust for Nature Conservation entered into an agreement with the Anglian Water Authority to manage the nature reserve and the following principles were declared:

a) To conserve the plant and animal life and to provide suitable conditions for wildfowl, whether resident or migratory, and for nesting birds.
b) To manage and maintain the woodland in a manner compatible with nature conservation, and to establish additional plantations and indigenous trees where suitable.
c) To observe and keep detailed records of changes in plant and animal life, particularly wildfowl, for scientific purposes.
d) To provide (so far as is compatible with the above mentioned objectives) educational facilities and to encourage the public to enjoy and appreciate the nature reserve.

A full-time Warden was appointed in April, 1975.

In the summer of 1976 the name of the reservoir was changed to Rutland Water and the nature reserve became Rutland Water Nature Reserve.

Design

From the outset it was envisaged that the nature reserve would be divided into zones so that access would be regularised and controlled. It was proposed that there should be three divisions - one section where there would be access for the general public, another designed specifically for the wildlife enthusiast and a third section that would remain comparatively undisturbed with access primarily for research (Fig. 1).

The nature reserve was to be located along the two arms of the western end of the reservoir. Here the water would range from extensive shallows to depths of ten metres, providing variable feeding conditions for all species of waterfowl. The shoreline would be irregular, with creeks and gullies, and once the tree planting programme was underway there would be cover for wildlife from the prevailing south-westerly winds. The area was as distant from the major recreational activities as possible, thereby reducing disturbance by other reservoir users.

Possible disturbance was a main concern for the conservation bodies. Evidence from other reservoirs showed that sailing craft in particular caused considerable disturbance to waterfowl (Batten 1977; Cooke 1977). Following close liaison with other prospective reservoir users, a zoning policy was introduced. Sailing would take place on an area of water of some 2037 acres. The sailing boats would not have access to the 1050 acres of water adjacent to the nature reserve. Marker buoys, indicating the sailing limits, would be set out (Fig. 2). Bank fishing would not be allowed from the nature reserve shoreline. The low profiled fishing boats would have access to the 1050 acres except for the bunded lagoons, the area set aside for research and a 50 metre margin along the length of the nature reserve shoreline. The fishing season on the reservoir would be confined to the summer months and the summer wildfowl populations would find adequate refuge within the lagoons. The 50 metre 'sanctuary margin' would provide further undisturbed areas for broods of water fowl and increase the breeding potential.

The lagoons were to be located in the southern arm between Lax Hill, Egleton Village and the Hambleton peninsula. Here the low-lying land above top water line presented a land drainage problem because of possible silting up of the field drains at and just below top water line. An ideal situation was seen to be the linking up of all the surrounding drains to one or two main channels. A further decision to divert the sewage effluent from the north arm to the south arm and into a lagoon

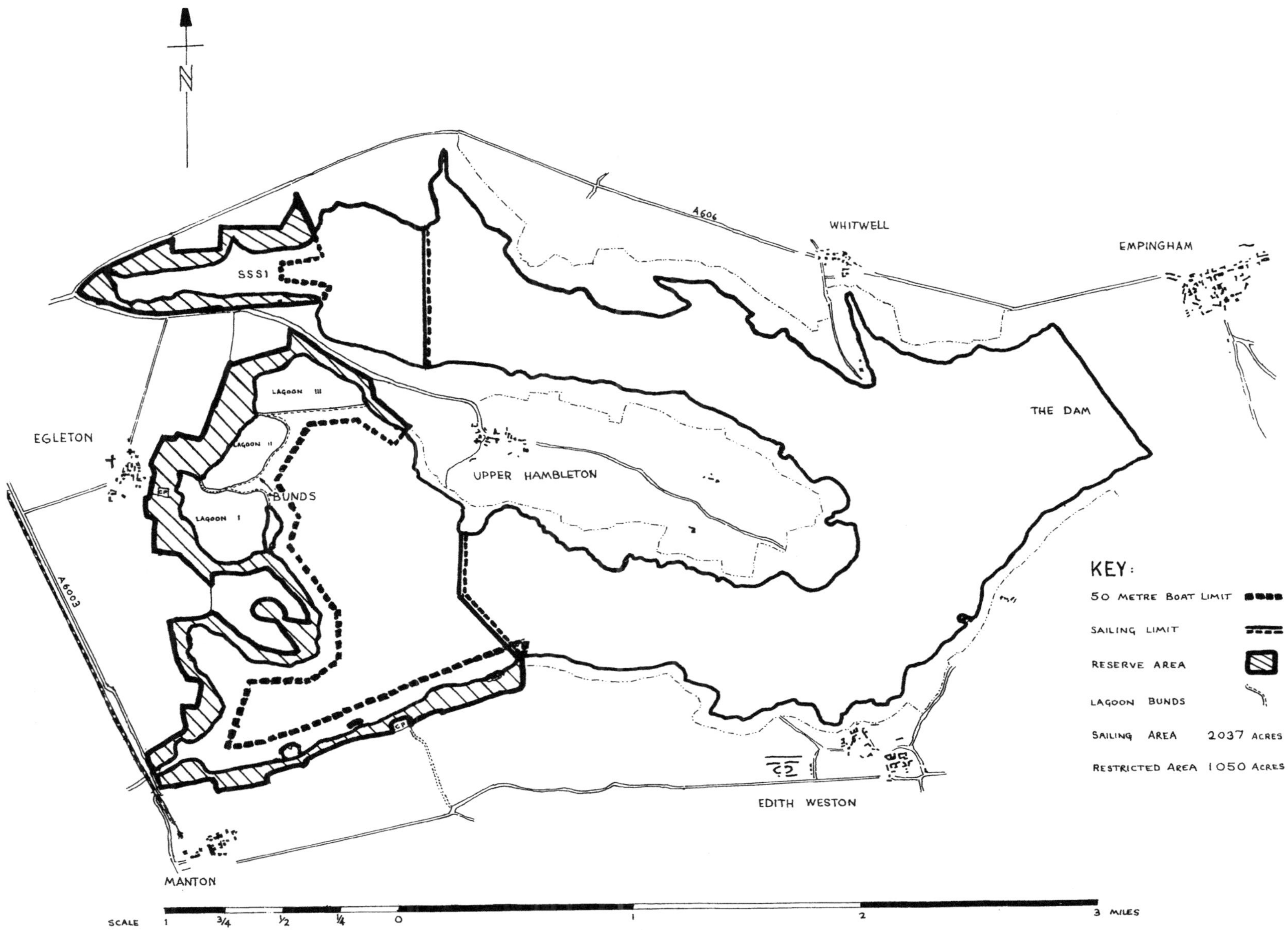

Fig. 1. Map of Rutland Water showing zonation of recreational activities.

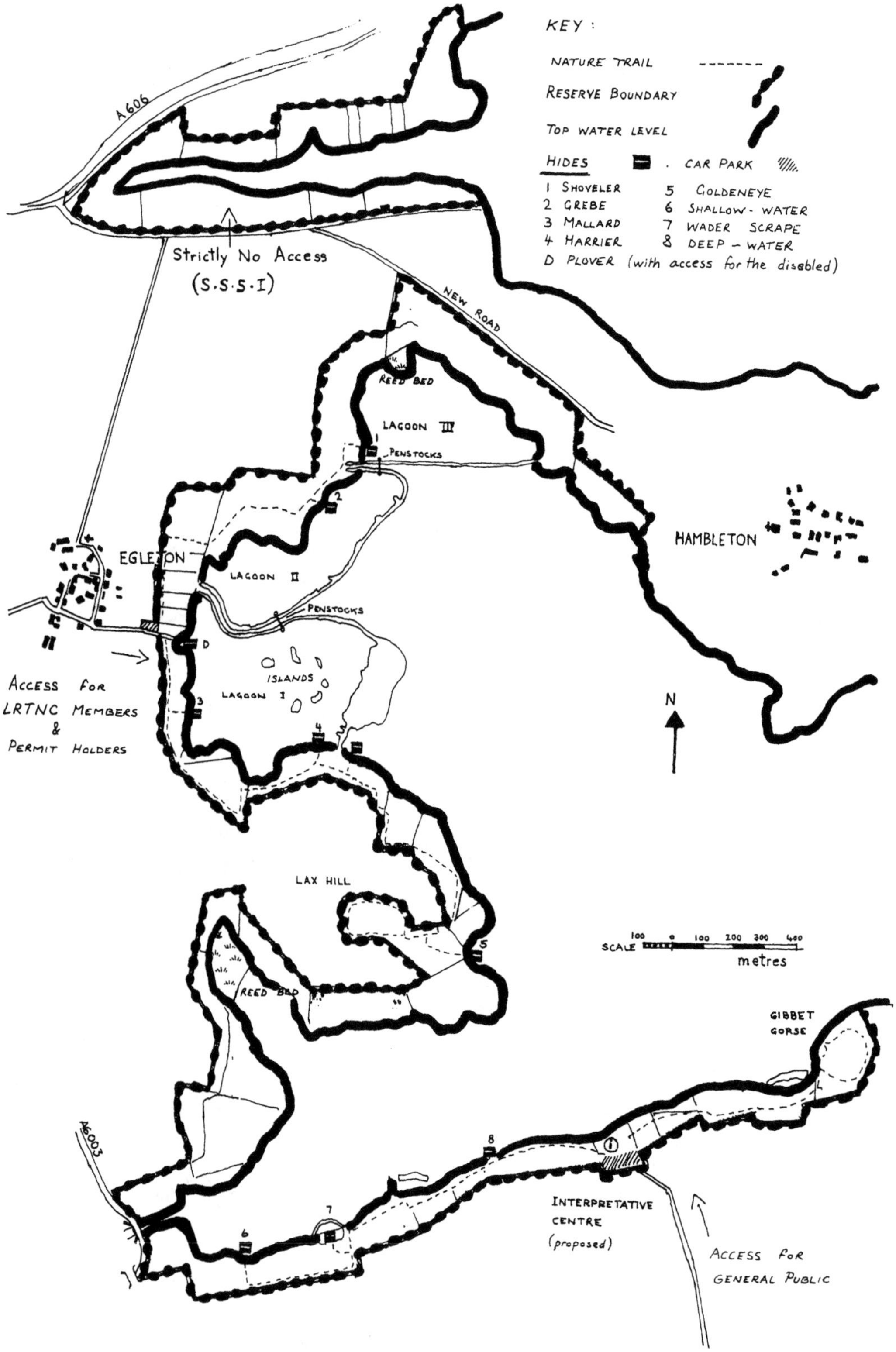

Fig. 2. Map of Rutland Water Nature Reserve.

made provision for additional tertiary treatment and gave a further safeguard to the reservoir in the event of a breakdown at the sewage works.

Plans were drawn up for a series of bunds which included a system of penstocks to control water levels in the lagoons thus created. This system would provide independent water controls for each lagoon and would be used to provide the best habitat conditions throughout the seasons. During the winter months the water levels would be increased, flooding areas colonised by plants during the summer months. This would produce a food source for winter wildfowl. The spring water levels would be maintained during the breeding season after which the gradual release of water would provide ideal feeding conditions on the exposed muddy littoral zone for migrant waders in the autumn.

The three sections of the reserve were now defined. It was planned to develop the public section along the southern shoreline where there was adequate vehicular access from the road between the villages of Manton and Edith Weston. Provision was made for the construction of a small car park, an information centre and the laying out of a nature trail. This section would incorporate Gibbet Gorse, a small woodland, thus providing additional habitat diversity. This section would be called the Lyndon Reserve.

It was proposed that the experienced naturalists would have access to the area around the lagoons near the village of Egleton. With careful planning the impact of the nature reserve would be greatly increased and the benefit to wildfowl in providing undisturbed waters during peak recreational periods would be enormous. It was thought that the lagoons would substantially increase the breeding potential as well as providing mid-summer moulting areas. Access to the Egleton Reserve would be limited to permit holders issued to naturalists such as members of the Trust. A small car park would be built and more nature trails created, one of which would lead to Lax Hill, an area of woodland within this section.

The third part of the nature reserve, known as Burley Fishponds, had been scheduled a Site of Special Scientific Interest by the Nature Conservancy Council in the early 1960s. Its values lay in the reed beds where *Acrocephalus scirpaceus* and other rare marsh birds bred. An earlier proposal had been put forward to construct a bund across the site when it was realised that the reed bed would be submerged at top water line. After this was abandoned it was realised that the effects of inundation on the habitat was potentially of great research interest and it was decided to restrict the area to research.

In June 1972 more detailed design plans were prepared. Items such as the siting and design of islands, cover planting, nature trails, hides and wardening were considered, together with the possibility of building a countryside interpretive centre as part of the overall plan.

A detailed tree planting programme was incorporated into the overall landscape plans (Appendix 1).

Low scrub planting was recommended for sections of the shoreline to provide a screening belt to minimise disturbance along the nature trails. The cover planting would consist of mainly *Salix* spp. and other water tolerant species. New tree plantations, consisting of local indigenous species such as *Quercus robur* and *Fraxinus excelsior,* would be diversified by the inclusion of a mixed shrub layer such as *Corylus avellana* and *Acer campestre.* A tree nursery was to be created as soon as possible. Bird watching hides would be sited in both the public and permit holders areas overlooking many of the proposed habitats. Designs would be based on existing hides in use elsewhere in the country.

Management

With the appointment of a Warden in April 1975, these plans began to be put into practice. The tree planting programme had, however, begun a little earlier in March 1975 when volunteers planted 2000 *Salix* spp. along a section of the Lyndon reserve. Water tolerant species such as *Salix viminalis* and *Salix cinerea* were planted along the top water line and in some sections a metre below it. It was intended that this would help to break up the shoreline and would provide nesting habitat for bird species at times when the willows were partially submerged at high water level. Above the top water line *Salix triandra* and *Salix fragilis* were the most numerous species planted.

During the summer of 1975 a shallow wader scrape was constructed. The site chosen for the scrape was relatively flat, close to the top water line and near to a natural spring. The top soil was removed to expose the underlying clay and the whole area graded to within a few centimetres. The excess soil was used to form a screened walkway to the hide site. A small clay bank was constructed around the perimeter of the scrape with a simple water control system built into the bank. The water from the spring was directed into the scrape to the desired level. The variable water depths caused by the grading would provide ideal feeding conditions for waders adapted to feeding at different depths.

Two islands were constructed along the Lyndon reserve designed to retain water during drawdown periods. The islands were constructed of waste materials from the reservoir site, a factor which substantially reduced their overall cost. Each island was designed to have a number of spits so that additional secluded bays were formed. It was also envisaged that the spits would provide nesting sites and loafing spots for wildfowl.

In September 1975 work began on the clay bund which was to enclose three lagoons on the Egleton reserve. The site was marked out and the top soil cleared to expose the underlying clay. Further clay was brought by motor scrapers from a borrow pit within the reservoir area and compacted by rollers to form a mile long bank. At various points on the bund the level of clay was raised so that at top water line the whole structure would appear as a series of islands and only during drawdown periods would the bund act as a dam. The original top soil was then replaced on top of the bund and the area seeded with grass. At certain points stone rip-rap was placed on the main reservoir side of the bund to prevent erosion from wave action. In order to divide each lagoon two banks were constructed forming canalised sections of water. Each section terminated in a set of penstocks that were linked to their respective lagoons and to the main bund.

The laying out of the nature trail in the small wood known as Gibbet Gorse in the Lyndon reserve was started in October 1975 in the form of a circular ride cut through the dense scrub. A number of glades were also formed to add a further diversity of habitats within the woodland. Also in 1975 a scheme to introduce nest boxes for specific species was started to compensate for the loss of many natural sites which resulted from the clearance work within the reservoir site.

A further 11 349 trees were planted during the winter of 1975/76 comprising cover belts of *Salix* spp., plantations of *Quercus robur* and *Fraxinus excelsior* and new hedgerows.

During 1976 work continued on the bunds, the construction of islands within the lagoons and the laying out and seeding of footpaths both in the public and permit holder sections of the reserve.

Following the severe drought during 1976 a large number of young trees died and these were replaced during the winter of 1976/77. *Betula pendula* and *Alnus glutinosa* were planted close to the top water line mark where the prevailing south-westerly wind would eventually help to disperse the seeds on the shallow water. Both species of tree produce an abundance of seed which is a valuable source of food for wildfowl.

By now the project was generating a great deal of interest and curiosity. It became necessary to set up a voluntary wardening scheme to help control the casual weekend visitors. The water levels in the reservoir were steadily rising and the wildfowl numbers were increasing as were the bird watchers. In March 1977, 31 Trust members joined the wardening scheme and this number has been maintained ever since. The voluntary wardens play an extremely important role on the nature reserve, assisting in wardening, recording and all forms of management work.

In order to maintain the interest shown by the public, the Lyndon reserve was opened for a series of Open Days during that summer. Hide construction had been delayed for as long as possible so that the proposed sites could be assessed for bird usage. Designs were drawn up and two 30 ft hides with concrete block sides and wooden top sections were built, one overlooking the wader scrape and the other an area of deep water on the Lyndon reserve. The siting of the third hide was delayed when the first site selected proved unsuitable. During the winter of 1977/78 two hides were built and erected by a team of voluntary wardens on the Egleton section of the reserve. These were all wooden structures, one sitting on a mound of clay and the other raised on stilts a metre from ground level.

Tree planting continued during that winter with a further 3 743 trees added to the reserve.

In the lagoons a series of small islands were raised near proposed hide sites. Some were planted with *Juncus* spp. to provide stability around the edges and cover for nesting birds; other were topped with shingle to attract species requiring that particular form of nesting habitat.

The Lyndon reserve was opened on a temporary basis during the summer of 1978. The full facilities were still incomplete but the shelter belt planted in 1975 had grown sufficiently to provide enough cover to allow public access. The visitor usage was monitored throughout the season and a full visitor survey conducted on two occasions. A small charge was made for adults while children entered free. 3012 adults visited the reserve on the 67 open days, although apart from a sign at the entrance to the reserve, no attempt was made to advertise the reserve. The potential of the reserve as an educational facility was recognised and to cater for this interest four Teachers' Courses were held during 1978 which resulted in a total of 120 teachers visiting the reserve.

Two more hides overlooking the lagoons were built in that summer and a fifth was completed in December. The nature trail around Lax Hill, a woodland on the Egleton reserve, was completed and linked up with the trails leading to the hides around the lagoons.

The Egleton reserve was opened officially on 6 January 1979. It was planned that the reserve would be open on Wednesdays, Saturdays and Sundays throughout the year. The number of visitors to this section was 1 807 in 1979 increasing to 3 227 in 1980. These figures represent day permit holders and do not include visits made by members of the Leicestershire and Rutland Trust.

The harsh winter of 1978/79 resulted in a much reduced planting season. Although the major planting programme had been completed there were still several hundred species, such as *Acer campestre, Corylus avellana, Viburnum opulus, Euonymus europaeus* and *Rhamnus catharticus,* which had been planted as seedlings in a tree nursery in 1975. It was intended that these species would provide an important shrub layer in the new woodlands on the reserve.

The reservoir finally reached top water level in March 1979. The water was maintained at a constant level for most of the following summer and only a slight drawdown occurred during the winter months. There had already been some erosion to the bunds and preventative measures were taken in the form of stone rip-rap on the reservoir facing side. Further erosion took place during the high water levels necessitating further repairs in 1980. To help combat future erosion, *Salix* spp. will be planted along the bund to help stabilise the bank.

In order to cope with the increase in visitors to both areas, the third hide on the public reserve was completed in 1980. Two of the existing hides on the Egleton reserve were extended from 15 to 30 ft. The siting of the hides made them inaccessible for disabled persons. A 45 ft hide was constructed during the summer of 1980. It was built facing the lagoon nearest to the car park with easy access for wheelchairs along the tarmacadam path leading to a ramp into the hide. A wheelchair is kept at the car park.

Adult visitors to the Lyndon reserve during the summers of 1979 and 1980 numbered 2911 and 3 238 respectively. The slight fall in 1979 was probably due to the rising costs of petrol.

The majority of the facilities for the general public visiting the Lyndon reserve were almost complete by 1980. What should have been the focal point of this area, the Interpretive Centre, was delayed due to problems of funding and management. However, it is envisaged that a centre, reduced both in size and facilities from the original design, will be operational by the end of the current year.

Wildlife recording

Daily ornithological records have been kept for six years. These include numbers and distribution of both common and rare species. A breeding bird census has been carried out annually since 1976. Mid-month wildfowl counts have been undertaken since May 1975. Table 1 represents the regular winter populations that have occurred at Rutland Water since flooding began in 1975. The 'regular' population for a water is the mean of the three highest counts in each winter for every winter in which at least five counts have been made (Atkinson-Willes 1963). Approximately 2 000 birds have been ringed annually since 1976. The total number of bird species recorded had reached 211 by May 1981 (Appendix 2).

The ornithological growth was related to the surface area of water. In the summer of 1975 the

Table 1. 'Regular' winter populations for 13 common waterfowl species at Rutland Water based on the mid-month Wildfowl Trust counts.

Species	75/76	76/77	Peak flooding period 77/78	78/79	79/80	80/81
Podiceps cristatus	46	32	27	134	99	54
Podiceps ruficollis	11	25	33	51	57	71
Cygnus olor	6	6	32	56	50	43
Anas platyrhynchos	795	2578	2145	1634	1318	1557
Anas strepera	0	15	49	118	231	122
Anas crecca	257	1553	993	921	803	942
Anas penelope	202	3106	2151	1612	1855	2935
Anas clypeata	92	189	326	395	335	207
Aythya fuligula	244	554	802	1711	1744	1289
Aythya ferina	238	1414	1069	408	672	498
Bucephala clangula	12	37	83	105	107	128
Mergus merganser	3	4	12	36	16	46
Fulica atra	513	1998	2709	2846	2883	1136

total surface area was 400 acres. The water area increased slowly during the winter of 1975/76. Wildfowl populations remained fairly static for the first part of that winter and only increased after January when water levels rose, providing more food for dabbling duck species (Table 2). The constant water levels during the 1976 summer proved ideal for breeding Podicipidae species. The gradual evaporation caused by a drought produced perfect feeding conditions for migrant waders in the same year.

In the autumn of 1976, water flowed into the reserve boundaries for the first time. Until then the only standing water had been the Site of Special Scientific Interest at Burley Fishponds. Between October 1976 and March 1977 the water surface area increased from 700 to 2 600 acres. The abundance of food associated with rapid flooding of arable land resulted in the highest numbers of wildfowl recorded at Rutland Water (Table 3).

Wildfowl bred in the lagoons for the first time in 1977 and seven species successfully raised young. The numbers and varieties of breeding waterfowl has steadily increased since 1977 (Table 4).

The recreational zoning policy was tested for the first time during that summer and autumn as the fishing season had started at Rutland Water; sailing had begun in 1976.

The 1977 mid-September waterfowl count produced a total of 4 333 birds. Ninety-three were

Table 2. 1975/76 Winter waterfowl counts at Rutland Water based on mid-month counts.

	September	October	November	December	January	February	March
Podiceps cristatus	67	43	8	5	8	9	28
Cygnus olor	6	7	6	0	2	5	0
Anas platyrhynchus	750	1038	590	268	599	596	547
Anas crecca	72	250	74	51	175	263	258
Anas penelope	3	7	21	83	65	293	230
Anas acuta	0	1	0	0	0	8	6
Anas clypeata	90	126	62	29	28	0	8
Aythya fuligula	180	246	236	252	198	83	124
Aythya ferina	189	65	147	164	195	331	170
Bucepyala clangula	0	0	4	3	15	11	11
Mergus merganser	0	0	0	0	2	1	8
Tadorna tadorna	4	2	0	0	0	0	0
Fulica atra	283	423	414	497	441	543	499

Table 3. Winter waterfowl counts at Rutland Water based on mid-month counts - 1976/77.

	September	October	November	December	January	February	March
Podiceps cristatus	27	31	12	12	25	38	29
Podiceps ruficollis	49	19	8	5	1	0	8
Cygnus olor	2	1	2	6	8	1	6
Cygnus bewickii	0	4	40	46	76	4	1
Anser albifrons	0	0	0	1	0	0	0
Anser fabalis	0	0	0	0	0	1	0
Anser brackyrhynchus	0	0	0	0	10	0	0
Anas platyrhynchos	1195	2207	2567	2961	2158	1652	1304
Anas crecca	161	877	720	1159	2038	1462	1106
Anas strepera	3	2	8	18	20	4	0
Anas penelope	5	94	333	656	4518	2352	2449
Anas acuta	0	46	40	2	10	16	16
Anas clypeata	143	231	194	105	130	64	93
Aythya fuligula	320	535	703	426	418	202	299
Aythya ferina	99	480	1444	1242	1556	563	247
Aythya marila	0	0	1	0	0	0	0
Fulica atra	1484	1145	1400	2569	1943	381	423
Total	3488	5672	7472	9208	12911	6740	5981

Table 4. Numbers of waterfowl breeding within the Nature Reserve waters at Rutland Water.

Species	1977 Year	1978 Year	1979 Year	1980 Year
Podiceps cristatus	6	18	23	18
Podiceps ruficollis	12	19	22	14
Cygnus olor	1	4	7	6
Branta canadensis	0	0	1	1
Anser anser	0	0	0	1
Tadorna tadorna	1	2	3	1
Anas platyrhynchos	19	38	38	35
Anas strepera	0	0	0	2
Anas crecca	1	1	0	0
Anas clypeata	4	2	1	5
Aythya fuligula	4	10	8	10
Aythya ferina	2	1	0	1
Oxyura jamaicensis	0	0	0	2

Table 5. Waterfowl distribution at Rutland Water 10 days before end of fishing season and 10 days after start of close season (sample year 1979).

Acreage	Recreation activity		Wildfowl numbers	
			Fishing	No fishing
2064	Sailing, boat/bank fishing, walking, riding		186	2092
850	Boat fishing only		1335	2827
150	Protected areas	Lagoons	6827	2130
50		S.S.S.I	1128	528

recorded within the sailing area where bank and boat anglers were also present. 995 birds were counted in the area behind the sailing limits used by boat anglers only. 3 245 birds were counted in the lagoons and Site of Special Scientific Interest. This seasonal waterfowl usage of the lagoons and Site of Special Scientific Interest has continued each year with up 7 955 birds recorded. At the end of each fishing season there is a decline in waterfowl usage of the lagoons as birds disperse over the whole water (Table 5). This trend continues throughout the winter months unless there is some weekend sailing activity.

Two bi-annual birds reports have been published (Appleton 1978, 1980). The reports comprise classified lists of all bird species recorded at the reservoir, the results of annual breeding bird censuses and records of all birds ringed on the nature reserve.

Discussion

The undoubted success of the Rutland Water Nature Reserve on an artificially constructed wetland is due to a variety of factors. The success in the improvement of existing habitats and the creation of new ones on a scale never before attempted in Great Britain, is confirmed by the significant number of wildfowl now wintering and breeding on the reserve. The numbers of *Anas clypeata, Anas strepera, Anas crecca, Anas platyrhynchos, Anas penelope, Aythya fuligula and Aythya ferina* recorded each winter makes the reserve of national importance. The numbers of *Bucephela clangula* and *Mergus merganser* are rapidly approaching the same status. Until recently *Anas clypeata, Anas strepera* and *Anas crecca* winter populations gave the reserve international status under the criteria adopted at Ramsar, Iran by the Convention on Wetlands of International Importance especially as Wildfowl Habitats. But with the new criteria (Scott D. A. 1980) in which the North West African wintering waterfowl populations are added to the West European population the reserve no longer achieves this status for these species.

The provision of lagoons and the zoning policies have alleviated the problems of disturbance to wildfowl, both in the summer and winter by other recreational users. This fact was endorsed in a report (Water Space Amenity Commission Research Report 5 1979) which described the Rutland Water reserve project as a prime example of its kind.

The future

The continued success of the reserve will depend on the alertness of the conservation organisations in safeguarding the wildlife interests. An increase in leisure activities will result in a demand for additional recreational facilities on reservoirs. Proposals such as the introduction of water skiing, motor boating and winter fishing can only reduce the wildlife potential.

Problems specific to Rutland Water, such as wave erosion of the bunds, will necessitate constant attention and resources. Normal annual drawdown can be tolerated and in most cases is beneficial to species using reservoir habitats. A prolonged drawdown, however, could have serious implications and reduce the wildfowl populations significantly. The future management of the reserve must be progressive and, to an extent, experimental, but the aims will remain, as ever, to conserve the plant and animal life and to provide a sanctuary for wildfowl.

Acknowledgements

It is to the Welland and Nene River Authority, later to become the Welland and Nene River Division of the Anglian Water Authority, that we owe our greatest debt. Their imaginative forward thinking and sympathetic understanding of the conservation requirements has seen the establishment of a nature reserve of great importance.

Leicestershire and Rutland Trust Honorary Officers Hugh Dixon, Honorary Treasurer and Rutland Water Reserve convener, Geoff Stansfield, Honorary Secretary, Pat Herbert, Administrative Officer and all members of the Reserve Committee have done much to bring the reserve to its present state of operation. Peter Langford, Divisional Manager of the Welland and Nene Division of the Anglian Water Authority and his fellow officers have supported the scheme from the outset.

Harry Crowe, the Reservoir Manager, Jonathan Inglesant, Recreation Officer and Frank Knights, Reservoir Engineer, have shown a tremendous understanding and willingness to help. Nature Conservancy Council officers Peter Nicholson and Edward Hammond have helped plan and advise on many aspects of design and management. Their involvement has been considerable.

Dame Sylvia Crowe, the Landscape Consultant for the reservoir project, showed enthusiasm and knowledge of conservation that proved a major factor in the success of the planting programme. The Anglian Water Authority's Head Forester, Tony Ford, has helped enormously with both advice and practical assistance.

I must also mention past assistant wardens Tim Pridmore and Graham Elliott and the current assistant warden, Chris Park, who have all contributed greatly to the creation of the reserve. The voluntary wardens, whose continuous support we enjoy, deserve special acknowledgement.

In preparing this paper, I am greatly indebted to Geoff Stansfield and David Harper for their advice and comments and to everyone who has helped type and re-type the manuscript.

References

Appleton, T. P., 1978. Birds of Rutland Water 1975–1976. Leicestershire and Rutland Trust for Nature Conservation.

Appleton, T. P., 1980. Birds of Rutland Water 1977–1978. Leicestershire and Rutland Trust for Nature Conservation.

Atkinson-Willes, G. L., 1963. Wildfowl in Great Britain. Monographs of the Nature Conservancy Number 3, H.M.S.O.

Batten, L. A., 1977. Sailing on Reservoirs and its Effects on Water Birds. Biological Conservation, Vol. 11, 1977, pp. 49–58.

Convention on Wetlands of International Importance Especially as Waterfowl Habitats. Cmd. 6465, Treaty series No. 34 (1976). H.M.S.O., London, 1976.

Cooke, A., 1977. The Birds of Grafham Water. Huntingdon Local Publications Group, Huntingdon, 1977.

Nicholson, P. B., 1972. Empingham Reservoir, Planning for Nature Conservation. Nature Conservancy Council 1972.

Scott, D. A., 1980. A Preliminary Inventory of Wetlands of International Importance for Waterfowl in West Europe and North West Africa. International Waterfowl Research Bureau.

Appendix 1. Number of trees and shrubs planted each winter at Rutland Water Nature Reserve.

Species	74/75	75/76	76/77	77/78	78/79	79/80	80/81
Larix decidua		261				208	6
Pinus sylvestris			100				
Populus nigra italica		58	185				
Populus trichocarpa		60					
Populus canescens				24		6	
Populus tremula		50					
Salix fragilis		311	80				
Salix alba		20	119				
Salix caledenaron					140		57
Salix cinerea		1626					
Salix daphnoides		127	50	156			
Salix triandra		205					
Salix viminalis		834	67	509		255	
Salix viminalis/cinerea		905	869				
Salix spp.	2000	336	2506	660		907	2005
Betula pendula		661	472	268		28	
Alnus glutinosa		954	700	155			
Carpinus betulus			50				7
Corylus avellana		203	100			28	56
Fagus sylvatica							4
Quercus robur		235	1109				50
Crataegus monogyna		2800	2840	1402	3580		2000
Sorbus aucuparia		40		47		16	
Sorbus terminalis				8			
Sorbus aria		11		27			
Malus sylvestris				125			
Pyrus pyraster				7			
Prunus avium		38					
Acer platanoides		31					
Ilex aquifolium			6	5			
Acer campestre		74	300	25		1	5
Acer pseudoplatanus		130					
Tilia cordata				11			
Fraxinus excelsior		1376	1974	20			
Cornus sanguinea		3		294	52	42	
Viburnum opulus						124	16
Euonymus europaeus						53	
Grand total 38265	2000	11349	11527	3743	3772	1668	4206

Appendix 2. Birds recorded at Rutland Water – April 1975 to May 1981.

Gavia arctica
Gavia immer
Gavia stellata
Podiceps cristatus
Podiceps griseigena
Podiceps auritus
Podiceps nigricollis
Tachybaptus ruficollis
Puffinus puffinus
Fulmarus glacialis
Sula bassana
Phalacrocorax carbo
Phalacrocorax aristotelis
Ardea cinerea
Botaurus stellaris
Ciconia ciconia
Anas platyrhynchos
Anas crecca
Anas querquedula
Anas strepera
Anas penelope
Anas acuta
Anas clypeata
Netta rufina
Aythya marila
Aythya fuligula
Aythya ferina
Bucephala clangula
Clangula hyemalis
Falco tinnunculus
Alectoris rufa
Perdix perdix
Coturnix coturnix
Phasianus colchicus
Rallus aquaticus
Porzana porzana
Gallinula chloropus
Fulica atra
Haematopus ostralegus
Vanellus vanellus
Charadrius hiaticula
Charadrius dubius
Pluvialis squatarola
Pluvialis apricaria
Arenaria interpres
Gallinago gallinago
Lymnocryptes minima
Scolopax rusticola
Numenius arquata
Numenius phaeopus
Limosa limosa
Limosa lapponica
Tringa ochropus
Tringa glareola
Melanitta fusca
Melanitta nigra
Mergus serrator
Mergus merganser
Mergus albellus
Oxyura jamaicensis
Tadorna tadorna
Tadorna ferruginea
Anser anser
Anser albifrons
Anser fabalis
Anser brachyrhynchus
Anser caerulescens
Branta bernicla
Branta leucopsis
Branta canadensis
Alopochen aegyptiata
Cygnus olor
Cygnus cygnus
Cygnus beweckii
Buteo buteo
Buteo lagopus
Accipiter nisus
Circus aeruginosus
Circus cyaneus
Pandion haliaetus
Falco subbuteo
Falco peregrinus
Falco columbarius
Calidris maritima
Calidris minuta
Calidris temminckii
Calidris malanotos
Calidris alpina
Calidris ferruginea
Calidris alba
Philomachus pugnax
Recurvirostra avosetta
Phalaropus lobatus
Glareola pratincola
Stercorarius parasiticus
Stercorarius skua
Stercorarius pomarinus
Larus marinus
Larus fuscus
Larus argentatus
Larus canus
Larus hyperboreus
Larus glaucoides
Larus melanocephalus
Larus minutus
Larus ridibundus
Larus sabini
Rissa tridactyla
Tringa hypoleucos
Tringa totanus
Tringa erythropus
Tringa nebularia
Limnodromus scolopaceus
Calidris canutus
Columba palumbus
Streptopelia turtur
Streptopelia decaocto
Cuculus canorus
Tyto alba
Athene noctua
Strix aluco
Asio otus
Asio flammeus
Apus apus
Apus melba
Alcedo atthis
Picus viridis
Dendrocopos major
Dendrocopos minor
Alauda arvensis
Hirundo rustica
Delichon urbica
Riparia riparia
Corvus corone corone
Corvus corone corvix
Corvus frugilegus
Corvus monedula
Pica pica
Garrulus glandarius
Parus major
Parus caeruleus
Parus ater
Parus palustris
Parus montanus
Aegithalos caudatus
Prunella modularis
Anthus pratensis
Anthus trivialis
Anthus cervinus
Anthus spinoletta
Motacilla alba
Motacilla alba alba
Motacilla cinerea
Motacilla flava
Motacilla flava flava
Bombycilla garrulus
Lanius excubitor
Sturnus vulgaris
Coccothraustes coccothraustes
Carduelis chloris
Chlidonias niger
Sterna hirundo
Sterna paradisaea
Sterna albifrons
Sterna sandvicensis
Columba oenas
Panurus biarmicus
Sitta europaea
Certhia familiaris
Troglodytes troglodytes
Turdus viscivorus
Turdus pilaris
Turdus philomelos
Turdus iliacus
Turdus torquatus
Turdus merula
Oenanthe oenanthe
Saxicola torquata
Saxicola rubetra
Phoenicurus phoenicurus
Phoenicurus ochruros
Luscinia megarhynchos
Erithacus rubecula
Cettia cetti
Locustella naevia
Acrocephalus scirpaceus
Acrocephalus schoenobaenus
Sylvia atricapilla
Sylvia borin
Sylvia communis
Sylvia curruca
Phylloscopus trochilus
Phylloscopus collybita
Regulus regulus
Regulus ignicapillus
Muscicape striata
Ficedula hypoleuca
Carduelus carduelis
Carduelis spinus
Acanthis cannabina
Acanthis flavirostris
Acanthis flammea
Pyrrhula pyrrhula
Fringilla coelebs
Fringilla montifringilla
Emberiza citrinella
Emberiza calandra
Emberiza schoeniclus
Plectrophenax nivalis
Passer domesticus
Passer montanus

Discussion

Mr. T. Appleton said that one of his aims was to demonstrate that, contrary to the opinions of some experts, a very effective wildfowl reserve could be constructed by manipulation of the shore line in shallow water areas of a reservoir. Such manipulations included the construction of scrapes, bunds, islands and bays to provide nesting sites, refuges; feeding areas, etc. The location of Rutland Water had proved ideal to attract birds using the valleys of the rivers Welland and Nene as migration routes.

He emphasised the importance of having both a warden on the site well before the reservoir began to fill and the willing cooperation of the responsible Water Authority as well as the local conservation trust. He then demonstrated that much of the modification had been achieved by utilising waste material from elsewhere in the reservoir construction, together with machinery already on site, and indicated the importance of volunteer groups in tree planting, hide construction and similar activities.

He closed by demonstrating the great importance of the nature reserve as a refuge for wildfowl from disturbance stemming from other activities on the reservoir.

Mr. E. Hammond, in opening, pointed out that not every reservoir could support the numbers of birds that had moved to Rutland Water, nor was every reservoir worth the effort. He also emphasised that, although the protection and encouragement of wildfowl and waders was the most outstanding feature of the reserve, there were plenty of other features of the reserve in which there was a vigorous conservation programme involving areas of old woodland and old pasture.

He had noted that the number of visitors was increasing and felt that in this, as in all else, there must be a clear management policy. In particular, a decision would have to be made as to when the numbers must be limited.

He considered it most important that the success of the nature reserve must be put into perspective. In Leicestershire, the gain of this reserve must be offset against the loss of some 170 ha of prime habitats since 1974, and he stated that there were now parishes in the county virtually devoid of any true wild life habitats, even those as simple as a pond or a hedge. He ended by asking Mr. Appleton what advice he would give to anyone setting up a reserve.

Mr. Appleton replied that, on the question of the number of visitors, birds rapidly became habituated to visitors so long as people were confined to the one route. Nonetheless, if numbers threatened the sanctuary status, limits must be applied.

He added that the most important advice he could give would be to ensure that conservation interests were represented effectively from the very beginning. Then, as in the Rutland Water case, a valuable working relationship could be initiated and maintained with all interested groups.

Mr. R. Fisher asked what mammals had been recorded in the area. He had noticed apparent damage by deer in one of the demonstration pictures. He further enquired about the occurrence of any unexpected species and any predators such as mink.

Mr. Appleton replied that fallow deer (*Dama dama*) had now disappeared from the immediate area, but would doubtless return. Neither mink (*Mustela vison*) nor coypu (*Myocastor coypus*) had been recorded, although there was some evidence for the presence of at least one otter (*Lutra lutra*). Small mammals were abundant and attracted many birds of prey such as short-eared owls. He had recorded many rare birds, including several N. American vagrants like the long-billed dowitcher (*Limnodromus scolopaceus*) and pectoral sandpiper (*Calidris melanotos*).

Dr. Huggins asked whether charging for admission could be used to limit numbers of visitors. Mr. Appleton said that he considered that people must be prepared to pay for facilities such as those provided at Rutland Water but that he did not believe that any but the most casual of visitors had been dissuaded from making a visit.

Mr. Davey asked what had been the basis for the selection of the reserve area. Had it been mainly a question of deciding the area and then modifying the habitat to attract the birds? He further asked if, with time, the birds would become fully accustomed to visitors and thus be unaffected by them. Mr. Appleton replied that the basic criterion had been shallowness of the water. This made it ideal for birds and less suitable for other activities such as yachting. Birds could become habituated to visitors, but wildfowl are quarry species and therefore would never become wholly at ease.

Dr. Bullock asked what other conservation activities, aimed at the terrestrial habitats, were currently in hand. Mr. Appleton said that old woodland management practices, such as coppicing, were being used to encourage diversity. New plantations had been established; meadows were left short, which encouraged widgeon, but also plant diversity was enhanced. Old hedgerows were properly laid, while ponds were being excavated for freshwater species.

The distribution of micro-organisms through a bunded lagoon in Rutland Water

J. T. Holah & J. D. McIver
Water Resources Unit, Trent Polytechnic, Nottingham, England

Keywords: micro-organism, pathogen, wildfowl, sedimentation, distribution, botulism

Abstract

The distribution of micro-organisms entering a bunded lagoon in Rutland Water from a sewage effluent outflow was assayed. The majority of micro-organisms were found to settle out in the immediate vicinity of the effluent outflow. In this area micro-organisms, especially spore formers, were present in numbers exceeding 1×10^6 organisms per gram of sediment. The lagoon lies within a nature reserve and a reduction in the reservoir water level could lead to contact between shore feeding waterfowl and pathogenic micro-organisms in the freshly exposed sediment.

Introduction

Oakham Sewage Works final effluent is discharged into a bunded lagoon in the north western corner of the lower arm of Rutland Water. This area is designated as a bird sanctuary and is of national importance for both endemic and migrant species of birds. Previous research at Trent Polytechnic, concerned with the removal of pathogens and the dispersal of effluents into watercourses, had shown the presence of a botulism indicator organism, *Clostridium perfringens*, in the sewage works final effluent. This investigation set out to follow the fate of micro-organisms entering the lagoon via the sewage effluent and to ascertain any potential health hazard to the birds.

It is thought that micro-organisms in sewage effluents tend to be attached to particulate matter (Weiss 1951; Hendricks 1971; Rittenberg *et al.* 1958) and hence sediment out of the liquid phase when flow rates are sufficiently reduced to allow deposition. Whilst few micro-organisms multiply within sediment, most are able to survive for significant lengths of time (Hendricks 1971; Nolongoski *et al.* 1975). The lagoon is very shallow and any reduction of the reservoir level such as occurs in summer drawdown would expose substantial amounts of lagoon sediment. Micro-organisms within this sediment would then be accessible to wading birds.

Of the diseases caused by micro-organisms in sediments one of the most pathogenic to wild fowl is botulism, a disease that since 1969 has become more widespread in this country (Blandford *et al.* 1969; Keymer *et al.* 1972). Botulism is a paralytic disease caused by ingestion of the lethal neurotoxin of *Clostridium botulinum*, an anaerobic spore-forming organism whose natural habitat is the soil. Mortality is frequently very heavy and it is not uncommon for a single outbreak to result in the death of many thousands of birds (Kalmback & Gunderson 1934).

The factors that favour the precipitation of an outbreak of botulism in waterfowl include a prolonged spell of warm weather, enlarged areas of shallow stagnant water, an abundance of aquatic invertebrates and oxygen depletion associated with large ammounts of rotting vegetation or other organic matter (Bell *et al.* 1955). With the shallow nature of the bunded lagoon and the normal

Hydrobiologia 88, 225–229 (1982). 0018-8158/82/0882-0225/$01.00.

summer drawdown of the reservoir, these factors could occur.

Methods

Samples of the lagoon water and sediment were taken twice weekly during April and May 1980. Water samples were taken just below the lagoon surface and corresponding sediment samples were taken directly below this point with a sterile drag sampler. Samples were then stored in a 'cool box' before analysis at Trent Polytechnic, usually within two hours.

Sampling strategy involved taking samples in an approximate straight line from a point close to the effluent inlet to a point on the lagoon perimeter. The strategy was adopted to take into account the variables of possible stream flow through the lagoon, the influence of surface currents, differences in organism and sediment load of the sewage effluent and differences in stream entry due to lowering of the reservoir level.

Both sediment and water samples were analysed for coliform organisms, the standard indicator organisms of water quality and *Clostridium perfringens,* an indicator of spore-forming pathogens such as *Clostridium botulinum.* Coliforms were identified using the 'Most Probable Number' method with analysis for both presumptive and differential coliforms. *Clostridium perfringens* was analysed on SPS agar (Difco) and black colonies formed after 24 hours were counted. All samples were filtered for dry weight analysis.

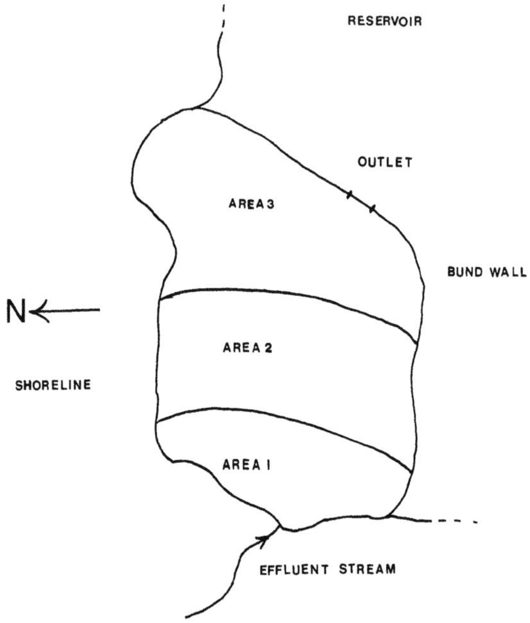

Fig. 1. Sketch map of the bundel lagoon showing the three arbitrary areas.

Results

The lagoon was divided into three arbitary areas in an attempt to analyse the effect of distance from the sewage effluent inlet on water and sediment sample counts. To produce these areas the lagoon was divided by two imaginary lines, all points on any one line being approximately equidistant from the effluent stream inlet (Fig. 1). When comparing counts from water and sediment samples, it was assumed that all micro-organisms were associated with particulate matter. The counts could then be correlated with the dry weight of the sample and be expressed as micro-organisms per gram solids. Mean organism counts were calculated for each of the areas of the lagoon and also for the effluent stream (Table 1). The correlation between micro-organisms counts and distance from the sewage

Table 1. Mean indicator organism counts per gram solids in each area of the lagoon.

Indicator organism	Sample	Stream	Area 1	Area 2	Area 3
Presumptive coliforms	Water	3.67×10^6	4.94×10^5	9.26×10^4	3.61×10^4
	Sediment	2.31×10^4	2.42×10^3	4.28×10^2	1.86×10^2
Differential coliforms	Water	3.23×10^4	3.73×10^3	1.88×10^3	1.65×10^3
	Sediment	2.85×10^2	6.07×10^1	2.67×10^1	1.05×10^1
Clostridium perfringens	Water	5.07×10^6	3.54×10^6	1.32×10^5	4.17×10^4
	Sediment	4.92×10^5	1.73×10^5	1.02×10^5	3.08×10^4

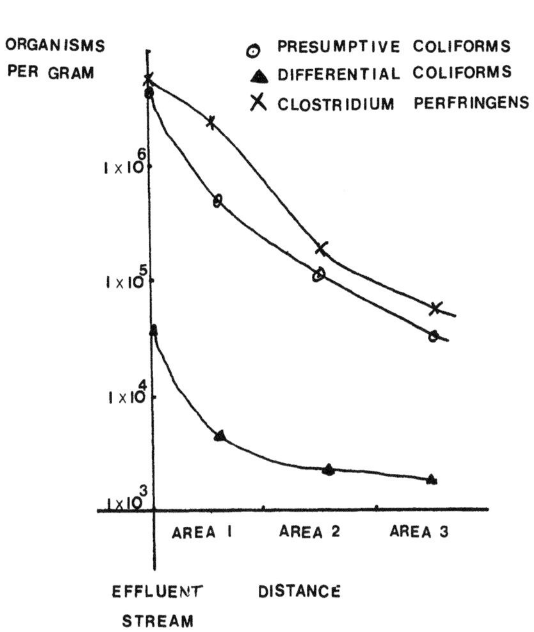

Fig. 2. Graph of mean indicator organism counts per area versus distance from the sewage effluent inlet for water samples.

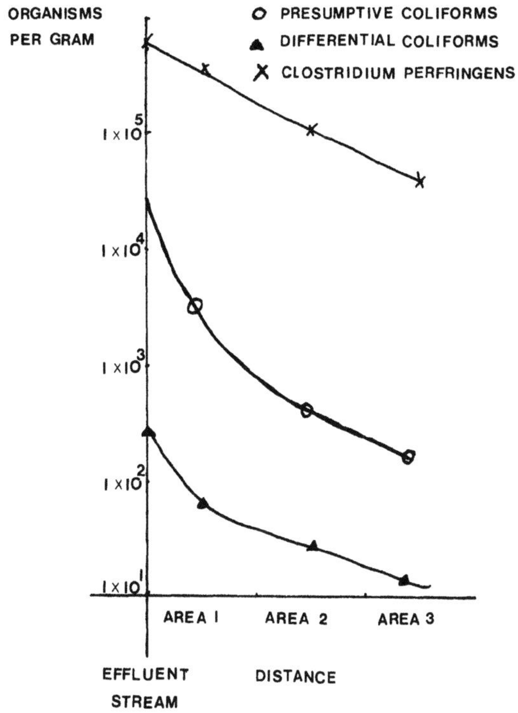

Fig. 3. Graph of mean indicator organism counts per area versus distance from the sewage effluent inlet for sediment samples.

effluent intake is shown in Figs. 2 and 3.

The assumption that micro-organisms are attached to particulate matter could be checked by comparing the effect of distance from the sewage effluent intake on micro-organism counts and on the dry weight content of the water samples. The relationship between mean dry weight of the water samples and distance from the sewage effluent intake is shown in Fig. 4.

A statistical analysis of the micro-organism mean counts for each area was undertaken using the 'Analysis of Variance' (ANOVA) test. As can be seen from Table 2, only the mean counts for sediment samples gave a significant difference between areas of the lagoon.

Discussion

The sewage final effluent flow forms the only inlet to the lagoon while the outlet under normal conditions is an adjustable gate to an adjacent bund. When the reservoir water level is high however, as during this project, the outlet may be extended by breaching the bund wall, making a continuous contact with the reservoir outside. In either case the water in the lagoon itself moves in a 'plug flow' fashion between inlet and outlet. Under these circumstances, if we were to ignore such

Table 2. ANOVA analysis of the indicator organism counts per gram solids between the three areas of the lagoon.

Indicator organism	Sample	F_{obs}	F 0.05	Significance
Presumptive coliforms	Water	1.41	3.46	No significant difference
	Sediment	6.59	3.49	Significant difference
Differential coliforms	Water	1.23	3.46	No significant difference
	Sediment	0.58	3.49	No significant difference
Clostridium perfringens	Water	4.10	4.35	No significant difference
	Sediment	10.17	4.12	Significant difference

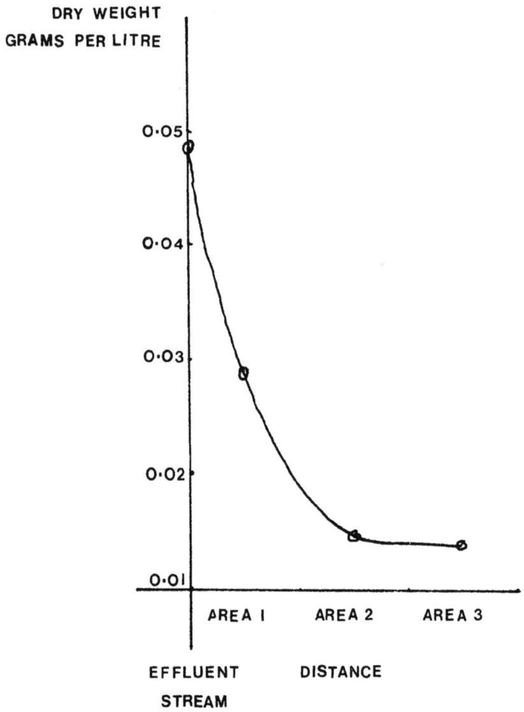

Fig. 4. Graph of mean suspended solids versus distance from the sewage effluent inlet.

effects as microbial mortality, sedimentation or growth, the number of micro-organisms leaving the lagoon should be that entering it. Figures 2 and 3 show however that there is a reduction in the concentration of micro-organism numbers as the water moves through the lagoon, with the greatest reduction within area 1 and a smaller reduction in areas 2 and 3. Dilution of the micro-organisms is not an explanation as in theory dilution does not occur in a plug flow system under steady state conditions.

If as assumed, the micro-organisms are attached to particulate matter, the distribution of the particles and hence the micro-organisms will depend on the mass of the particle and the velocity of the water. Figure 4 shows the distribution of suspended particles through the lagoon. The major reduction in organism numbers in area 1 is due to heavy sedimentation after the fast flowing effluent stream has entered the sluggish lagoon water. The particulate concentration tends to stabilize in areas 2 and 3, probably because the velocity of the lagoon water will remain fairly constant over this zone. The superficial shape of the plots in Figs. 2 and 3 is similar to that for the plot in Fig. 4, indicating that a large proportion of the micro-organisms entering the lagoon from the effluent stream do settle out in area 1. However it must be noted that Fig. 4 is arithmetically plotted whereas Figs. 2 and 3 have logarithmic Y axes. This means that the majority of micro-organisms attached to particles in the water phase are not apparent in the sediment phase resulting from that water. As such it must be concluded that these organisms have died. It will also be noted that the mortality rate is lower for *Clostridium perfringens,* an organism which, being spore-forming, is better able to survive the conditions in the lagoon.

The statistical ANOVA test was inconclusive due to three factors. Firstly arbitrary lines were chosen to divide the lagoon into areas of approximately the same size and any change in the positioning of these lines may alter the significance of the analysis. Secondly a large variation in sample counts, not uncommon when analysing bacteria in water, resulted in high count standard deviations and hence low significance. Thirdly the assumption has been made that micro-organisms are solid bound and suffer the fate of the particle to which they are attached. If, as is likely, only a proportion of the effluent organisms are so related, then the analysis, including as it does death and growth as well as deposition, is unlikely to show any measurable correlation between the water phase and the resulting sediment. This view is enhanced by the results from Table 2 where significant differences are found only for organisms within the sediment.

Few numerical relationships between indicator organisms and pathogens are quoted in the literature and hence it is difficult to judge pathogen numbers in the lagoon water and sediment. However Van Donsel & Geldreick (1971) quoted a relationship of one salmonella in the sediment to 150 differential coliforms in the overlying water. This would indicate a level of 10–25 salmonellae per gram of sediment in the lagoon. With *Clostridium perfringens* counts of the order of 3×10^6 in area 1 there is a reasonable chance however of finding similar pathogenic organisms, including *Clostridium botulinum.*

From the distribution of micro-organisms through the lagoon and general observations made whilst undergoing this investigation a number of points can be concluded:

1. Micro-organisms entering a shallow reservoir from a sewage effluent are likely to settle out in the immediate area of the effluent discharge.

2. Micro-organisms, especially spore formers, are likely to be present in high numbers within the discharge area.

3. Reduction of the reservoir water level will expose sediments in the discharge area, making them available to wading birds.

4. If the reservoir is bunded to produce lagoons, a reduction in reservoir water level is likely to create areas of stagnant water, a condition that aids the development of many pathogens including *Clostridium botulinum.*

References

Bell, J. F., Sciple, G. W. & Hubert, A. A., 1955. A micro-environment concept of the epizoology of avian botulism. J. Wildlife Management 19: 352–357.

Blanford, T. B., Roberts, T. A. & Ashton, W. L. G., 1969. Veterinary Record 85: 541.

Hendricks, C. W., 1971. Increased recovery of salmonellae from bottom sediments versus surface waters. Appl. Microbiol. 21: 379–380.

Kalmback, E. R. & Gunderson, M. F., 1934. Western duck sickness: a form of botulism. USDA Tech. Bull. 411: 1–81.

Keymer, I. F., Smith, G. R., Roberts, T. A., Heaney, S. I. & Hibbard, D. J., 1972. Veterinary Record 10: 111.

Nolongoski, Klug & Kellog, 1975. The metabolism of organic compounds in fresh water lake sediments by Clostridia species. Abstracts of the Annual Meeting of the American Society for Microbiology, Vol. 75: p. 184.

Rittenberg, S. C., Mittwer, T. & Ivler, O., 1958. Coliform bacteria in sediments around three marine sewage outfalls. Limnol. Oceanogr. 3: 101–108.

Van Donsel, D. J. & Geldriech, E. E., 1971. Relationships of salmonellae to faecal coliforms in bottom sediments. Water Res. (NY) 5: 1079–1087.

Weiss, C. M., 1951. Adsorption of Escherichia coli on river and estuarine silts. Sewage and Industrial Wastes 23: 227–237.

Wildlife surveys of the reservoir site prior to inundation 1971–1976

I. M. Evans
Leicestershire Museums Service, 96 New Walk, Leicester, England

Keywords: wildlife surveys, inundation, birds, mammals, insects, plants

Abstract

When it became certain in 1970 that construction of the reservoir now known as Rutland Water would take place, members of the Rutland Natural History Society planned a survey of the natural history of its site. Fieldwork spanned the six years 1971–76 and focussed on the flowering plants, insects, birds and mammals. The results of the survey are contained in a 70 page report to be published shortly on behalf of the Society by the Leicestershire Museums Service (RNHS 1981).

Two introductory chapters set out the organisation of the survey and the timetable of the construction programme to which it was tailored.

The third chapter consists of a detailed description of the vegetation of the watercourses and adjacent farmland based on a field-by-field survey carried out in 1971–73. It also encompasses other botanically interesting areas, such as marshes, spinneys and woods, that were affected by reservoir construction. The narrative is divided into 13 sections corresponding to sub-divisions of the survey area and appended to it is a list of the 413 species of flowering plants and ferns identified. Two of the species, panicled sedge *Carex paniculata,* and lesser reedmace *Typha angustifolia,* had not been recently recorded elsewhere in Rutland and attempts were made to transplant both to safer sites. However the fate of the complex assemblage of willows, *Salix* spp., upstream of Burley Fishponds, at the inflow to the north arm, which could not be conserved in this way, has still to be assessed.

The fourth chapter is based on surveys of the insect life of the site extending over three two-year periods 1971–72, 1973–74 and 1975–76, this arrangement reflecting the need to adapt recording techniques as site clearance proceeded. A narrative account of the available habitat and changes in progress in each two-year period is followed by lists of all species identified and the localities in which they were found. Lepidoptera dominate these lists, with a total of 20 species of butterflies and over 150 species of moths. However, there are useful records of grasshoppers and crickets, dragonflies and seven other orders of insects, although coverage of these last is acknowledged to be patchy. One of the most striking features of the entomological report is the effect on populations of butterflies, for example, of the cessation of agricultural operations in areas that were destined to be flooded. Species such as common blue and small copper, which had been very scarce in the area during the 1960s, appeared early in the course of the survey on abandoned pasture and in successive years spread until they had colonised much of the reservoir site. Dragonflies, of which eight species were recorded, showed an apparent increase in numbers and grasshoppers colonised the large areas of temporarily abandoned grassland. The survey provided an opportunity for the investigation of the butterfly fauna of woodlands in the vicinity of the reservoir site such as those at Hambleton and Burley, but there was an inexplicable absence of species characteristic of this habitat such as the hairstreaks.

The fifth chapter, on the birds of the site, is centred on a tabulated summary of observations made at fortnightly intervals throughout the six years 1971–76, a remarkable achievement by any standards. Field

Hydrobiologia 88, 231–232 (1982). 0018-8158/82/0882-0231/$00.40.

records were originally summarised by parish and month and then totalled for the whole area for the 'winter' and 'summer' months. The 'winter' tables record the occurrence between October and March in the five seasons 1971/72 – 1975/76 of 99 species of birds, the 'summer' tables the occurrence between April and September of the five years 1972–1976 of 112 species. The lists overlap of course and the total count for the period was 121 species. A brief commentary on changes in status of particular groups of birds follows the tables. Predictably, species of open water and its margins increased in numbers from the summer of 1973 when water first began to accumulate in shallow lagoons upstream of the dam. Equally predictably, as hedgerows and woodland were cleared from the site, certain associated species, such as game birds, berry-feeding thrushes, tits and some warblers, declined in numbers. However the creation of a huge open area with a vigorous 'weed' flora suited small mammal predators such as kestrel and short-eared owl and large flocks of finches were seen in the winter months. The tables and summary, inadequately summarised here, bear eloquent witness to the transformation of the avifauna of an area of Midlands farmland. They put on record, for example, the first known occurrence in the area of birds such as shelduck, in 1974, and cormorant, in 1975, the first now breeding at Rutland Water, the second now present almost all year round.

The report concludes with a brief account of the mammals of the reservoir site, based on lists of sightings made at fortnightly intervals throughout the five seasons 1972–76. Again, attention was concentrated on the watercourses and immediately adjacent land, although records of bats were obtained when buildings were demolished. The records are tabulated, but the table differs from those for the birds in that the survey area was subdivided into five, permitting comparisons between parts of the site at differing stages of development. A general table for the whole site shows general trends in numbers. In all, 22 species of mammals were recorded, including pipistrelle, noctule and long-eared bats, fallow deer and harvest mouse. There is a detailed account of the fate of five badger setts within the site perimeter. The occupants of two of these were persuaded to vacate them in advance of engineering work and in the other three voluntary evacuation was followed by successful reoccupation. At the conclusion of operations at least three occupied setts remained within the boundary of the Water Authority's land and their future seems to be assured.

The report is illustrated with some 24 photographs of the reservoir site. These include a number providentially taken prior to any work in the area, both broad landscapes and closer views of typical areas of habitat. There are a number of contrasting pairs of the same view before and after clearance and some of the destruction of habitat as it took place. Taken with the systematic records they constitute a unique record of a now vanished landscape and its wildlife.

Reference

Rutland Natural History Society, 1981. Before Rutland Water. Report of a survey of the wildlife of the site before and during the construction of the reservoir, 1971–76. Leicestershire Museums, Art Galleries and Records Service. ISBN 0 85022 098 X. Obtainable from the Leicestershire Museums, Art Galleries and Records Service, 96 New Walk, Leicester, England, price £3.00 including postage and packing.

GPSR Compliance

The European Union's (EU) General Product Safety Regulation (GPSR) is a set of rules that requires consumer products to be safe and our obligations to ensure this.

If you have any concerns about our products, you can contact us on ProductSafety@springernature.com

In case Publisher is established outside the EU, the EU authorized representative is:

Springer Nature Customer Service Center GmbH
Europaplatz 3
69115 Heidelberg, Germany

Batch number: 10593311

Printed by Printforce, the Netherlands